RF Power Amplifiers for Wireless Communications

Second Edition

DISCLAIMER OF WARRANTY

For a listing of recent titles in the *Artech House Microwave Library,* turn to the back of this book.

RF Power Amplifiers for Wireless Communications

Second Edition

Steve C. Cripps

**ARTECH
HOUSE**

BOSTON | LONDON
artechhouse.com

Library of Congress Cataloging-in-Publication Data
A catalog record for this book is available from the Library of Congress.

British Library Cataloguing in Publication Data
Cripps, Steve C.
  RF power amplifiers for wireless communications.—2nd ed.—(Artech House microwave
library)
  1. Amplifiers, Radio frequency 2. Power amplifiers
  I. Title
  621.3'81535
  ISBN-10: 1-59693-018-7

ISBN 10: 1-59693-018-7
ISBN 13: 978-1-59693-018-6

Cover design by Igor Valdman

International Standard Book Number: 1-59693-018-7

10 9 8 7 6 5 4 3 2 1

Contents

CHAPTER 11

Power Amplifier Bias Circuit Design   337

CHAPTER 12

Load-Pull Techniques   359

CHAPTER 13

Power Amplifier Architecture   371

CHAPTER 14

Power Amplifier Linearization Techniques   397

Preface to the Second Edition

When I started writing the first edition of *RF Power Amplifiers for Wireless Communications*, some time back in 1997, it seemed that I was roaming a largely uninhabited landscape. For reasons still not clear to me there were few, if any, other books dedicated to the subject of RF power amplifiers. Right at the same time, however, hundreds of engineers were being assigned projects to design PAs for wireless communications products. It was not, therefore, especially difficult to be successful with a book that was fortuitously at the right place and the right time.

As the original manuscript progressed both Artech and I, not to mention the bemused reviewers, became aware that the book was developing into something of an experiment; a not entirely premeditated attempt to write a technical book that was readable. It even had jokes. My word, if that's not *really* radical; isn't engineering supposed to be a serious business? Well, my own experiences in the engineering labs around the world is that engineers don't seem to be all that serious, at least not all of the time. Then there was the issue of equations; I happen to like symbolic analysis, but only the kind that I can understand. Once the Σ's get into double-nesting I tend to switch off, and I have a sneaking suspicion I'm not alone.

Well, times have changed. PA design and their high-volume manufacture have advanced enormously, and there now seem to be thousands of engineers and technologists who know more about the subject than I do. I like to think I played a part in getting some of them started, but it makes the current task rather more challenging than the initial carefree romp. I suppose I should just get more serious, cut the jokes and fill many more pages with Σ's. But that's not my style, for better or for worse.

Scanning the chapters, the general flow of the original book is the same. There are several new chapters, and nearly all of the original material has been updated and extended as appropriate. A new Chapter 4 digs a bit deeper into all of the things that make PA devices behave differently at GHz frequencies than at audio. Some of this analysis leads into another new chapter, Chapter 8, where I make an undisguised attempt to persuade RF designers that transistors do not behave as switches anymore, and the classical "switch" modes can be approximated using more conventional RF thinking on impedance and harmonic matching. I hope this sits OK with the two potential sides of a debate which I hope to promote. I am not in any way suggesting that the switch modes and their associated voluminous literature are in any way wrong, but simply trying to build a more solid bridge between the real world of GHz PA design and the elegant but idealized world where transistors are modeled as switches. In so doing I am not claiming any kind of priority either, but trying to persuade people to think a bit differently.

Modulation and nonlinear effects in PAs are closely interwoven, and so I have decided to cover both in a single chapter (Chapter 9). As modulation systems get more complicated, RF designers have a more difficult time understanding them, and the numerous excellent books that have appeared on the subject tend to cover areas of specific interest to the PA designer rather briefly and in some cases barely at all. With a little trepidation, therefore, I have tried to cover modulation systems from the PA designer's viewpoint. To do so, I place heavy reliance on actual measurements and observations of signals coming out of commercial generators, rather than delving into the challenging depths of symbolic analysis. This is something of a novel approach to the subject, and I hope it will not get me into too much trouble.

The design of bias networks for PAs has been belatedly promoted to chapter status, Chapter 11. The problems of maintaining stability and simultaneously minimizing supply modulation effects can cause as much difficulty for a PA designer as does the design of the RF matching networks, and hopefully this new chapter will provide some useful new insights for the PA practitioner. In further recognition and promotion of the school of pragmatic PA design, there is a new chapter on load-pull techniques, covering both passive and active methods. I am particularly indebted to Professor Paul Tasker and his group at Cardiff University for allowing me extensive access to their facilities, which include active harmonic load-pull and accurate measurement of RF waveforms. I have always held the view that unless you are able to measure RF voltage and current waveforms you can't say with any certainty what the "mode" of PA operation is, and it was my own measurements using this facility that formed the starting point for some of the less conventional views on harmonic matching to be found in Chapters 4 and 8.

PA linearization has become a huge subject, and several books devoted to the subject (including my own, see bibliography) have appeared in the interim. Chapter 14 has therefore been carefully limited to an overview, as in the original edition. But it is nevertheless intended as a bit more than an introductory overview. Its retention is also relevant, in that several of the chapters make reference to the issues of PA design for specific linearization environments. In an RF world ever more intruded by digital techniques, such considerations are having a major impact on the traditional design objectives of RF designers and will continue to do so.

Efficiency enhancement techniques, as in the first edition, are considered as a separate subject from linearization, and Chapter 10 is a much extended and updated treatment of this important topic. In both Chapters 10 and 14 I have finally been forced into quoting patents as references. I have always been reluctant to enter this minefield, as much as anything because I can only quote a few representative patents among so many that appear to be very similar. I don't want my selections to be viewed as any kind of priority judgment; they are quoted entirely as bibliographic aids, nothing more.

One of the features of this second edition is the inclusion of a CD which contains most of the simulation files and some Excel spreadsheets that are used in the text. The PA_waves program, which I call a "P.A. postulator" runs in Excel VBA, and is something of a new concept and may hopefully find some uses outside the confines of this book. In Chapter 9, I have made quite extensive use of Excel spreadsheets to illustrate and analyze various aspects of relevant modulation systems. These files are also included on the CD and comprise mainly a 4096 data point "representation" of IQ streams for various flavors of IS95, IS54, EDGE, and 802.11. The 4096 restriction has been selected mainly because the Excel FFT routine will only handle this number of points. In fact, the main text uses the data primarily for drawing IQ diagrams and doing some statistical analysis, particularly on efficiency in Chapter 13. Appendix B shows how the files can be used to perform some spectral analysis, with various models for PA nonlinearity, but this has not been used in the main text because most readers will have access to more comprehensive CAD analysis tools that will perform this kind of analysis more satisfactorily. But I hope this additional feature will be found useful, especially perhaps for students who (like myself) cannot afford to purchase the latest CAD tools.

This revision represents my own ongoing experiences as a consultant in the RFPA field. I would like to make a special acknowledgment to one of my clients, Filtronic (UK) and in particular the chairman of Filtronic, Professor David Rhodes, for many stimulating discussions on the subject over the last three years. In particular, these discussions have had an important impact on much of the material presented in Chapters 4 and 8. I am also indebted to Richard Ranson for his contributions to these discussions and Chris Potter (Cambridge RF) for much enlightenment on digital modulation techniques and test methods.

Steve C. Cripps
Somerset, England
May 2006

CHAPTER 1

Introduction

1.1 Introduction

A logical starting point for a book on RF power amplifier design would be to define what a power amplifier (PA) actually is. A technical definition might be an amplifier which is designed to deliver the maximum power output for a given selection of active device. Such a definition is useful in that it emphasizes that many of the techniques described in this book will be of interest to "small signal" amplifier designers. For example, the problem of obtaining low noise performance and also maximum dynamic range from an amplifier used in the front end of a receiver could be considered to be a PA design problem, and the techniques described in Chapter 2 would be applicable.

But in practice, by a "PA," we really mean an amplifier that can do damage to something. We are talking about a world of uncharacteristically high excitement in the RF lab, where attenuators and terminations get hot, and the life expectancy of expensive test equipment is greatly reduced. This means we are talking about RF outputs anywhere from the 1 watt level upwards. Above this level, we start to "feel" our amplifier RF output, even when safely dissipated in an attenuator. PAs are amplifiers whose outputs either directly or indirectly make an impact on the human sensory system.

Pyrotechnics notwithstanding, many power amplifier designs are simple extensions or modifications of linear designs. In the first place it will therefore pay to look at amplifiers in general, and to recall some of the classical results of linear RF amplifier theory. PAs do however usually operate with the active device displaying some, or maybe gross, nonlinear behavior, and we run straight into the problems of nonlinear modeling and characterization. This is a big subject and has been the focus of hundreds of papers and several books in recent years. The focus in this book is the development of useful, practical *a priori* design methods, and we will focus on developing such methods using some very basic models, which will be used throughout this book. These models will be introduced in Sections 1.3 and 1.4 of this chapter, where the important distinction between weak and strong nonlinear behavior will be defined.

One of the principal differences between linear RF amplifier design and PA design is that for optimum power, the output of the device is not presented with the impedance required for a linear conjugate match. This causes much consternation and has been the subject of an extended debate about the meaning and nature of conjugate matching. It is therefore necessary to swallow this apparently unpalatable result as early as possible (Section 1.6), before going on to give it more extended interpretation and analysis in Chapter 2.

1.2   Linear RF Amplifier Theory

The basic results of matched two-port linear RF amplifiers were derived originally by Mason [1] and Rollett [2], although most microwave engineers first encounter these results in a famous application note ("AN-95") written by Bodway under the old Hewlett Packard label and still available from Agilent [3]. There are also many books on basic RF design techniques which cover this material in greater detail [4].

The key results can be illustrated by examining the schematic shown in Figure 1.1. A transistor is represented as a two-port s-parameter matrix. The input and output impedance, or reflection coefficient, presented to the transistor, can be adjusted using conceptual tuning devices; since these tuners are realized using passive circuitry (e.g., transmission lines and shunt capacitive slugs or stubs), the reflection coefficients Γ_s and Γ_L at the input and output device reference planes are restricted to the range of $0 < |\Gamma_{s,L}| < 1$ in magnitude.

The use of tuners as a necessary means for obtaining useful gain from a microwave device is often an immediate source of puzzlement for "analog" electronic designers who have learned their trade at much lower frequencies. Much the same could also be said for the statutory microwave assumption of 50Ω terminations in Figure 1.1. This, more than anything else, is the target for much factionalism between the microwave and analog camps. At microwave frequencies, active devices will typically give little useful gain due to the domination of their parasitic reactances. Much attention needs to be paid to resonating these out in order that useful voltage levels can be established at the device active nodes. As the size and power capability of a microwave device increases, this problem becomes more acute.

The reasons for the 50Ω standard are sound and well proven, and concern the parasitic effects of the interconnecting wires between devices themselves, and also between devices and test instruments. At GHz frequencies, any interconnecting wire will have a significant electrical length in comparison to a wavelength at the application frequency, so that "connections" can no longer be assumed to display the same voltage at each end. Thus the interconnections have to be regarded as part of the circuit design, and some form of physical uniformity needs to be introduced. Uniform transmission lines, in their various physical forms, provide a solution to this requirement, and the selection of a standard characteristic impedance is necessary. It is a matter of the fundamental physical constants in our universe that transmission lines which have characteristic impedances below about 20Ω, and above about 100Ω, are

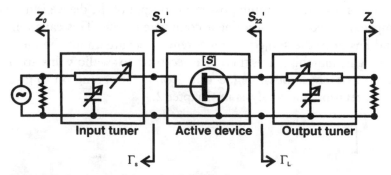

Figure 1.1   Schematic for 2-port gain and stability analysis.

difficult to manufacture. Very low, or very high, characteristic impedance transmission lines have either very narrow or very wide spacing of the two conductors. Just about anywhere in between these limits would therefore be an acceptable choice, and in the pre–World War II pioneering days of microwave engineering, both 40Ω and 80Ω were used. It was the emergence of a microwave test equipment industry that really sealed the 50Ω standard, with companies such as General Radio and Hewlett Packard then calling most of the shots. For better or for worse, the microwave world revolves around the 50Ω standard, and there seems no end in sight. Take a transistor, match its input and output to 50Ω, and, presto, you are a microwave designer. But there are some hazards on the way, which largely stem from the fact transistors are not entirely unilateral; there is reverse, as well as forward, transmission.

Returning to Figure 1.1, much of the complexity of the behavior of this system is characterized by the relative magnitude of the reverse transmission parameter s_{12}; however, a more concise parameter emerges after some extensive analysis, and this is called k, or Rollett's stability factor. The key equations are quite straightforward, and represent the change in the input reflection s_{11} to s'_{11}, due to the output plane being presented with the output load reflection Γ_L:

Input match:

$$s'_{11} = s_{11} + \frac{s_{21} \cdot s_{12} \cdot \Gamma_L}{1 - s_{22} \cdot \Gamma_L} \qquad (1.1)$$

Output match:

$$s'_{22} = s_{22} + \frac{s_{21} \cdot s_{12} \cdot \Gamma_S}{1 - s_{11} \cdot \Gamma_S} \qquad (1.2)$$

So for a conjugate match, one can set

$$s'_{22} = \Gamma_L{}^* \text{ and } s'_{11} = \Gamma_S{}^*$$

giving two equations which, in principle, can be solved for Γ_S and Γ_L.

It should now be noted that *in general*, this solution always exists; however, there is an additional constraint that the magnitudes of Γ_S and Γ_L must be less than unity for the solution to be realizable using passive external circuitry. This is where the k-factor makes its appearance. After some lengthy but admirable manipulations of complex algebra, Rollett et al. showed that the $0 < |\Gamma_{S,L}| < 1$ condition corresponds to a parameter known as the "k-factor" being greater than unity, where

$$k = \frac{1 - |s_{11}|^2 - |s_{22}|^2 + |D|^2}{2 \cdot |s_{21}| \cdot |s_{12}|}$$

and

$$D = s_{11} \cdot s_{22} - s_{12} \cdot s_{21}$$

So in practical terms, if $k > 1$ the device will never display an input or output reflection coefficient magnitude which is greater than unity, no matter what passive matching may be placed at its input or output. This condition therefore makes an additional statement about the stability of the device. Unfortunately, formulating the problem as a 2-port introduces some restrictions which are not general enough for some applications, and there has been an ongoing quest in the literature to find a "best and final" set of conditions for absolute stability under all conditions [5, 6]. But the simple $k > 1$ condition is undoubtedly a good practical guideline to follow.

It should not, however, be stretched too far. In the world of PA design, one is often struggling to obtain adequate signal gain, as well as extracting optimum power from a device. This is an inevitable consequence of cost-driven designs; large periphery transistors have lower gain and one is usually constrained to use the lowest cost technology. We will see later how, when the signal gain drops below 10 dB, the extra RF drive power required will often cancel out any efficiency advantage that had been carefully designed. The upshot of this is that one is often looking for an optimum situation where the k-factor is greater than unity, but not too much greater. Devices with high k-factors also tend to have low gain, and some extra gain can be retrieved by allowing positive feedback around the device, while keeping the k-factor above unity. This trick is often used, albeit sometimes inadvertently, by the microwave ECM and mm-wave amplifier community; and was a mainstay of the pre–World War II tube radio industry, "reaction" controls and all.

If $k > 1$, expressions can be found in the literature for the conjugate match and corresponding "maximum available gain" (MAG). These expressions are well known and are not repeated here [3, 4]. But there are a few ramifications which are worth noting:

1. Any device which has a k-factor greater than, but not much greater than, unity displays a more aggressive gain/match characteristic than a theoretical unilateral device. In particular, the final MAG may be considerably higher, in a nearly matched condition, than a simple voltage standing wave ratio (VSWR) mismatch calculation would indicate. For example, such a device displaying a 10 dB return loss may show more than the calculated 0.7 dB increase in gain when finally matched to −20 dB return loss.

2. Circuit losses can play havoc with the k-factor, and especially the frequency where it crosses unity. In practice, devices can be safely used some way below the unity k-factor point if the k-factor is based on fully de-embedded s-parameter measurements.

3. The circuit environment in which a transistor is placed can modify significantly its effective s-parameters, and especially the critical reverse transmission parameter, s_{12}. This is probably the main cause of unexpected, or unsimulatable, stability problems.

4. k-factor analysis, as presented here in its classical form, is only applicable to a single stage amplifier. In a multistage environment, the condition $0 < |\Gamma_{s,L}| < 1$ no longer applies, because the input and/or output planes of an intermediate stage are terminated with active networks. So taking a multistage amplifier as a single two-port and analyzing its k-factor is a necessary, but by no means sufficient, condition for overall stability. This problem is often bypassed in

multistage RF amplifier designs by the use of some form of isolation between stages, although multistage stability analysis and design strategies have been published [7].

1.3   Weakly Nonlinear Effects: Power and Volterra Series

The possibility, or reality, of nonlinear effects in linear amplifiers are usually first introduced in the form of a power series expression, as shown in Figure 1.2.

The amplifier symbol here represents a transistor with its associated input and output matching circuitry, and the lower case voltages are the small RF signal variations about the transistor operating point. The amplifier output consists of an infinite series of nonlinear products which are added on to the linear gain represented by the first term:

$$v_o = a_1 v_i + a_2 v_i^2 + a_3 v_i^3 + a_4 v_i^4 + \dots \text{(etc.)} \tag{1.3}$$

This "power series" is trotted out in many elementary textbooks as a generalized formulation for nonlinear behavior; yet it clearly has limitations. Most notably, there is no allowance for the timing of the various output components. We know, for sure, that even if the contents of the amplifier icon in Figure 1.2 were a passive R-C network, the output voltage would show both amplitude and time, or phase, changes from the input signal. A much stronger formulation of the power series, called the Volterra series, includes phase effects.[1] The power series is, nevertheless, useful for characterizing the behavior of nonlinear products in a small operation zone in the immediate vicinity of the DC operating point. This usefulness is limited, additionally, by the problem that the a_n coefficients are very sensitive to changes in the input and output tuning, and also the bias levels at input and output.

Such a formulation, with some enhancements, is useful for analyzing the "weakly nonlinear" properties of an amplifier or device. Weak nonlinearities may,

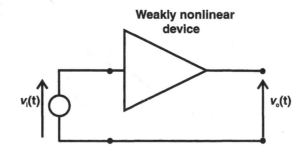

$$v_o(t) = a_1 v_i(t) + a_2 v_i^2(t) + a_3 v_i^3(t) + a_4 v_i^4(t) + \dots \text{(etc.)}$$

Figure 1.2   Weakly nonlinear device, and power series.

---

1.   The Volterra series in this book is regarded as a power series with complex coefficients; however, the formulation is much more general than this.

for example, be intermodulation distortion at levels lower than, say −30 dBc.[2] Such low level nonlinearities are very difficult to predict using device models based on I-V curve-fitted equations, and direct measurement of the parameters in the power series is one step toward obtaining a method of predicting low level, or weakly, nonlinear effects. This approach, using the stronger Volterra series formulation, has been pursued in depth by Maas [8] and should be of much interest to the designer of receiver components, where low level products down to −100 dBc will still be of importance in system specification and performance.

Generally speaking, PAs operating at or beyond the 1 dB compression point require more careful treatment, since the nonlinearities become "strong" and arise through the cutoff and clipping behavior of the transistor. Beware, especially, of analyses which truncate the power series to include only second- and third-order effects. The third-order nonlinearity is, undoubtedly, an important contributor to compression and saturation effects in small signal amplifiers, but in PAs the fifth- and seventh-degree terms become significant as the 1 dB compression point is approached and can dominate at higher drive levels.

1.4   Strongly Nonlinear Effects

Strongly nonlinear effects refer to the distortion of the signal waveform which is caused by the limiting behavior of the transistor. This is best illustrated by a simple idealized field effect transistor (FET) transfer characteristic, shown in Figure 1.3. The drain current exhibits cutoff, or pinchoff, when the channel is completely closed by the gate-source voltage, and reaches a maximum, or open-channel, condition

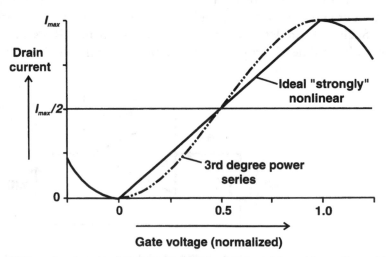

Figure 1.3   FET transfer characteristic, showing attempts to model weakly nonlinear (dotted) and strongly nonlinear (solid) effects.

2.   The dB and dBc notation will be used throughout this book. The reader is assumed to be familiar with the decibel, or "dB," as a logarithmic power ratio; the "dBc" term refers to the relative power level of a spectral distortion product to the main signal, expressed in dB.

where further increase of gate-source voltage results in little or no further increase in drain current. The characteristic shown in Figure 1.3 is highly idealized, both in the abruptness of the onset of both limiting conditions, and also in the perfect linearity of the intermediate region. This is what we will term an *ideal strongly nonlinear model*, and despite its rather conspicuous ideality, is widely used in textbooks to derive formulae for PA power and efficiency.

A more realistic model for the transfer characteristic of a FET is shown in the dotted curve of Figure 1.3. Here we see a more gradual cutoff, with the transconductance "fading" as the final cutoff point is reached, and a similarly gradual approach to the open channel condition. It is instructive to attempt to model this curve mathematically, since one quickly sees the difference between strong and weak nonlinear effects, and experiences the difficulty of satisfying both with the same model.

The fading transconductance is very typical of many kinds of FET and often represents approximately square-law behavior. The limiting characteristic, in turn, can be recognized as a third-order effect. So an equation of the form

$$I_d = g_0 + g_1 \cdot V_g + g_2 \cdot V_g^2 + g_3 \cdot V_g^3 \qquad (1.4)$$

will have some of the desired characteristics. If the gate voltage between cutoff and saturation is normalized to a range of 0 to 1, it can be shown that a suitable equation is

$$I_d = \quad + 3 \cdot V_g^2 - 2 \cdot V_g^3 \qquad (1.5)$$

Note that this equation is not a power series as such; it is an attempt to model the transistor transfer characteristic over a full operating range. The apparent absence of a linear term is disconcerting but will be explained shortly. This, in fact, is the equation plotted in the dotted curve of Figure 1.3, between the cutoff and saturation points. Outside this range, the simple third degree equation does not show the hard saturation and cutoff behavior and any simulator has to be made aware of this. To model these strongly nonlinear regions, many more higher order terms would need to be added to (1.4). An alternative, of course, is to impose the hard-saturation and full-cutoff conditions on to the weakly nonlinear model, with suitable "if" statements. The result of this is shown in Figure 1.4. This is easy enough to implement when using a computer to draw the curves and compute the equations, but represents a departure from simple analytical techniques. We will nevertheless use the resulting "strong-weak" model from time to time in this book, to bring greater reality to the overly idealized, but analytically tractable, ideal strongly nonlinear model of Figure 1.3.

It is also instructive to close the loop here with the discussion on weakly nonlinear effects, by taking the characteristic of (1.5) and computing the power series coefficients at a selected bias point.

Substituting $V_g = V_b + v_s(t)$ in (1.4) gives:

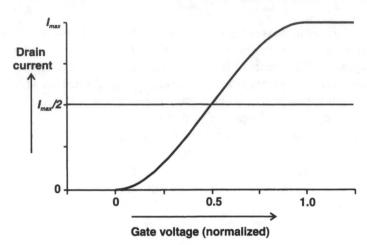

Figure 1.4 Combined weak and strong nonlinear model.

$$i_D(t) = 3\left(V_b + v_s(t)\right)^2 - 2\left(V_b + v_s(t)\right)^3$$
$$= \left(3V_b^2 - 2V_b^3\right)$$
$$+\left(6V_b - 6V_b^2\right) \cdot v_s(t) \tag{1.6}$$
$$+\left(3 - 6V_b\right) \cdot v_s^2(t)$$
$$+\left(-2\right) \cdot v_s^3(t)$$

So, by comparing the coefficients of $v_s^n(t)$ in (1.6) with the original power series formulation in (1.3), we obtain:

$$a_1 = \left(6V_b - 6V_b^2\right)$$
$$a_2 = \left(3 - 6V_b\right)$$
$$a_3 = \left(-2\right)$$

Several useful and more general observations can be made on these expressions for the a_n power series coefficients:

1. The "small signal gain", a_1, is a fairly strong function of the DC bias point, V_b.
2. The second degree nonlinearity, a_2, crosses zero and changes sign, giving a "null" at $V_b = 0.5$. So at this bias point, there would be no second order products. In higher degree systems, such as would represent a more accurate model for a real device, these nulls can occur for all degrees of nonlinearity. Such null points, or "sweet spots," are commonly observed in real devices. They are manifestations of the simple geometric properties of curves and inflexion points, and in this sense do not need to be explained in physical terms.
3. The third degree nonlinearity, a_3, is constant due to the selection of only a third order power series. In practice it would be more variable with bias

point, but the negative sign shows that limiting action will occur and serves to model the saturation of the device as hard saturation is approached.

1.5   Nonlinear Device Models for CAD

In order to devise a comprehensive model for a device, it is clearly necessary to characterize both the weak and strong nonlinear behavior. Unfortunately, each of these nonlinearity traits, in a particular device, arises from quite different aspects of the device physics. This poses a challenge to the device modeler, not to mention the end user, since final solutions still seem to be elusive. Unfortunately, the PA designer is much more sensitive to some of the shortcomings of widely used CAD models than designers of many other kinds of RF devices. This would seem to be equally true for RF power bipolar junction transistors (BJTs) and metal oxide semiconductor field effect transistors (MOSFETs) as it is for the more recent metal semiconductor FET (MESFET) and heterojunction bipolar transistor (HBT) devices, although the latter seem to catch most of the attention in technical journals and symposia.

Attempts to model devices for CAD applications fall into three general categories:

- Physical models, where the device physics and fabrication geometries are modeled from the "bottom-up";
- Equivalent circuit models, where the device physics is translated into analogous circuit elements;
- Behavioral models, where an essentially abstract set of equations is fitted to the measured three-terminal DC characteristics.

Generally speaking, most CAD device models for RF applications use the second category, with some additional help from either the first or the third. The BJT modeling community has, historically, used the physical ("bottom-up") model path, while the available FET models have more of the "top-down" flavor. This is probably explained by differences in the physical operation of the two devices. The physical process of a field effect device, even in its most simplified formulation, presents an immediately intractable problem of electostatics, in determining the exact profile of the depletion region. Conversely, and ironically, the more esoteric quantum mechanics of BJTs and semiconductor junctions leads to tractable mathemetical expressions for their I-V characteristics, which still form the basis of most BJT models used in today's advanced CAD packages. Most models used in commercial CAD simulators end up in the form of an equivalent circuit, with many linear elements and a number of nonlinear ones. These elements in some cases have physical counterparts, but in many cases are there because it has been determined empirically that they are needed to get a better fit to measured data. One of the biggest outstanding issues with these models is that elements may have "coupled" values, due to their empirical, rather than physical, basis. This makes yield and process variations difficult to simulate, since any physical process variation may require several, or even in some cases all, of the equivalent circuit elements to be changed.

The third category, behavioral modeling, has received more attention recently. There was a time when it was thought that it would never be possible to formulate a mathematical function that would describe the behavior of a nonlinear device entirely in terms of its terminal voltages and currents. There appear however to be some recent developments in behavioral modeling that could have a major impact on modeling techniques in the future [9].

The central issue in modeling an RF power transistor is "scaling." Almost always, the detailed modeling is done on a much smaller sample device than the device being used. Such models may be quite accurate. Designers of analog ICs and receiver components at higher frequencies typically have models they use confidently and which generate products with acceptable, or even impeccable, measured performance when compared to simulated results. The power amplifier designer has to take this small cell and scale it by tens, or even hundreds, in order to "build" a power transistor. This scaling is not, unfortunately, a simple set of electrical nodal connections, which can be handled easily enough on a modern circuit simulator. The large periphery device will display a range of secondary phenomena, which may have been quite negligible in the model cell. Nonuniform thermal effects, for example, can play havoc with the customary assumptions made about equal currents and voltages across an array of "identical" circuit elements. The typically very low impedances obtained by multiple parallel connections can also cause other effects to come into play that would normally be neglected: current spreading at bondwire contacts, electro-acoustic coupling in the semiconductor crystal, mutual coupling between bondwires, and many more.

Perhaps the most insidious aspect of the inadequacy of scaled-up RF power transistor models is the difficulty of putting model and device through simple comparative tests. Even a basic I-V measurement can pose serious difficulties for an RF power transistor. Most laboratory I-V curve tracers work at speeds which are several orders of magnitude slower than the RF signal for which the model is required and can be slow enough that transient junction heating effects, which will not occur to any significant extent during an RF cycle, intrude into the measurement. This assumes that the curve tracer is even capable of supplying several amps of current in the first place and that the voltage and current sensors have been thoroughly corrected for voltage drops. An additional problem frequently arises in the form of oscillation. Low frequency stability can be a serious problem with large periphery RF transistors and more often than not will show itself at some point on the I-V characteristic, either as "kinks" or more dramatically as chaotic-attractor spirals. Unfortunately, the oscillation remedies which will be discussed later in this book, for amplifier designs, are less effective for I-V curvetracing, since the inevitable inclusion of high value decoupling capacitors at the device terminals modifies the I-V curves. "Looping" I-V characteristics can sometimes be inherent device behavior, but they can also be caused by fixture decoupling circuitry.

Accurate I-V curves are difficult to obtain for RF power transistors; this has led many to develop custom-built test rigs, usually incorporating a pulsed measurement scheme [10]. An alternative approach is to build a curvetracer that sweeps the through the I-V characteristics at an RF speed [11]. Both approaches pose challenging problems of fast, accurate voltage and current sensing. Usually the I-V data from these setups look quite different from those obtained using a conventional low frequency curvetracer and asks important questions about any curvefitting efforts based on simple DC I-V measurements. This "dispersion" between DC and RF I-V trajectories has been thoroughly researched for microwave GaAs MESFETs, but it seems to have attracted little attention in the lower RF power transistor world of Si BJT and MOSFET. There may be some physical reasoning to suggest that this effect would be less troublesome in a bipolar device, but it would certainly seem prudent to have a closer look.

The measurement problems for RF power transistors are not limited to I-V curves. The RF impedances are typically so low, compared to a 50 ohm reference, that even simple linear s-parameter measurement is fraught with calibration problems. Once again, the standard, and dubious, way around this is to measure much smaller cells and scale up. Alternatively, prematching networks can be placed close to the large transistor die to transform the impedances up to a more measurable range, but this poses equally challenging calibration problems in de-embedding the matching networks. More and more, one is driven to the conclusion that the best way of deriving accurate models for RF power transistors is to build amplifiers, even nonoptimized ones, and fit the combined circuit and device models to the measured results.

The literature on these subjects is extensive, and outside the scope of this book. References [12–16] provide a starter bibliography, but the research continues, and papers on modeling still occupy multiple sessions in major international conferences. Undoubtedly, some individual groups of workers with the time and the facilities have managed to get on top of most of the modeling issues for the successful simulation of specific PA products, but their full results are not always available to the general public for use in commercially available CAD products. Although the GaAs RFIC foundry customer may be able to access good CAD models for the process in question, packaged RF power transistors seem to be poorly represented in commercial model databases. One is thus faced with having to make best use of what is available, and this can mean deriving your own models using generic templates. It is undoubtedly this unsatisfactory situation that causes many workers in the field to resort to older, traditional cut-and-try methods. It is hoped to show in this book that CAD tools do have much utility in RF power amplifier design and to encourage the use of both older and newer design methods in a complementary manner.

1.6   Conjugate Match

The concept of conjugate match has already been used in Section 1.2, and would seem to require no further explanation. A generator, as shown in Figure 1.5, delivers maximum power into an external load when the load resistance is set equal to the real part of the generator impedance, assuming also that any reactive component has been resonated out. Such an elementary result would seem unnecessary to revisit, but there is in fact a trap which has to be carefully avoided if the ideal generator is replaced by a real-world device, and especially if it happens to be the output current generator of a transistor.

The simple proof, whereby the generator is shown to deliver maximum power to the variable load resistance, takes no account of the possibility that the generator

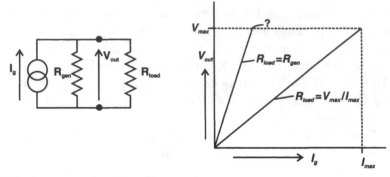

Figure 1.5 Conjugate match and loadline match.

will have physical limits, both in terms of the current it can supply, and (especially) the voltage it can sustain across its terminals. Take, for example, a case where the current generator can supply a maximum limiting current of 1 Amp, and has an output resistance of 100 ohms. Applying the conjugate match theorem, a load of 100 ohms would be selected for maximum power transfer. But the voltage appearing across the generator terminals would be 50V. If the current generator were the output of a transistor, it is very likely that this would exceed the voltage rating of the device. There is an additional restriction in that the transistor voltage swing is limited by the available DC supply.

To an outside observer, only able to observe the power (but not the voltage or current waveforms) in the load resistor, the device would therefore show limiting action at a current considerably lower than its physical maximum of I_{max} (see Figure 1.5). This is clearly an undesirable situation; the transistor is not being used to its full capacity, or "efficacy." In order to utilize the maximum current and voltage swing of the transistor, a lower value of load resistance would need to be selected; the value is commonly referred to as the "loadline" match, R_{opt}, and in its simplest form would be simply the ratio

$$R_{opt} = V_{max}/I_{max}$$

where it has been assumed that $R_{gen} \gg R_{opt}$; if R_{gen} is taken into account, it would be necessary to solve the equation:

$$\frac{R_{gen} \cdot R_{opt}}{R_{gen} + R_{opt}} = \frac{V_{max}}{I_{max}}$$

The two results discussed in this section, conjugate match and loadline match, are both elementary results of electronics. Yet when encountered together in the RF environment they appear to many as contradictory. Discussions on this "contradiction," and possible ways to resolve it, have been popular subjects in the correspondence sections of some trade and hobbyist journals. The key issue appears to be the statement of the problem. The basic conjugate match theorem only applies in a completely unrestricted case where currents and voltages at the generator terminals are unbounded by physical constraints. The loadline match is a real-world compromise, which is necessary to extract the maximum power from RF transistors and at the same time keep the RF voltage swing within specified limits and/or the available DC supply.

The controversy, nevertheless, continues. It is frequently asserted, for example, that a transistor which has a loadline match, rather than a conjugate match, will cause reflections and VSWR problems in a system to which it is connected. This is only a half-truth, however, in the sense that if the PA is presented with an appropriate load, it will operate perfectly well regardless of how the load environment is configured in the system context. The flip side of the coin is that whatever component the PA output connects into will very likely be experiencing an input termination which is different from its specified condition. The question then arises, and here we come perhaps to the core of the issue: what output impedance does a PA present to the subsequent device in the chain? As explained above, the reason for the loadline match is to accommodate the maximum permissible current and voltage swings at the transistor output. This says nothing about the impedance of the device, which can be assumed to remain approximately constant throughout the linear range.

A device which operates in Class A, with no excursions into the clipping or cut-off regimes, will essentially present something close to its small signal output impedance, represented by the S_{22} parameter, to the external world. In such conditions, the PA will more or less follow the rules of linear cascading, traceable back to Thevenin's theorem. But once a device starts to operate in any significantly nonlinear fashion, even for only a small portion of each RF cycle, the whole *concept* of output impedance starts to break down, due to the fact the waveforms are no longer sinusoidal. For example, in Chapter 3 we will introduce "reduced conduction angle," or Class AB, operation. In these very common modes for PA design and implementation, the input RF voltage swings the device into a region where it is completely cut off and draws no current. In such a condition, by definition the device must now present an open circuit to the outside world. In such an amplifier the instantaneous output impedance takes on the form of a switching characteristic, as shown in Figure 1.6. Thevenin did not consider such a device when formulating

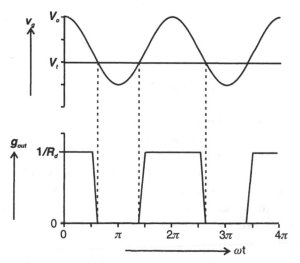

Figure 1.6   Output conductance variation of device in Class AB (output parasitic reactances ignored).

his law and would surely be concerned if he knew the firebrand zeal with which electrical engineering undergraduates have his law rammed down their throats. There is a large class of situations in practical electronics where Thevenin's theorem is, simply, inapplicable.

One is tempted to recall the words of Nobel Prize–winning physicist Richard Feynman on a similar topic, that of the use of complex impedances in electrical engineering[3]:

> ...the quantity R+jωL+1/jωC is a complex number and is used so much in electrical engineering that it has a name: it is called the complex impedance, Z. Thus we can write V=ZI. The reason [electrical engineers] like to do this is that they learned something when they were young: V=RI for resistances, when they only knew about resistances and DC. Now they have become more educated and have AC circuits, so they want the equations to look the same. Thus they write V=ZI, the only difference being the resistance is replaced by a more complicated thing, a complex quantity...

But Feynman was offended, it seems, only by the relatively minor extension of Ohm's law to include reactive impedance under sinusoidal excitation. He must surely then be quite horrified at the much bigger stretch which is asked by those who seek to represent a palpably nonlinear device such as depicted in Figure 1.6 using Thevenin's theorem. Such idealized symbolism seems certainly to have much of the same elements, in its attempt to extend elementary rules learned when young, beyond the point where they are useful. We learn to call the Thevenin conductance the "output conductance" of the device. Even though it is a behavioral, rather than a physical, element, we seem to develop a relationship with this entity which, over the years, metamorphoses from conceptual to physical. A device such as that shown in Figure 1.6 simply has to be measured under whatever external environment it is expected to experience, and perhaps then a more global model for its behavior can be determined, but such a model will not be a simple Thevenin equivalent.

An engineer has limited time, and probably also attention span, to deal with such matters of philosophy and semantics. Highly nonlinear PA output stages are usually interfaced with the outside world through the use of an isolator or a balanced configuration (see Chapter 13), which are simple and effective ways of dealing with this problem.

1.7   RF Power Device Technology

In the beginning, there was silicon (Si), and along came Gallium Arsenide (GaAs). Si was cheap, GaAs was expensive, but above 2 GHz GaAs soon hijacked the microwave world. This happened initially in low noise receiving applications. Somehow, despite some serious material and reliability issues, which took a decade or two to solve, high power GaAs, up to tens of watts, became available as well. The new millennium saw things get more complicated. The low cost requirements of high volume consumer products in the wireless communications sector demanded cheaper

---

3.   Unfortunately, there is no reference for this quotation; the style however seems authentic.

semiconductor technologies, and silicon started on a comeback trail. High power, driven by cellular basestation requirements at 830 MHz and 2 GHz, was an easy target for a new Si technology, which could not compete with GaAs in high GHz military and satellite communications applications but could give comparable performance at 2 GHz at lower cost. This technology, pioneered by Motorola, was a derivative of the long established RF MOS (Metal Oxide Semiconductor) process, and termed laterally diffused metal oxide silicon, or LDMOS. LDMOS has become the default choice for any high power PA application below 2 GHz and has essentially annihilated the RF power Si bipolar industry, other than a few small pockets that still manufacture such devices for specialist applications and long-term supply contracts.

GaAs technology has meanwhile taken on an unpredicted role in the form of a bipolar device, known as the heterojunction bipolar transistor (HBT), which has become the default technology for low power mobile handset PAs. This has been another unexpected development, based on at least two decades' worth of reliability-plagued research into HBTs. The mobile application favors a device which requires only a single supply voltage and which can be completely shut off. This made the HBT a "must-have" technology, although its diffusion into other higher power applications remains slow.

The most recent development in GHz RF power technology is the emergence of high bandgap semiconductors, from the research labs to commercial availability. Although Silicon Carbide (SiC) has long been the most widely touted candidate, it has been overtaken by the new wonder-material of the 00's, Gallium Nitride (GaN). The key property such technology brings to the RF power party is high voltage operation. GaN devices capable of running at 100 volts are becoming available, which can offer as much as five times the peripheral power density, at impedance levels which are an order of magnitude higher than their GaAs counterparts. Although the most logical application for such technology, and the source of much of the development funding, is broadband military electronic counter measures (ECM), the burgeoning power requirements of third generation (3G) mobile communications systems are making GaN look more attractive as the months go by. As a minimum, it seems that projections for a vacuum electronics renaissance[4] may have been a little off target.

References

[1]   Mason, S., "Power Gain in Feedback Amplifiers," *IRE Trans. Circuit Theory*, CT-1, June 1954, pp. 20–25.
[2]   Rollett, J. M., "Stability and Power-Gain Invariants of Linear Twoports," *IRE Trans. on Circuit Theory*, March 1962, pp. 29–32.
[3]   Bodway, G., "S-Parameters, Circuit Analysis and Design," Hewlett Packard Application Note AN95.
[4]   Vendelin, G., A. Pavio, and U. Rhode, *Microwave Circuit Design*, New York: Wiley, 1990.
[5]   Woods, D., "Reappraisal of the Unconditional Stability Criteria for Active 2-Port Networks in Terms of S-Parameters," *IEEE Trans. Circuits Syst.*, CAS-23, February 1976, pp. 73–81.

---

4.   Thank goodness for second editions..."Klystrinos" are still a nice concept, though.

[6] Gupta, M.S., "Power Gain in Feedback Amplifiers, A Classic Revisited," *IEEE Trans. Microwave Theory & Tech.*, MTT-40, May 1992, pp. 864–879.

[7] Macchiarella, G., et al., "Design Criteria for Multistage Microwave Amplifiers with Match Requirements at Input and Output," *IEEE Trans. Microwave Theory & Tech.*, MTT-41, August 1993, pp. 1294–1298.

[8] Maas, S., *Nonlinear Microwave Circuits*, Norwood, MA: Artech House, 1988.

[9] Wood, J., and D. E. Root, (eds.), *Fundamentals of Nonlinear Behavioral Modeling for RF and Microwave Design*, Norwood, MA: Artech House, 2005.

[10] Barton, T., et al., "Narrow Pulse Measurement of Drain Characteristics of GaAs MESFETs," *Elect. Lett.*, Vol. 23, No. 13, June 1987, pp. 686–687.

[11] Smith, M., et al., "RF Nonlinear Device Characterization Yields Improved Modelling Accuracy," *Proc. IEEE Intl. Microw. Symp.*, MTT-S, 1986, pp. 381–384.

[12] Snowden, C., and R. Miles, (eds.), *Compound Semiconductor Device Modelling*, New York: Springer-Verlag, 1993.

[13] Seitchik, J. A., C. Machala, and P. Yang, "The Determination of Spice Gummel-Poon Parameters by a Merged Optimization-Extraction Process," *Bipolar and Circuits Technology Meeting*, 1989, pp. 275–278.

[14] Statz, H., et al., "GaAsFET Device and Circuit Simulation in Spice," *IEEE Trans. Electron Devices*, ED-34, February 1987, pp. 160–169.

[15] McCamant, A., G. McCormack, and D. H. Smith, "An Improved GaAs MESFET Model for Spice," *IEEE Trans. Microwave Theory & Tech.*, MTT-38, June 1990, pp. 822–824.

[16] Root, D., and B. Hughes, "Principles of Nonlinear Active Device Modelling for Circuit Simulation," *Proc. IEEE MTT ARFTG Conf.*, Tempe, AZ, 1988, pp. 3–26.

CHAPTER 2

Linear Power Amplifier Design

This chapter will show how some of the concepts in Chapter 1 can be further developed into a simple but complete strategy for the design of linear RF power amplifiers. Essentially, a linear power amplifier can be designed using the same basic matching principles used for small signal designs, but with a power-matched output which will not appear to be conjugately matched. In some respects, power amplifier design becomes an analogous process to low noise amplifier (LNA) design, where in order to achieve the best possible noise performance from the device being used, its input will need to be presented with a reflection coefficient which will differ significantly from the conjugate match of the input impedance. In the linear PA case, the device has to be presented with a power match on the output in order to extract the maximum power from the device in question.

In older times, the exact value of the power match impedance at RF and microwave frequencies was regarded as something which could only be measured experimentally. Much like noise match data, this was something which designers had to measure for themselves, or was (preferably) available from the device manufacturer. Thus the art and science of "load-pull" was founded, and remains with us to this day, in the form of not inexpensive but very powerful computer-controlled measurement systems. This chapter introduces load-pull techniques and their application in linear PA design. Actual load-pull systems and more advanced harmonic load-pull techniques will be discussed further in Chapter 12. The main theme in this chapter is to show that there is a simple underlying theory which can do a fairly decent job of predicting the experimentally generated load-pull data for Class A type amplifiers, as defined in the next section.

2.1 Class A Amplifiers and Linear Amplifiers

It is widely assumed that the two terms "Class A" and "Linear" are almost synonymous, certainly as far as RF power amplifiers are concerned. In fact, Class A amplifiers are often by no means linear, and highly linear amplifiers are not necessarily, or even frequently, of the Class A type.

Class A is easy to define in its classical manner. Figure 2.1 recalls the ideal "strongly nonlinear" device transfer characteristic from Chapter 1 (Figure 1.3).

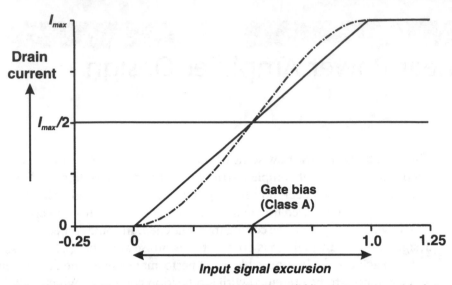

Figure 2.1 FET Class A bias point (see also Figure 1.3). The solid line represents an ideal strongly nonlinear response, used for simplified analysis; the dotted line shows the more realistic weakly nonlinear response, with low-level harmonic generation even at low signal levels.

Such a device is assumed to be perfectly linear in the region between cutoff and saturation. If an amplifier is constructed using the exact mid-point of the linear range as the bias point, perfectly linear operation will result, providing the RF drive signal never exceeds the boundary values. If the RF drive signal is perfectly sinusoidal, then the output current will also be sinusoidal with no harmonic content. This is a classical Class A amplifier.

The amplifier of Figure 2.1 is clearly also a linear amplifier, provided the drive signal does not exceed the stated limits. So in this ideal case, a Class A amplifier will be linear for this limited range of drive. In practice, of course, the linear region will contain weak nonlinearities, as discussed in Section 1.3 and indicated in Figure 2.1. The "weakness" of these nonlinearities will become less evident as the signal drive level is increased; clearly in order to make a Class A power amplifier, it is necessary to swing the device through the increasingly nonlinear perturbations of the transfer characteristic in order to reach the hard-clipping limits. In this case, the current in the output generator will have a significant harmonic content. In practice, the output current generator will be working into a reactive matching network which transforms the 50 ohm environment to the required load resistance value. Such networks are typically low-pass, and the harmonics generated by the transconductive nonlinearities will be greatly attenuated. This gives such an amplifier an unfairly "clean" image, which fades somewhat when a modulated signal is substituted, and quite high levels of spectral spreading or intermodulation are typically seen. As will be discussed in more detail in Chapter 9, the distortion processes that generate harmonics will also cause spectral distortion of an amplitude-modulated signal. Class A amplifiers are, in general, "cleaner" than the Class AB amplifiers, which will be introduced in Chapter 3, which make deliberate use of the strongly nonlinear regions of the device characteristic in pursuit of higher efficiency.

2.2   Gain Match and Power Match

Figure 2.2 shows the power transfer characteristic of a Class A amplifier with two different output matching conditions. The solid line shows the response for an amplifier which has been conjugately matched at much lower drive levels. The two points A and B refer to the maximum linear power and the 1 dB compression power, respectively. It has become something of a standard in the RF amplifier community to use the 1 dB compression point as a general reference point for specifying the power capability of an amplifier or an amplifying device (transistor). It also represents a practical limit for "linear" operation. We will see in due course that the 1 dB compression point actually represents a moderate, rather than a weakly, nonlinear point, and in this chapter we will focus primarily on point A, which represents the point at which nonlinear behavior (gain compression) can initially be detected.

In a typical situation, the conjugate match would yield a 1 dB compression power significantly lower than that which can be obtained by the correct power tuning, shown by the dashed line in Figure 2.2. At both points A and B, the device would be delivering 2 dB lower power than the device manufacturers specifications. Unfortunately, power transistors are often the most expensive individual components in a system, and such "wastage" of performance can be translated directly into unnecessary cost. So the power matched condition has to be taken seriously, despite the fact that the gain at lower signal levels (i.e., the lower left-hand corner of Figure 2.2) may be 1 dB or so less than the conjugate matched condition. The two curves in Figure 2.2 should be compared with the pair of lines in Figure 1.5. The behavior has the same root cause in each case, although in the RF world one only has the circumstantial evidence of RF power, rather than direct current and voltage readings, to go by.

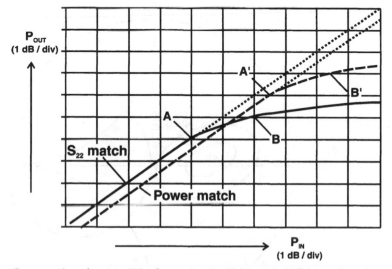

Figure 2.2 Compression characteristics for conjugate (S_{22}) match (solid curve) and power match (dashed curve). The 1 dB compression points (B, B′) and maximum linear power points (A, A′) show improvements under power-matched conditions.

It is important to note in Figure 2.2 that no matter which criterion is used for RF power, the power match gives about the same 2 dB improvement. That is to say, the maximum linear power (A, A′) increases with power tuning by about 2 dB as well as the 1 dB compression power (B, B′). This is a typical observation, albeit a strictly qualitative one. Across a wide range of devices and technologies, the actual difference may vary over a 0.5 to 3 or 4 dB range, but the improvement with power match will be fairly constant over a range of gain compression.

2.3   Introduction to Load-Pull Measurements

The two power sweep measurements of Figure 2.2 indicate that there is some kind of functional relationship between output power and output match. The logical next step is to measure more than two data points. Such a measurement is termed a load-pull measurement, presumably originating from analogous measurements performed on RF oscillators. In its simplest form, a load-pull test setup consists of the device under test with some form of calibrated tuning on its output. The input will probably also be tunable, but this is mainly to boost the power gain of the device, and the input match will typically be fixed close to a good match at each frequency. Some kinds of RF transistor, particularly bipolar transistors, show significant dependency between output power and input load. It is in practice quite difficult to differentiate between true "source-pull" effects and the changes in gain caused by input matching. Devices which show the most prominent source-pulling effects are usually operating close to their maximum useable frequency, and this situation is best avoided by using a higher frequency technology.

The hardware itself will be considered some more in Chapter 12, but a typical set of load-pull data is shown in Figure 2.3. Such a set of data may take days, hours,

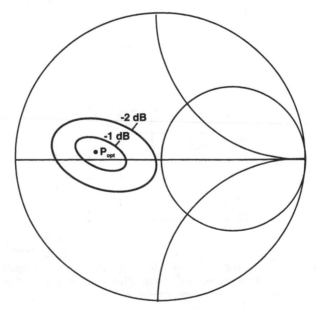

Figure 2.3   Typical load-pull data.

or minutes to compile, depending on the degree of complexity, expense, and time invested in the equipment. The results show closed "contours," marking the boundaries of specified output power levels. For most practical purposes, the PA designer is mainly concerned with the 1 dB and 2 dB contours, these representing levels relative to the maximum or optimum power output of the device at the test frequency.

The most obvious observation in looking at the data in Figure 2.3 is that the constant power contours, plotted on a Smith chart, are not circles. Unlike noise and linear mismatch (gain) circles, they resolutely refuse to display a circular profile, no matter how carefully the equipment is calibrated. Over the years, they have been variously compared with eggs, potatoes, and rugby balls, but they are definitely not oranges or soccer balls. This was assumed to be a manifestation of nonlinear behavior, but there was a curious twist on this that was largely unobserved; the contours are still roughly the same shape, even if the criterion for power measurement is shifted to the *maximum linear power* (i.e., points A, A' in Figure 2.2). This leads to a straightforward explanation for the noncircular shape of the contours, which will be considered in detail in Section 2.5.

Load pull data has been the mainstay of RF and (especially) microwave power amplifier design for many years. It gives the designer a well-defined impedance design target, on which to base the strategy for suitable matching network design. It apparently converts an intractable nonlinear problem into one which can be attacked and solved using linear techniques and even linear simulators. With the recent availability of good, fast, nonlinear simulators, and slowly improving large signal models, it could fairly be speculated that load-pull equipment may go down the same road to nostalgic oblivion as the slide rule and the slotted line. This is certainly not evident yet. Indeed, the most stringent verification test that can be applied to a simulator is to perform a "virtual" load-pull measurement. The results from such comparisons (reluctantly performed, it seems, requiring substantial cooperation between antagonistic parties) are at best only fair. The load-pull test seems likely to continue to provide the basic measurements required to derive and fine-tune nonlinear models for RF power devices.

2.4 Loadline Theory

As long ago as 1983, well before the widespread availability of nonlinear simulation CAD tools, it was shown the most basic and elementary loadline principles could be extended to predict load-pull contours at microwave frequencies, for a device kept in its linear range [1]. This novel and simplistic viewpoint had a major impact on the design of PAs at GHz frequencies. The biggest surprise was that the resulting contours, plotted on a Smith chart, were areas of intersection between constant resistance and constant conductance circles, displaying a "pointed" version of the familiar flattened circles obtained from load-pull measurements. Figure 2.4 shows a typical result, showing a direct comparison between the theoretical prediction and actual measured data. Over the intervening years, several attempts have been made to refine this theory, removing some of the idealizing assumptions made in the original analysis [2]. By and large, however, the simplicity of the original technique has

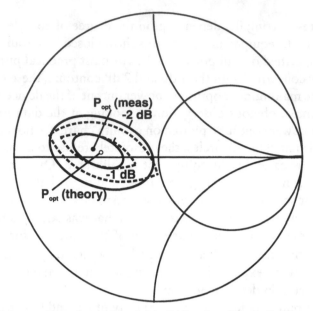

Figure 2.4 Direct comparison of experimental (solid lines) and theoretically generated power contours (dotted), using the loadline analysis and equations in this section [1].

survived attempts to refine it, and it remains a very useful *a priori* design method to use as a starting point for a PA design iteration.

The starting point for this analysis of an RF power amplifier is a heavily idealized device model, shown in Figure 2.5. This is an ideal strongly nonlinear transconductive device, represented here as a voltage controlled current source with zero output conductance and zero turn-on (or "knee") voltage. The transconductance is linear except for its strong nonlinearities represented by pinchoff (for input voltages below V_t) and hard saturation at I_{max}. A key feature of this analysis is that the device is never allowed to breach these limits of linear operation; in this sense the analysis is valid up to, but not beyond, the onset of gain compression.

Figure 2.6 shows the RF circuit in which the ideal device is analyzed. The output (drain) is AC coupled to an RF load. The DC bias is fed through a separate "choke,"

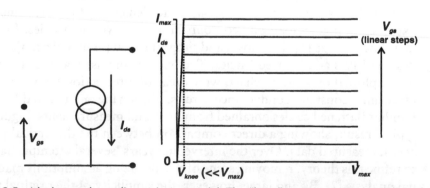

Figure 2.5 Ideal strongly nonlinear device model. The "knee" turn-on region is neglected, and between 0 and I_{max}, the output current generator is linearly controlled by the input generator.

which is assumed to have a very high reactance at the RF frequency.[1] An important detail is that although the transistor is being considered as an ideal voltage-controlled current generator, the RF load may have a reactive component, but any output parasitics of the transistor will be considered to be part of the external load.

Assuming that the reader is familiar with the concept of a loadline match, we can progress immediately to look at the RF waveforms in Figure 2.6, which show the device under conditions of sinusoidal excitation and optimum loading. The current swings over its maximum linear range (zero to I_{max}), that is, with an amplitude of $I_{max}/2$. The voltage swings over its maximum range of zero to $2V_{dc}$, an amplitude of V_{dc}. Clearly, the loadline resistor in this optimum power-matched condition has a value of

$$R_{opt} = V_{dc}/(I_{max}/2) = V_{dc}/I_{dc} \qquad (2.1)$$

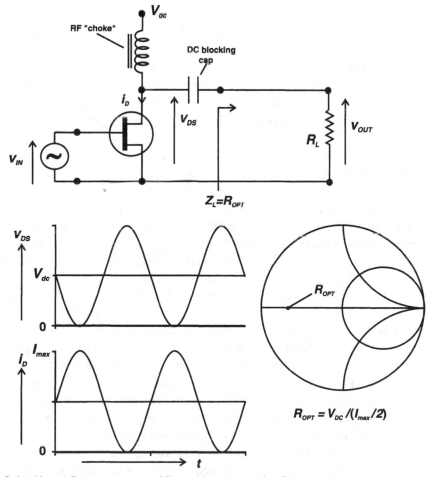

Figure 2.6 Class A linear power amplifier, with optimum loadline match.

---

1. Throughout this book, I intend to overrule the grammatical tautology contained in such terms as "RF frequency" and "DC current" in favor of the technical clarity these terms convey.

This is one of the most basic and elementary results of electronics, and there is no suggestion that we are here indulging in any attempt at reinvention. We are, however, making an important assumption that currents and voltages at RF frequencies follow the same trajectories as they do on a DC curve tracer. This issue has been the subject of much debate [3], and has been extensively addressed by semiconductor equipment manufacturers [4], but at this juncture we make this quasi-static assumption along with the other idealizations. It is also worth noting that the peak RF voltage of $2V_{dc}$ arises primarily from the symmetry of the assumed sinusoidal waveforms. An RF waveform that is symmetrical about its mean level has to rise to a peak voltage of twice the DC supply in order to remain above zero on the downward part of the cycle. We will see later that when the waveforms become asymmetrical due to harmonic content, the peak voltages can be greater (or less) than $2V_{dc}$.

Figure 2.6 shows the device in its optimally matched condition, where it will deliver an RF power P_{opt}, where

$$P_{opt} = (1/2) \cdot V_{dc} \cdot I_{dc}$$

This represents a classical Class A linear amplifier, with a drain efficiency of 50%.

It is now possible to use this simple model and circuit to trace out a load-pull contour by examining the effect of moving the resistive and reactive component of the RF load away from the optimum value of $R_{opt} + j.0$. In particular, we will determine the range of load terminations which will give a power of P_{opt} / p, with $p = 2$ for initial clarity.

Figure 2.7 shows that there are two resistive terminations that result in a maximum linear power of P_{opt}/p, $p \cdot R_{opt}$, and R_{opt}/p. In the case of the lower resistive load R_{opt}/p, the device can swing over the full current range of I_{max}, but the corresponding voltage swing is only $2 \cdot V_{dc}/p$, which corresponds to an RF power output of P_{opt}/p. In the case of the higher resistive load $p \cdot R_{opt}$, the drive level has to be backed off to reduce the current swing and keep the voltage swing at the maximum linear peak value of $2 \cdot V_{dc}$. It can be clearly seen that the current amplitude has to be reduced by the ratio of p, corresponding to a reduced power of P_{opt}/p.

We thus have two points on a load-pull contour for a power output level of P_{opt}/p. Figure 2.8 shows how the R_{opt}/p point can be extended into a continuous arc segment of constant power if some series reactance is added to the load resistance. The key point is that while the current swing remains at its maximum permissible value, the low voltage swing can be increased by adding series reactance, without affecting the power. This series reactance has a maximum value $\pm X_m$, whose value is such that the series combination of X_m and R_{opt}/p gives a complex magnitude of R_{opt}. In a similar manner, Figure 2.9 shows that the $p \cdot R_{opt}$ point can be extended into another arc segment of constant power if some shunt susceptance is added to the load conductance. In this case, the maximum voltage swing can be kept constant while the RF current swing is increased, provided that a corresponding shunt susceptance is added. The limiting case here is that the complex admittance magnitude of the load is equal to $1/R_{opt}$ ($= G_{opt}$).

The two arcs of constant power conveniently follow the printed Smith chart circles of constant resistance ($R = R_{opt}/p$) and constant conductance ($G = 1/(p \cdot R_{opt})$). It

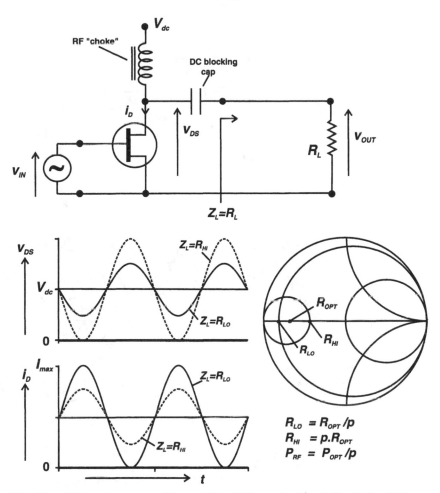

Figure 2.7   Class A linear power amplifier case 1: resistive output loads for $R_L<R_{OPT}$; RF power output is P_{OPT}/p ($I_{pk}=I_{max}$). For $R_L>R_{OPT}$, RF power output is P_{OPT}/p ($I_{pk}=I_{max}/p$).

remains to be shown that the contour is indeed closed, and that the limiting points of constant power, as derived, coincide at the apex of the intersection of the separate arcs. This is the simple exercise of showing that the impedance $R_{opt}/p + j. X_m$ is the same as the admittance $1/(p. R_{opt}) + j. B_m$, where

$$X_m^2 = R_{opt}^2 \left(1-1/p^2\right) \quad \text{and} \quad B_m^2 = G_{opt}^2 \left(1-1/p^2\right)$$

The final result, drawn for the case of $p = 2$ (the 3 dB power contour), is shown in Figure 2.10. The noncircular nature of constant power contours is clearly indicated. Figure 2.11 shows a family of contours, easily constructed as described, for a range of p values which correspond to 1 dB steps. It is especially interesting to note the small target area represented by the 1 dB contour and the much larger size of a conventional 1 dB mismatch circle centered on R_{opt} (shown dotted in Figure 2.13). This is evidence against the notion that power match is a simple manifestation of a "large-signal" movement of s_{22}.

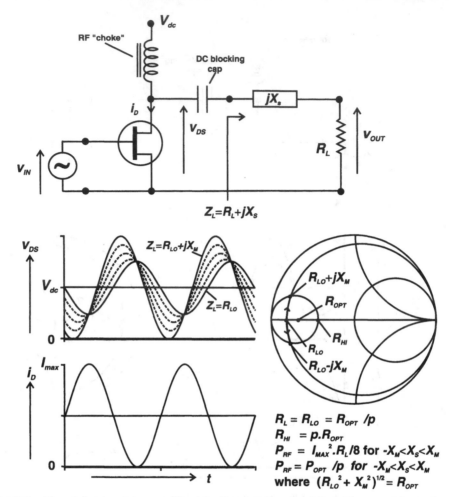

Figure 2.8 Class A linear power amplifier case 2: output load resistive component lower than R_{OPT}. For $R_L=R_{OPT}/p$, RF power output is P_{OPT}/p, over a range of series reactance—$X_M<0<X_M$.

The contours shown in Figures 2.10 and 2.11 will apply at any frequency. This comes as something of a surprise to load-pull stalwarts who are accustomed to seeing substantial movement of the measured contours as a function of frequency. The key point is the choice of reference plane for impedance measurement. In this section the reference plane has been the terminals of the transistor output current generator, whereas in practice the closest physical point for measurement purposes lies outside of the die and package parasitic reactances. Another way of looking at this is to recognize that the impedances represented by the contours as analyzed here are measured in absolute units of ohms, resistive and reactive. Consequently, the external circuit can be analyzed at any frequency and plotted on the same chart as the load-pull contours as described. If the reference plane is shifted to a point where parasitic reactances with specific values are included in between the current generator and the reference point, then the contours will become frequency dependent, inasmuch as capacitors, inductances, and transmission lines have frequency-dependent reactances. This "de-embedding" procedure is necessary to compare the theoretical results derived in this section with actual measured data. It is, nevertheless, a very

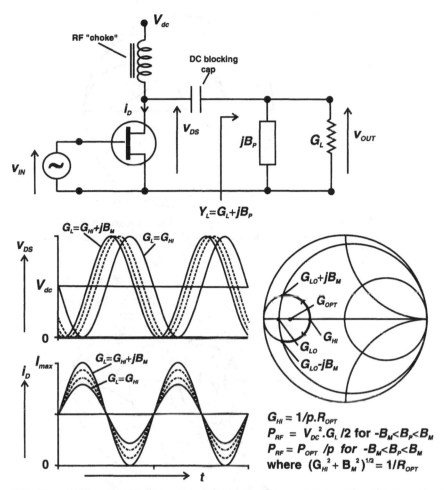

Figure 2.9   Class A linear amplifier case 3: output load resistive component higher than R_{OPT}. For $R_L = p.R_{OPT}$, RF power output is P_{OPT}/p, over a range of shunt susceptance—$B_M < 0 < B_M$.

useful feature to be able to remove the frequency dependency of load-pull contours by suitable choice of reference plane location. This will be illustrated in Section 2.6 as a particularly useful aid for broader bandwidth designs.

2.5   Package Effects and Refinements to Load-Pull Theory

The basic load-pull theory developed in the last section needs some extension in order that direct comparisons with measured data can be made. Figure 2.12 defines the problem in a typical case. The load-pull contours have been derived at plane A, the terminals of the transistor current generator. We need to transform this result to plane B, which might represent the output soldering tab of the packaged die. The device output capacitance, bondwire inductance, and the package parasitics have to be taken into account at plane B in order to present an impedance which maps onto the original contour at plane A. This is a simple task for a linear circuit simulator, although care is needed in applying the necessary transformations. For example, Figure 2.12 shows the transformation at plane B where the only significant parasitic

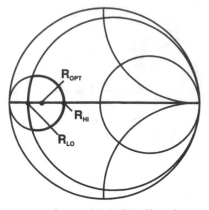

Figure 2.10 Constant power contour for $p = 2$ (−3 dB). Closed contour is formed by intersecting circles of constant resistance ($R = R_{LO} = R_{OPT}/p$) and constant conductance ($G = 1/R_{HI} = 1/pR_{OPT}$).

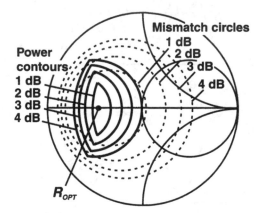

Figure 2.11 Power-match contours (solid, 1 dB to 4 dB contours shown), compared with gain mismatch circles (dotted) based on a source impedance equal to the optimum power match ($R_{OPT} = 20\Omega$).

element is the shunt capacitance. The contour appears to rotate, or slide, around the circle of shunt conductance, in going from plane A to plane B. But the direction is opposite to that of an impedance transformation, due to the fact that here we are compensating, or de-embedding, the shunt element. Clearly, the transformation will be frequency dependent if the capacitor has a fixed value. Figure 2.13 shows the effect, over frequency, of applying the necessary transformation to a typical transistor and package.

Referring back to Figure 2.4, it can be seen that the theoretical contour (dotted) has been transformed in the manner described to allow for a series bondwire and a device output capacitance, thus appearing to topple over backwards from its original upright position. The agreement in Figure 2.4 is good, and it is general experience that the loadline theory tends to set slightly tighter design target zones, an acceptable situation for practical purposes.

The comment has been made that the removal of frequency dependency by transforming the design impedance reference to plane A is unnecessarily cumbersome, given that device and package parasitics may not be accurately known. This

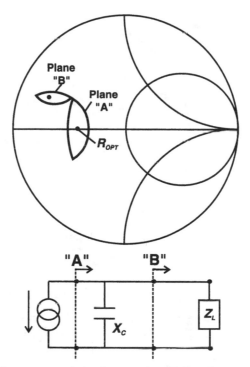

Figure 2.12   Power contour measured at reference plane A (transistor current generator) and plane B, showing the appropriate compensation (de-embedding) required at B due to the shunt capacitance. Contours plotted for $R_{OPT} = 20\Omega$, $X_c = 35\Omega$ (5 pF at 2 GHz).

would apply particularly for narrowband designs, which are common in wireless communication bands. A single frequency, measured load-pull data set would in principle enable a design to be done, but then the lack of any ability to analyze the performance of different matching networks over frequency is a serious limitation, even for narrowband designs. It will also be shown in the design example of Section 2.7 that the theoretical load-pull contour approach enables some useful tolerancing analysis to be done, something that a single data set would not permit. For numerous reasons, it is good RF practice to have some idea of the package parasitics. Reputable device vendors should be able to supply package models, and in the limit simple package models are not difficult to derive for frequencies up to 2 GHz. One approach is to take the *s*-parameters of the die by itself (which may be available as a stand-alone product) and fit a simple package model to the packaged device data.

Although the general agreement between theory and measured data in Figure 2.4 is good, there are several refinements which are advisable in using this method for the design of practical power amplifiers. Most notably, the calculation of the value of R_{opt} needs to be re-examined. Referring back to Figure 2.6, the device RF current is shown at a peak value of I_{max} at the same instant that the drain voltage is at zero. This is clearly a direct result of the ideal turn-on, or knee, characteristic shown in Figure 2.5. In order to sustain gate-controlled current, it is necessary to reduce the value of the load resistor so that the drain (or collector) voltage is kept above the appropriate knee voltage for the device being used. For BJTs, HBTs, and GaAs FETs, a suitable value for this reduction, denoted by V_k, would be 1 volt. LDMOS

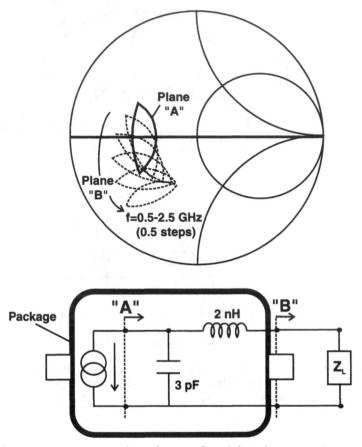

Figure 2.13 Power contour measured at reference plane A (transistor current generator) and plane B for a typical set of package parasitic reactances.

devices would require a larger correction, although will also be running at higher supply voltages.

Some comment is also needed on the peak current I_{max}, which is not a commonly specified parameter on data sheets. It turns out that I_{max} is not an easy parameter to measure directly. The best way of estimating its value is to build an optimized Class A amplifier and observe the DC supply current. Assuming that the RF current is sinusoidal, the value of I_{dc} will be equal to $I_{max}/2$. Most data sheets will give a typical Class A operating condition, including values for RF power input and output, and DC bias, so that a representative value for I_{max} can be obtained. This approach is particularly useful for bipolar transistors, which compared to FETs do not have such an easily identifiable current saturation characteristic. So the modified value for R_{opt} is:

$$R_{opt} = (V_{dc} - V_k)/I_{dc} \tag{2.2}$$

Some FETs, particularly MESFETs, show significant slope on their I-V characteristics, corresponding to a DC output conductance. It is tempting, therefore, to include this as a physical conductance which acts in parallel with the external load resistor. In practice, and as can be confirmed by running nonlinear time-domain simulations, the slope of the DC characteristics tends to be cancelled as the device sweeps over the loadline range of current and voltage, and the assumption of zero AC output conductance usually gives a better approximation than the simple DC conductance correction. This approach represents an approximation and a useful starting point. A nonlinear simulator, with an accurate curve-fit to the whole IV function, is the only way to obtain a more accurate answer to the precise optimum value for R_{opt}, and the difference between this value and the approximate one may often lie inside the global variation of transistor current/threshold specifications.

2.6 Drawing the Load-Pull Contours on CAD Programs

Although the simplicity of the loadline approach to PA design has found wide appeal in the microwave amplifier community, CAD package developers have not embraced the concept so readily. It would be a very useful addition to any linear RF or microwave circuit analysis program to have a simple routine which plots out the load-pull contours for a simple set of device parameter inputs, in much the same way that noise circles are usually offered.[2] Fortunately, it is a very simple matter to trick a simple circuit analysis program into plotting the contours.

In Figure 2.10, it is clear that the two arcs of the load-pull contour could be modeled as series and parallel LCR circuits. The values can be chosen such that over the frequency sweep range in use for the design in question, the impedances of these "dummy" circuits sweep precisely over the area of intersection, defining the actual limits of each contour. Figure 2.14 shows the construction of such circuits, and gives equations for the value of the components in terms of the required contour (in dB) and the frequency sweep range. Obviously, the values have to be changed when the sweep range is changed, but most CAD packages include options for defining element values in terms of expressions involving the frequency limits and other component values in the circuit, so this would enable the contours to be plotted essentially independently of frequency sweep settings.

2.7 Class A Design Example

The basic steps for designing a Class A amplifier using the techniques developed in the previous sections will now be illustrated using a design example. As stated at the beginning of this chapter, the goal here is to transform a challenging nonlinear design problem into one which can be solved using the most basic design tools. The procedure is analogous to the classical design of a low noise amplifier, where the input impedance transformation is based on knowledge of the optimum noise match and derivative noise circle family. Here the output matching network is based entirely on the loadline approach and the resulting Smith chart construction of 1 dB load-pull contours. The input match is performed using the linear s-parameters; the design is based on a power level which is on the threshold, but not beyond,

---

2. It is indeed ironic that such noise circle constructions have been given virtually "axiomatic" status by the RF and microwave community, despite being based on highly idealized models, in a spirit of simplification which is very comparable to the idealizations used in the loadline theory for load-pull contours.

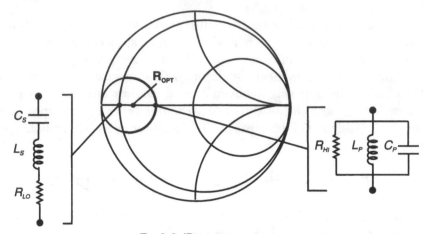

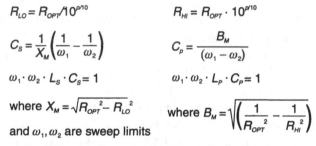

For "p" dB contour:

$$R_{LO} = R_{OPT}/10^{p/10} \qquad\qquad R_{HI} = R_{OPT} \cdot 10^{p/10}$$

$$C_S = \frac{1}{X_M}\left(\frac{1}{\omega_1} - \frac{1}{\omega_2}\right) \qquad\qquad C_p = \frac{B_M}{(\omega_1 - \omega_2)}$$

$$\omega_1 \cdot \omega_2 \cdot L_S \cdot C_S = 1 \qquad\qquad \omega_1 \cdot \omega_2 \cdot L_p \cdot C_p = 1$$

where $X_M = \sqrt{R_{OPT}^2 - R_{LO}^2}$ $\qquad$ where $B_M = \sqrt{\left(\frac{1}{R_{OPT}^2} - \frac{1}{R_{HI}^2}\right)}$

and ω_1, ω_2 are sweep limits

Figure 2.14 Construction of load-pull contours on linear circuit analysis programs by use of dummy series and parallel resonant circuits for series and shunt arcs.

the point of measurable gain compression, where the s-parameters still have values close to those measured at much lower signal levels. It has already been shown that a good power match at this point is likely to be close to the optimum for higher levels of compression, and in any case represents a good compromise for linear performance in the presence of variable-envelope signals.

The design is based on a typical 1-watt GaAs MESFET at a frequency of 1.9 GHz. This is a Class A design; in this respect the final design cannot be expected to meet the efficiency requirements of a cellular handset PA but serves as an illustration of the design principles so far discussed.

Step 1: Define target specs, select device, determine R_{opt}.
In this case, a transistor specified to give 29 dBm typical power at its 1 dB compression point is selected. The manufacturer's data show that this power is obtained at a Class A DC bias of 4.8v and 375 mA. The design frequency band is 1.75–1.85 GHz.

Allowing 1V for the turn-on of a typical GaAs MESFET, (2.2) gives

$$R_{opt} = (4.8 - 1)/0.375$$
$$= 10.1\ \Omega$$

Step 2: Set up schematic and output matching topology to give R_{opt} at plane A.
Referring to Figure 2.15, it is assumed that we know the values of the transistor output capacitance, and the package parasitics. The matching problem is therefore to present an impedance of 10.1Ω at plane A, a point inside the package and the device output capacitance. The initial matching strategy is to present the optimum impedance at the midband frequency and then look at a swept frequency analysis to determine the bandwidth of specified operation. A simple low-pass section is shown, consisting of a length of 50Ω microstrip line and a shunt capacitor. These values can be adjusted to give a suitable mid-band match, as shown in Figure 2.16. Note that the 1 dB contour has been plotted out as described in Section 2.6, using dummy series and parallel resonant circuits in the analysis file.

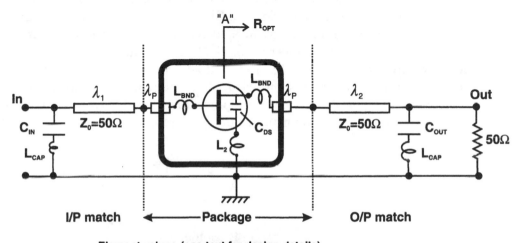

Element values (see text for design details):
C_{OUT}=2.7 pF, C_{IN}=4.7 pF, L_{CAP}=0.6 nH, C_{DS}=1.5 pF
L_{S}=0.08 nH, L_{BND}=0.2 nH, λ_{P}= 40 mil (ε_{r}=10, h=10)
λ_{1}= 150 mil (ε_{r}=4.5, h=31 mil), λ_{2}= 180 mil (ε_{r}=4.5, h=31mil)

Figure 2.15   Schematic of 1.9 GHz 1-watt linear PA design.

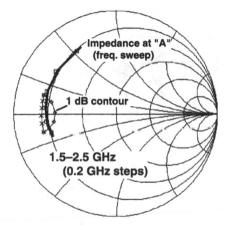

Figure 2.16   Plane A impedance sweep, with 1 dB power contour (single-section matching network).

Looking at the broadband sweep, it can be seen that although the matching topology is narrowband in nature, the small percentage bandwidth is satisfactory for this particular application. Figure 2.17 shows how much greater bandwidth could be obtained, if necessary, by using a double section matching network; the whole band from 1.5 to 2.5 GHz shows a good power match, comfortably nestled inside the 1 dB contour. This kind of match uses extra components and represents a possible overkill for typical wireless communication bandwidths. Nevertheless, it is a dramatic demonstration of the improved bandwidth of a two-section network, and also the utility of the "Plane-A" load-pull reference plane technique. These matching networks will be considered in more detail in Chapter 5.

The single section matching network can be realized using a length of microstrip line and a surface mount capacitor. Typical surface mount (SMT) capacitors, even smaller types such as 0402 and 0201, will have substantial parasitics which need to be included in the circuit simulation at 2 GHz. In this case, a series inductance of 0.6 nH has been included. For a narrowband design of this kind, the effect of such a series parasitic is simply to lower the original design value of capacitance from 4 pF to an effective value of 2.7 pF. Such a simple accommodation of parasitics will cease to be valid for wider band designs, where the solution may be to use lower parasitic components. At this point in the design procedure it is worthwhile to take a preliminary look at component and manufacturing tolerances.

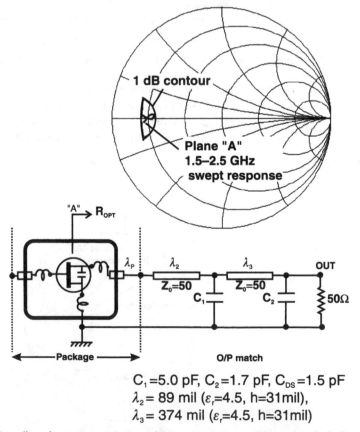

$C_1 = 5.0$ pF, $C_2 = 1.7$ pF, $C_{DS} = 1.5$ pF
$\lambda_2 = 89$ mil ($\varepsilon_r = 4.5$, h=31mil),
$\lambda_3 = 374$ mil ($\varepsilon_r = 4.5$, h=31mil)

Figure 2.17 Broadband power match (matching capacitor parasitics not included).

Since this is a power amplifier, one of the most important issues is how much the power will vary across a range of manufacturing tolerances for the two key matching elements on the output, the capacitor and the microstrip line. Assuming that the width of the line and the dielectric constant of the material will be held to very tight tolerances, the capacitor value and the length of the line (represented by its exact placement) may vary by as much as 10%. Obviously, the larger this percentage, the lower the cost and the higher the yield. Figure 2.18 shows the impedance variation at plane A for a ±10% variation in the output matching capacitor (C_{out}) and the microstrip matching section length (λ_2). Although the range of impedances (shaded area) still clings to the inside of the 1 dB contour in this specific example, it shows that a better centering of the design values exists. This ability to view the projected output power variation for both frequency and element tolerances is a powerful, and not widely recognized, feature of this simple design technique.

Step 3: Design input match using linear (s-parameter) methods.
Having designed the optimum power match on the output side, it is now permissible to complete the design using standard linear techniques. This results in a similar single section lowpass input matching network, as shown in Figure 2.15. The whole amplifier linear response can be plotted out, along with the input and output VSWR, as shown in Figure 2.19. Note that the output VSWR measured in terms of

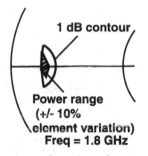

Figure 2.18   Power variation due to manufacturing tolerances on output matching elements.

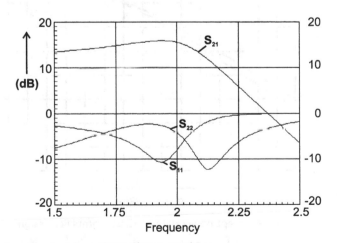

Figure 2.19   Gain and return loss of 1W Class A amplifier.

the overall s_{22} is poor, showing around 3 dB return loss. This is due in part to the reactive input match where k is close to unity, but is mainly due to the power match on the output. Even if the input Q-factor was reduced using a lossy element, the output would remain mismatched to 50 ohms. This indeed is the main difference between the design now completed and a simple linear conjugate matched design; the latter will give good output VSWR but maybe 2–3 dB lower maximum power.

Step 4: Build it!

A linear simulator cannot, of course, give a sweep of the power performance. In general, this would be the point at which the whole design would be analyzed using a nonlinear simulator to remove the idealizations and fine-tune the design values, as a final step in the design procedure, prior to actual board layout. Figure 2.20 shows some measured data taken on an amplifier based on the above design. The GaAs MESFET device is a low-cost packaged part typical of several on the market. The power sweeps at 1.8 and 1.9 GHz show 1 dB compression of at least 29 dBm. The power performance at 1.9 GHz seems to be somewhat better than at the 1.8 GHz design frequency, but the performance is within expectations.

2.8   Conclusions

This chapter has shown that Class A power amplifier design can be reduced to a linear design problem through the use of loadline techniques to determine the optimum power match. Although the material in this chapter may be considered by some to be of an elementary and approximate nature, the basic methods described have been used to design a wide range of RF and microwave amplifiers, from narrowband to

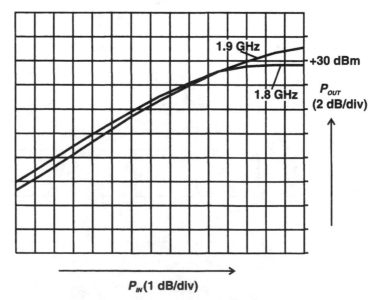

Figure 2.20   Measured swept power data on Class A amplifier design.

broadband, from tens of milliwatts to hundreds of watts, and from low frequencies to above 100 GHz. Those wishing to implement these methods are encouraged to do so without being over-awed by the more advanced concepts introduced in Chapter 3 onward.

References

[1] Cripps, S., "A Method for the Prediction of Load-Pull Power Contours in GaAs MESFETs," *Proc. IEEE Intl. Microw Symp.*, MTT-S, 1983, pp. 221–223.

[2] Kondoh, H., "FET Power Performance Prediction Using a Linearized Device Model," *Proc. IEEE Intl. Microw Symp.*, MTT-S, 1989, pp. 569–572.

[3] Barton, T., et al., "Narrow Pulse Measurement of Drain Characteristics of GaAs MESFETs," *Elect. Lett.*, Vol. 23, No. 13, June 1987, pp. 686–687.

[4] Ladbrooke, P., "Pulsed IV Measurement of Semiconductor Devices," available from Accent Inc., sales@accentopto.com.

CHAPTER 3

Conventional High Efficiency Amplifier Modes

3.1 Introduction

This chapter will introduce what will be termed conventional, or reduced conduction angle, high efficiency amplifier modes. These are the familiar Class AB, Class B, and Class C configurations.

The concept of making a more highly efficient RF amplifier by biasing the active device to a low quiescent current and allowing the RF drive signal to swing the device into conduction is very old, dating back to the earliest days of vacuum tubes. For this reason, it is often considered to be an elementary subject, not worthy of extensive discussion. In fact, we will see that there are many issues that come out of some elementary analysis, based on ideal device models, that are of great significance in the context of modern wireless communications systems.

Straightforward but more detailed analysis than is often presented in elementary textbooks will show that merely reducing the conduction angle of an RF power device is necessary, but frequently not sufficient to obtain a useful improvement in efficiency. In general, it is necessary to increase the drive level substantially from the Class A condition and to provide suitable impedance terminations at harmonics of the signal frequency. Most older textbooks assume that all higher harmonics will be shorted at the output of the PA device. This simplifies the analysis and was a much easier condition to realize in the days of tube amplifiers. This has led to some confusion regarding the kinds of matching topologies which should be used for today's transistor counterparts, and this subject will be discussed in more detail in Chapter 4.

Today's wireless systems impose challenging requirements on the PA designer. A mobile phone handset PA has to be as efficient as possible in order to conserve battery power. Base stations also have efficiency specifications due to power and cooling limitations. One of the most important concepts in comparing different PA configurations is what will be termed "power utilization factor" (PUF). This term will be defined in this chapter, but is basically the ratio of RF power delivered by a device in a particular mode under consideration to the power it would deliver as a simple Class A amplifier. PUF is a statement on cost efficacy, or "watts per dollar." There is limited use in the solid state world for a high efficiency configuration which delivers half of the RF power that the device in question can supply in a Class A mode. Even if the efficiency is 95%, two chips, or one chip of twice the periphery, have to be used to meet a given output power specification. This is not just an issue of spending twice as much either; the larger chip will have lower RF gain and less matching bandwidth.

Wireless systems also typically use modulation schemes that have variable envelope amplitude. This subject will be considered in some depth in later chapters, but it will be important in analyzing conventional high efficiency modes. For this reason, the behavior of Class AB modes over a range of input power levels will be considered. This is the subject of Section 3.5 and carries one or two surprises.

3.2 Reduced Conduction Angle—Waveform Analysis

The basic process of reducing the conduction angle is illustrated in Figure 3.1. The device is biased to a quiescent point beyond the Class A condition, toward cutoff. It is clear that a sufficiently large amplitude of RF drive will swing the device beyond its cutoff point V_t, on the negative portion of the RF cycle. It is also clear that in order for the current to swing up to the idealized saturation point, I_{max}, the drive level has to be increased from the Class A condition. In quantitative terms, the required signal voltage amplitude will be

$$V_s = \left(1 - V_q\right) \tag{3.1}$$

where V_q is the normalized quiescent bias point, defined according to

$$V_t = 0, V_o = 1$$

Initially, it will be assumed that as the quiescent point is varied, the signal voltage will be increased, according to (3.1), so that a peak current of I_{max} is maintained. It will also be assumed, for the time being, that the device is ideally transconductive, and that the output voltage is being maintained above zero to keep the device turned on. Considerations of output RF loading will be considered in detail in later sections.

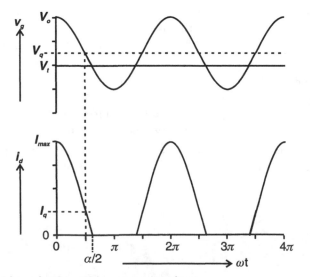

Figure 3.1   Reduced conduction angle current waveform.

So the current in the device has the familiar looking, truncated sinewave appearance. The conduction angle, α, indicates the proportion of the RF cycle for which conduction occurs. In this chapter we will elect to use a cosine function to describe the current waveform.[1] Due to the symmetry of a cosine function about zero on the time axis, there has been some confusion about the definition of the conduction angle. In this book, we will adopt a possibly nonstandard, but more mathematically sound, definition that α represents the entire angle of conduction, including the equal contributions either side of the zero time point. So the current cutoff points are at $\omega t = \pm \alpha/2$. We can now define the classical modes of operation, in terms of quiescent bias point and conduction angle, as shown in Table 3.1.

Looking at the current waveform in Figure 3.1, it is quite intuitive that the mean component, or DC supply, will decrease as the conduction angle is reduced. It is less obvious what happens to the fundamental component, and clearly there will be harmonics generated as well. The answers can be found by Fourier analysis of the waveforms.

The RF current waveform can be written as:

$$i_d(\theta) = I_q + I_{pk} \cdot \cos\theta, \quad -\alpha/2 < \theta < \alpha/2;$$
$$= 0, \quad -\pi < \theta < -\alpha/2; -\alpha/2 < \theta < \pi$$

where $\cos(\alpha/2) = -\left(\dfrac{I_q}{I_{pk}}\right)$, and $I_{pk} = I_{max} - I_q$,

so $i_d(\theta) = \cdot \dfrac{I_{max}}{1 - \cos(\alpha/2)} \cdot (\cos\theta - \cos(\alpha/2))$.

The mean current, or DC component, is given by

$$I_{dc} = \frac{1}{2\pi} \cdot \int_{-\alpha/2}^{\alpha/2} \frac{I_{max}}{1 - \cos(\alpha/2)} \cdot (\cos\theta - \cos(\alpha/2)) \cdot d\theta$$

and the magnitude of the nth harmonic is

$$I_n = \frac{1}{\pi} \cdot \int_{-\alpha/2}^{\alpha/2} \frac{I_{max}}{1 - \cos(\alpha/2)} \cdot (\cos\theta - \cos(\alpha/2)) \cdot \cos n\theta d\theta$$

Table 3.1   Classical Reduced Conduction Angle Modes

Mode	Bias Point (Vq)	Quiescent Current	Conduction Angle
A	0.5	0.5	2π
AB	0–0.5	0–0.5	$\pi-2\pi$
B	0	0	π
C	< 0	0	$0-\pi$

---

1.   We will, however, still use "sinusoidal" as a generic qualitative description; consinusoidal isn't really an English word.

there being no quadrature components due to the selection of an even function for the RF current waveform.

The results from the evaluation of these integrals up to $n = 5$ are shown in Figure 3.2, but the all-important cases of I_{dc} and I_1 are:

$$I_{dc} = \frac{I_{max}}{2\pi} \cdot \frac{2 \cdot \sin(\alpha/2) - \alpha \cdot \cos(\alpha/2)}{1 - \cos(\alpha/2)} \tag{3.2}$$

$$I_1 = \frac{I_{max}}{2\pi} \cdot \frac{\alpha - \sin\alpha}{1 - \cos(\alpha/2)} \tag{3.3}$$

Examining the curves of Figure 3.2 more closely, it is clear that the DC component decreases monotonically as the conduction angle is reduced. In particular, the class B condition, $\alpha = \pi$ in (3.2), gives

$$I_{dc}(\text{Class B}) = I_{max}/\pi;$$

This can be compared to the Class A condition, where

$$I_{dc}(\text{Class A}) = I_{max}/2;$$

Furthermore, the fundamental component for the Class B condition, $\alpha = \pi$ in (3.3), gives

$$I_1(\text{Class B}) = I_{max}/2;$$

which is, of course, the same as the fundamental component in the Class A condition. So from the viewpoint of current waveforms only, there appears to be a possi-

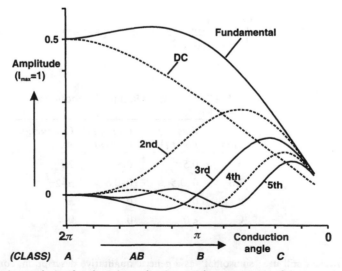

Figure 3.2 Fourier analysis of reduced conduction angle current waveforms.

bility of decreasing the DC supply power by a factor of $\pi/2$, without changing the RF fundamental component. In other words, the efficiency should increase from 1/2 in the Class A mode, to $\pi/4$ (about 78.5%) in Class B. But before these possibilities can be realized, the output termination and voltage waveform has to be considered.

For conduction angles lower than π, corresponding to Class C operation, the DC component continues to drop, but the fundamental component of current also starts to drop below its Class A level. This will result in a mixed blessing of yet higher efficiency, but lower PUF; the fundamental power will be lower than the Class A rating of the transistor. This will be discussed in more detail in a later section.

The first few harmonic amplitudes are also plotted in Figure 3.2. Note that throughout the Class AB range, and up to the midway Class B condition, the largest harmonic, other than the fundamental, is the second. It is positive, or in-phase, with the fundamental, and from a frequency domain viewpoint, the action of the partially cutoff transistor is to generate a substantial amount of second harmonic, which reduces the dips of the fundamental sinewave and sharpens the peaks, result-ing in a lower mean level for approximately the same overall peak-to-peak ampli-tude. We can note, in passing, that using the cutoff characteristic of a transistor is not the only possible method for generating such a desirable second harmonic com-ponent. It may be possible, for example, to tailor the frequency response of the input matching network such that the drive voltage has some appropriate in-phase second harmonic enhancement, assuming it is coming from another, possibly nonlinear, amplifier driver stage. This has some interesting possibilities (see Section 3.7), since we will see shortly that the conventional method of using the transistor cutoff char-acteristic is rather wasteful in terms of drive power requirements. This is an espe-cially important consideration at GHz frequencies, where power transistor gain is usually at a premium.

The odd harmonics (Figure 3.2) can be seen to pass through zero at the Class B point, but in AB mode, the third harmonic is certainly not negligible. In general, however, at levels lower than about 0.1 (magnitude) there will be weakly nonlinear components of the transconductance characteristic, which may substantially change the overall picture in terms of linearity. But it can be noted here that a small bias adjustment around the Class B ($\alpha = \pi$) point can be considered as a viable method of controlling the precise level of the third harmonic component; this will be significant in later work on Class F amplifiers.

3.3 Output Termination

It is clear that the transistor current contains some significant harmonic compo-nents, especially second harmonic. It is therefore a more complicated problem to analyze and draw the resulting voltage for a given load condition. In the first instance, the work can be greatly simplified by assuming that all harmonics are pre-sented with a short circuit; indeed, this condition forms part of the definition of clas-sical, or conventional, high efficiency amplifier modes. We will see that for modern solid state designs, this assumption will be somewhat idealistic. Most older text-books, however, make the assumption for three reasons:

1. A harmonic short, along with an optimum resistive termination at the fundamental, will always represent a close approach in performance to more optimum solutions involving multiple harmonic terminations (some of these alternative solutions will be considered in later sections and Chapters 4 and 5).

2. A harmonic "short" was fairly easy to realize in older tube PAs, where the loadline resistance, even for a kilowatt device, was measured in kΩ; a simple parallel resonant circuit would do the job easily.

3. The analysis becomes trivial, due to the reduction of the output RF waveform to a sinusoid.

We will therefore proceed with the analysis of the conventional modes using this classical simplifying assumption, but we will need to scrutinize its implications and ultimately modify it.

Figure 3.3 shows the circuit for further analysis. The reduced conduction angle current waveform flows through an AC-coupled load, which consists of a funda-

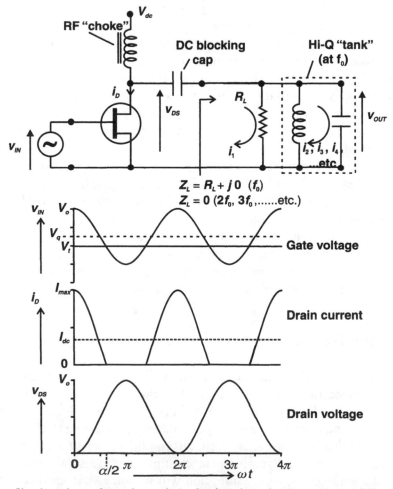

Figure 3.3 Circuit and waveforms for analysis of reduced conduction angle amplifier modes.

mental load resistor, and a conceptual harmonic short, indicated here by a shunt connected parallel resonant circuit at the fundamental. All harmonics of the load are shorted and generate no voltage, so the drain voltage is a sinewave whose magnitude will be set by the load resistor value to generate the maximum permissible voltage swing, in a similar fashion to the previous consideration of a Class A amplifier. In this analysis, we will assume that the maximum voltage swing will be generated at the input drive level corresponding to a peak current of I_{max}. With these assumptions, and an appropriate choice of load resistor, the power output and efficiency can be calculated directly from the curves of Figure 3.2, given that the RF output voltage at the transistor will in every case be a sinewave of amplitude V_{dc}.

So the RF fundamental output power is given by:

$$P_1 = \frac{V_{dc}}{\sqrt{2}} \cdot \frac{I_1}{\sqrt{2}} \tag{3.4}$$

where I_1 is given by (3.3), and the DC supply is given by:

$$P_{dc} = V_{dc} \cdot I_{dc} \tag{3.5}$$

where I_{dc} is given by (3.2).

The output efficiency is defined by:

$$\eta = \frac{P_1}{P_{dc}} \tag{3.6}$$

Unless stated otherwise, this definition of efficiency will be used in this book. This usually offends system designers, who prefer to take account of the drive power required, which in an RFPA is quite substantial. This leads to an alternative definition, the so-called "Power Added Efficiency" (PAE):

$$PAE = \frac{P_1 - P_{IN}}{P_{dc}} \tag{3.7}$$

where P_{IN} is the RF drive power. Generally speaking, if the RF power gain is less than 10 dB, then the drive power requirements will start to take a serious bite out of the drain, or output, efficiency of a PA stage, and the higher the efficiency the more serious the effect. On the other hand, a designer often has options for increasing gain, either by narrowing bandwidth, possibly judicious application of some positive feedback, better packaging, or even use of a "hotter" technology. In this respect, it is just as well to keep gain and output efficiency as separate entities, with the full knowledge that at system level, or even multistage PA level, they will behave interactively on the overall efficiency.

So the results of (3.2) to (3.6) can be plotted to show the overall variation of output power and efficiency as a function of conduction angle. These results are plotted out in Figure 3.4. Ideal conditions have been assumed: a perfect harmonic short, maximum linear current swing up to I_{max}, and maximum voltage swing of $2 \cdot V_{dc}$, with no knee region. Under these conditions, the optimum value of load resistor will be:

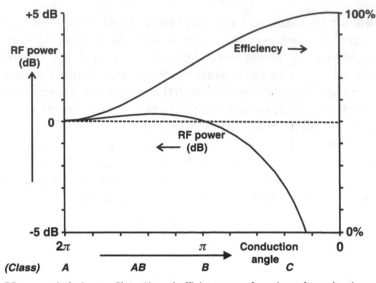

Figure 3.4   RF power (relative to Class A) and efficiency as a function of conduction angle; optimum load and harmonic short assumed.

$$R_{OPT} = V_{dc}/I_1 \qquad\qquad (3.8)$$

where I_1 is the fundamental Fourier component of current, given by (3.3) and shown plotted in Figure 3.2. We have already noted that the Class B condition has the same fundamental component as Class A, provided that the drive voltage magnitude is increased to obtain the same RF peak current value (I_{max}). The load resistor value for Class B is therefore, conveniently, the same as the linear loadline value discussed extensively in the last chapter. Theoretically, this value should be reduced somewhat for Class AB operation: about 10% lower in the mid-AB range to allow for the small increase in the fundamental amplitude from the Class A value. In practice, considerations of weakly nonlinear effects and a more realistic knee turn-on region will modify the actual optimum by at least 10%, so the theoretical variation over the AB region is often ignored initially.

The main features of Class AB, B, and C operation can be determined from the curves shown in Figure 3.4, and they will be discussed in some more detail in the following sections. But in a preemptive summary, the following points can be made:

1. Between Class A and Class B operation, the fundamental RF output power is approximately constant, showing a few tenths dB increase in the mid-AB range over the Class A power output.

2. The Class B condition delivers the same power as Class A, but with a DC supply reduced by a factor of $\pi/2$ compared to Class A, giving an ideal efficiency of $\pi/4$.

3. The Class C condition shows an ever-increasing efficiency as the conduction angle is reduced to low values; this is, however, accompanied by a substantial reduction in RF output power.

3.4   Reduced Conduction Angle Mode Analysis—FET Model

This section will further illustrate and analyze the various conventional reduced conduction angle modes by considering some specific cases. The cases are analyzed using an Excel spreadsheet (**AB_Waves**), which computes the Fourier components of the RF current waveforms for a given set of conditions for selected conditions of bias and input signal amplitude. Initially, idealized device characteristics are assumed.

The program requires three input parameters, all of which are normalized voltages:

- V_q is the input DC bias point, normalized (as throughout this book) between 0 (cutoff, threshold, or pinchoff) and 1 (saturation, or open-channel). The program assumes an ideal strongly nonlinear transconductance model, and the range of input voltage can exceed these limits, where cutoff or hard saturation, respectively, are assumed.
- V_s is the amplitude of the input signal, assumed to be sinusoidal. In order to obtain maximum current swing (i.e., to swing up to I_{max}), the value of V_s should be chosen such that:

$$V_s = 1 - V_q$$

- R_L is the normalized value of fundamental load resistance. This value is selected by the user R_L being normalized to the Class A loadline value of unity.

So the device current i_d is defined to be

$$i_d = g_m \cdot \frac{I_{max}}{2} \cdot \left(V_q + V_s \cos\theta\right), \left(V_q + V_s \cos\theta\right) > 0,$$

$$= 0, \left(V_q + V_s \cos\theta\right) < 0$$

where the transconductance g_m is included only to avoid a dimensional conflict and is assumed to have a normalized value of unity. For the purposes of this computation, the peak current I_{max} is also normalized to unity.

For a selected value of V_q, selections of V_s values can be made between zero and the maximum drive level $(1-V_q)$. The current waveform is constructed, and the DC and fundamental components, I_{dc} and I_1 are computed. If an output harmonic short is assumed, then the RF output power is simply

$$P_{RF} = \frac{I_1^2 R_L}{2} \tag{3.9}$$

which can be compared directly with the Class A power, which will have a normalized value of ½. The corresponding DC supply current, I_{dc}, can also be compared directly with the Class A normalized value of $I_{max}/2$, or ½. Thus the efficiency at the maximum drive level for each value of V_q can be computed relative to the Class A condition,

$$\eta = \frac{1}{2} \cdot \frac{I_1}{I_{dc}} \qquad (3.10)$$

which assumes that the full rail-to-rail voltage will be permitted, and the load resistor will have a value

$$R_L = R_A \cdot \frac{I_1}{\frac{1}{2}} \qquad (3.11)$$

where R_A is the Class A loadline resistor value.

When computing power and efficiency for "backed-off" signal drive levels, it is normal to assume the load resistor remains constant, at its optimum value for the peak power condition. Thus the power in the backed-off condition can be computed using (3.9) and the efficiency will be given by

$$\eta_{pbo} = \eta_{max} \frac{I_{dc\,max}}{I_{dc}} \frac{I_1}{I_{1\,max}} \qquad (3.12)$$

where the "max" suffix refers to the respective values of I_{dc} and I_1 at the maximum drive level. Note that the backed-off efficiency is *not* given by (3.10), which gives the efficiency for the impractical case where the load resistor can be increased during power backoff to maintain a rail-to-rail voltage swing.

Case 1: Class A

The Class A condition provides a starting point for checking the output display and also a reference point for power and efficiency. The required inputs for the ideal Class A condition are:

$$V_q = 0.5; V_s = 0.5; R_L = 1.0$$

and the resulting output is shown in Figure 3.5.

This display shows no surprises and corresponds to an ideal loadline-matched linear amplifier, showing 50% output efficiency. At full drive, the power meter shows 0 dB, indicating a PUF of unity. Other subsequent cases will measure power relative to this condition. With the sinusoidal output voltage amplitude of V_{dc}, due to the optimum selection of load resistor, the output efficiency is 50%.

Figure 3.5(b) shows the same amplifier with a drive level reduced by 3 dB. For the present sequence, it is assumed that the input impedance of the transistor remains constant and so a 3 dB decrease of input power corresponds to a factor of √2 reduction in voltage swing amplitude. It is also assumed that the output load conditions (and in particular, the value of load resistor) will remain constant as the input power is varied. So in this case, the output voltage and current swing will each be reduced by the same factor of √2. The resulting output power therefore drops by 3 dB, as expected for a linear amplifier. Note however, that the efficiency drops to an alarming 25% value in the 3 dB backoff condition. We note here, in passing, that the maximum voltage swing can be restored, with a corresponding increase in RF

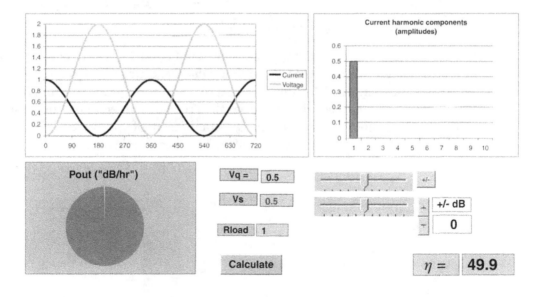

(a)

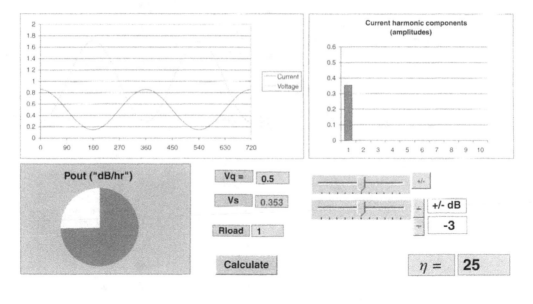

(b)

Figure 3.5   Class A mode: (a) full drive, and (b) 3 dB backoff conditions.

power and efficiency, if the load resistor is increased by a factor of √2 for the reduced drive condition. But this would then represent a new and lower maximum linear power; the load resistor cannot simply be changed as the drive level varies.

Case 2: Class AB

The input parameters for a mid Class AB condition are:

$$V_q = 0.3; \; V_s = 0.3; \; R_L = 0.94$$

and the resulting output waveforms are shown in Figure 3.6. Note that the value of R_L is now reduced from the Class A loadline value, reflecting the higher fundamental current component. The RF output shows a small increase (0.25 dB) from the Class A condition (indicated by a shading reversal on the power meter box) and is accompanied by a significant reduction in DC component of the device current. This causes a useful increase in efficiency (68%). Such beneficial effects come only at the

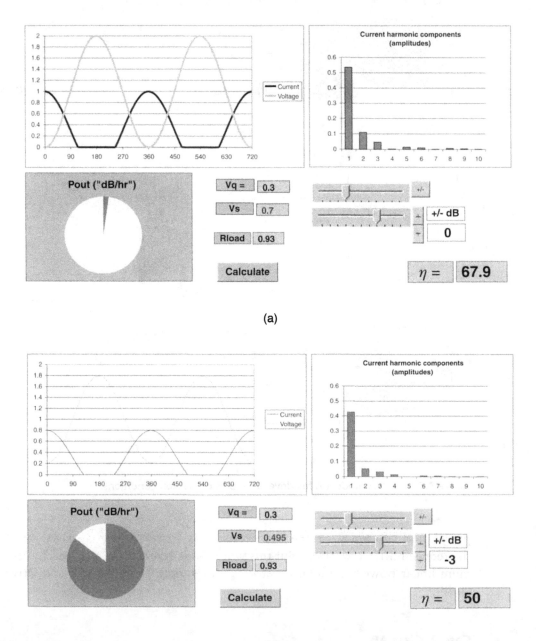

(a)

(b)

Figure 3.6　Class AB operation, $I_q = 0.3$: (a) full drive, and (b) 3 dB backoff.

expense of drive power. The increase in V_s from 0.5 to 0.7, as compared to Class A operation, translates ideally into about 3 dB extra drive power, which can be considered as a reduction in overall power gain from the Class A mode. If the linear gain starts off below 10 dB, then the Power Added Efficiency (PAE) will start to show a markedly less attractive increase.

There is an important issue in ideal Class AB operation that is often overlooked. Figure 3.6(b) shows the effect of a 3 dB reduction in drive level, and unlike the Class A case, *the output power does not show a corresponding 3 dB drop.* In other words, this is not a linear amplifier; a signal with an amplitude-modulated envelope will be distorted significantly at this peak power level. The reason for this is the simple mathematical fact that in Class AB operation, the conduction angle is a function of drive level, as well as the quiescent bias point. This may come as a surprise to PA stalwarts who often find experimentally that Class AB can offer a useful linear dynamic range, similar to Class A operation. This apparent conflict with practice will be explained later (see Section 3.5), but will not in any way invalidate the present observation.

Case 3: Class B

The input parameters for a Class B condition are:

$$V_q = 0; V_s = 1.0; R_L = 1.0$$

and the resulting waveforms are shown in Figure 3.7. The RF power has returned to its original (Class A) value, corresponding to a PUF of unity. The DC supply is reduced by a factor of $2/\pi$ compared to the Class A condition, resulting in an efficiency of $\pi/4$, or about 78.5%; a classical result.

The downside is that in theory 6 dB more drive power is needed to achieve the Class B conditions shown. This is a large reduction in power gain at RF and microwave frequencies, where even linear gains much above 10 dB are quite a luxury. On the one hand, it points toward the use of a higher gain device technology for the frequency in use. The widespread use of GaAs HBT and PHEMT technology for PAs in cellular phone handsets below 2 GHz is an illustration of this problem. On the other hand, it should also be fairly stated that the BJT device fares somewhat better in power gain reduction than the FET devices so far considered. In order to generate the necessary current waveform, it has already been noted that nonlinearities of the right kind on the input might be harnessed in order to reduce the heavy drive requirements of sinusoidal signals. The BJT does this, to some extent, through the I-V characteristic of its input diode junction, and the reduction in Class B power gain may be as little as 2 dB. Even with FET devices, it may be possible to contrive a way of adding some of the second harmonic, which is generated in the output, to the input drive signal. This would also have the effect of reducing the amplitude of fundamental drive signal required or Class B operation. This issue will be considered in more detail in Section 3.7.

The Class B mode shows a welcome benefit, however, in its linearity under reduced drive conditions, as shown by the 3 dB backoff condition in Figure 3.7(b). Due to the symmetry of the drive signal about the pinchoff level, the conduction

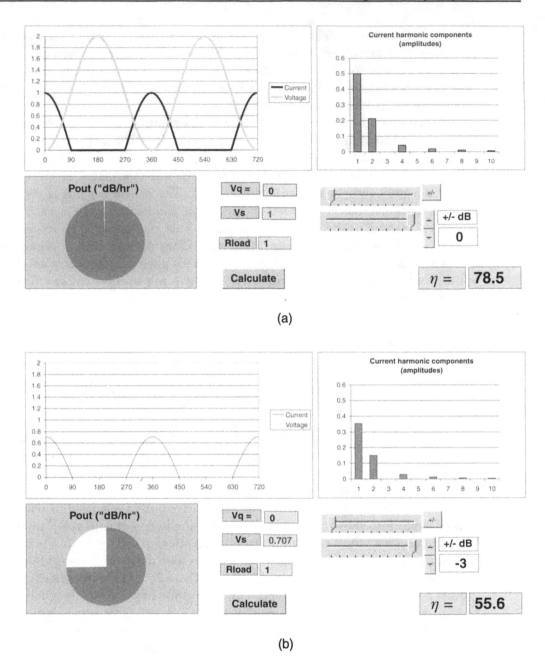

Figure 3.7   Class B operation, $V_q = 0$: (a) full drive, and (b) 3 dB PBO.

angle remains constant for varying drive levels. So a 3 dB reduction in drive power results in a corresponding 3 dB reduction in output power, and the Class B amplifier appears to behave in a linear manner. With a fixed value of load resistor, the efficiency at the 3 dB backoff condition is reduced to 55%. The output power and efficiency could be increased, but at this drive level only, by increasing the load resistor value and allowing a larger voltage swing. In fact, if the voltage swing is allowed to increase all the way back to its maximum value (using a value of $\sqrt{2}R_{opt}$), the efficiency will be back at 78.5%. Although this may appear impractical at first sight, it

opens up some possibilities for improving efficiency over a variable signal amplitude range, such as is encountered with amplitude modulated signals. This was first
recognized in a classical paper by Doherty in 1936 [1], which will be discussed in
detail in Chapter 10.

It should be noted that "linearity" is being defined here in terms of the final
sinusoidal output signal level, with the assumption that all harmonic voltages are
short-circuited. At first sight it may seem inappropriate to describe a device operating in Class B as an amplifier at all, let alone a linear one. Clearly, highly nonlinear
processes are involved internally, but the power input-output characteristic shows
linear behavior.

Case 4: Class C

The input parameters for a "mid" Class C condition are:

$$V_q = -0.6; \; V_s = 1.6; \; V_k = 0; \; R_L = 1.2$$

and the resulting output, at maximum drive and the −3 dB condition, is shown in
Figure 3.8.

The current waveform starts to take on the appearance of a train of short pulses,
which have low DC component, but also have a lower fundamental RF component
than in the Class AB cases. Consequently, very high efficiencies can be obtained, but
at the expense of lower RF output power and very heavy input drive requirements.
The 3 dB backoff condition, Figure 3.8(b), also shows a highly nonlinear response
with a large gain expansion. One of the major problems in utilizing Class C modes is
the large negative swing of input voltage, which coincides with the drain/collector
output voltage peaks. This is precisely the worst condition for reverse breakdown in
any kind of transistor, and even small amounts of leakage current flowing at this
point of the cycle will have a detrimental effect on the efficiency. For this reason,
along with the rapidly decreasing PUF and power gain, true Class C operation has
not often been used in solid state amplification at higher RF and microwave frequencies. Some renewed interest is currently apparent, mainly due to the possibility
of driving the RF power device using a digital signal. This topic will be considered
further in Chapters 7 and 10, however Figure 3.9 shows a comparison between the
theoretical power and efficiency for classical Class AB and C modes, and a conceptual device driven such that the current waveform is a rectangular pulse, rather than
the cap of a sinewave. Clearly, this concept appears to have useful potential.

Cases (1) through (4), and the accompanying 3 dB PBO linearity tests, give a
somewhat restricted view of the linearity characteristics of reduced conduction
angle modes. Figure 3.10 shows a more generalized set of characteristics for a representative range of quiescent bias settings and a 20 dB input power backoff range.
These plots have been computed using the procedure defined in (3.9) through
(3.12). With an ideal FET model, the Class A ($V_q = 0.5$) and Class B ($V_q = 0$) conditions give perfectly linear responses, but the intermediate Class AB settings show
substantial gain compression. Thus we see that the ideal FET model is not an optimum characteristic for Class AB linearity. This is an important observation which is
usually overlooked. In practice the transconductance of the device will often show

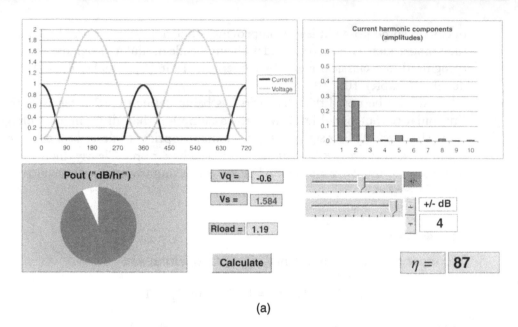

(a)

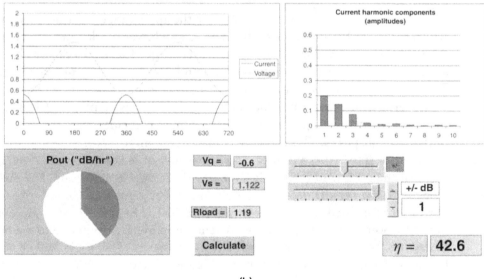

(b)

Figure 3.8   Class C operation, $V_q = -0.6$: (a) full drive, and (b) 3 dB backoff.

some gain expansion in the active region, which will tend to reduce, or in some cases actually cancel, the gain compression caused by the reduced conduction angle process. This subject will be considered in more detail in Chapter 4.

Figure 3.11 shows the corresponding output power versus efficiency characteristics for the same set of quiescent bias settings. The Class A PBO shows an efficiency which decreases in inverse proportion to the power, and the Class B case shows a roll-off which is proportional to the square root of the power. Intermediate cases transition between these two characteristics and most notably revert to the inverse power law when the signal drops below the level where truncation starts.

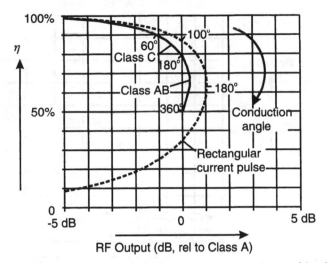

Figure 3.9 Power/efficiency sweeps for ideal Class AB/C PAs at maximum drive (solid); equivalent plot for rectangular pulses also shown (dotted).

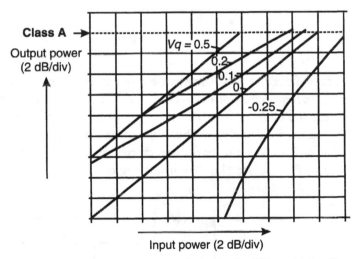

Figure 3.10 Linearity of reduced conduction angle modes for FET model; loadline match at the peak power level and ideal output harmonic short-circuit are assumed.

3.5 Reduced Conduction Angle Mode Analysis—BJT Model

The analysis of reduced conduction angle modes has thus far assumed an idealized FET characteristic. The BJT is a very different device in physical terms but has had much the same Class AB/C concepts applied to it, and experience seems to indicate that similar results can be obtained. This section therefore analyzes BJT reduced conduction angle modes in much the same spirit of idealization that was used for the FETs.

Although the BJT is widely regarded as a current amplifier in linear IC design, it is more appropriate to use a transconductive model for PA analysis. If the output current in a BJT is represented by the relation

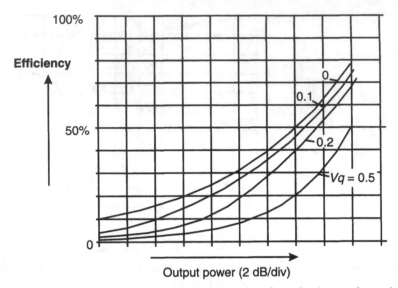

Figure 3.11 Efficiency versus power backoff (PBO) for reduced conduction angle modes, FET model.

$$i_c = I_{max} e^{k(v_{be}-1)}$$

(3.13)

the collector current i_c will show an exponential dependency on the base-emitter voltage v_{be}, as required by physical considerations. The value of k can be chosen such that the transfer characteristic can be plotted on the same normalized input voltage scale used in the FET analysis. This is illustrated in Figure 3.12, which shows that a value of $k = 2$ makes the two characteristics look at least functionally similar. Whereas the FET device shows a linear region after it turns on abruptly, the BJT

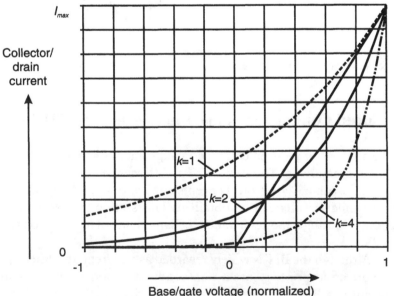

Figure 3.12 Exponential BJT transfer characteristic, compared to ideal FET.

device turns on more smoothly but has no such linear region. Higher values will make the turn-on more abrupt, but will give a more nonlinear active region. Clearly, all three k values plotted in Figure 3.12 could be the same device if the normalization of the input voltage scale was changed appropriately.

Reduced conduction angle modes for a BJT can be examined for various cases of quiescent bias and sinusoidal input voltage amplitude,

$$v_{be} = V_q + V_s \cos \omega t \tag{3.14}$$

where as before in the FET analysis, the quiescent bias V_q and the signal amplitude V_s are measured on a scale of zero to unity, the unity value of v_{be} giving the maximum permissible collector current, I_{max}, in accordance with (3.13).

Unfortunately, the ongoing analysis of a BJT in Class AB mode is more questionable in that some additional assumptions have to be made concerning the device input voltage, v_{be}. In the case of a FET device, it has been assumed thus far that the input impedance is essentially an open circuit, and more especially that the input voltage is a linearly scaled version of whatever external signal is being applied. This may not be such a good assumption in the BJT case, since the device input presents a nonlinear resistance to the external load. Ironically, this extra nonlinear effect can turn out to be helpful in linearizing the response (see Chapter 4 and [2]), but it does add significant complexity, which in this initial analysis will be ignored; v_{be} will for now be assumed to represent a linear scaling of the sinusoidal input signal.

Substituting the expression for v_{be} (3.10) into (3.9), we obtain

$$i_c = I_{max} e^{k(V_q - 1)} \cdot e^{kV_s \cos \theta}, (\theta = \omega t)$$

and expanding the second exponential, this becomes

$$i_c = I_{max} e^{k(V_q - 1)} \left(\begin{array}{l} 1 + (kV_s) \cos \theta + \dfrac{(kV_s)^2}{2} \cos^2 \theta + \dfrac{(kV_s)^3}{6} \cos^3 \theta + \\[2mm] \dfrac{(kV_s)^4}{24} \cos^4 \theta + \dfrac{(kV_s)^5}{120} \cos^5 \theta + \ldots \end{array} \right) \tag{3.15}$$

So as kV_s decreases below unity, the device characteristic will appear increasingly linear. For $kV_s > 1$, an increasing number of higher degree terms in (3.15) take on significant values, although for $kV_s = 2$ it can be noted that the seventh degree coefficient is 8/315, or about 1 percent of the fundamental component coefficient. The value $kV_s = 2$ is significant in the $k = 2$ case shown in Figure 3.12, since it represents the BJT equivalent of a Class B condition.

It would be possible to expand the cosine powers in (3.15) and obtain power series expressions for the output current in terms of the DC, fundamental, and harmonic components. With statutory reluctance, however, the direct computation option will be taken here. Figure 3.13 shows the power transfer characteristics for a selection of V_q values. As with the corresponding computation for the FET model (Figure 3.10), it has been assumed that the output load resistor has been selected to give a full rail-to-rail swing of voltage at the maximum drive point for each V_q setting, and that all harmonic voltages are short-circuited. The corresponding PBO

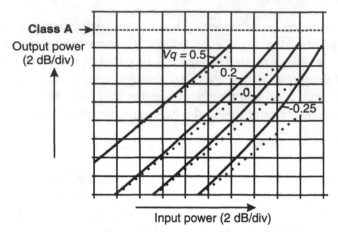

Figure 3.13 BJT Class AB power transfer characteristics (same scales and equivalent normalization as in the FET case, Figure 3.10).

efficiency plots are shown in Figure 3.14. The quasi–Class A condition $V_q = V_s = 0.5$ shows reasonable, although not perfect, linearity, with a peak efficiency of about 45%. Lower quiescent settings show an increasing gain expansion characteristic at higher drive levels. The quasi–Class B condition $V_q = 0$, $V_s = 1$ shows a peak efficiency of 70%, along with about 4 dB gain expansion in the upper 6 dB power range.

Although these results can be characterized as being generically similar to the comparable FET computations, there are some clear distinctions. In particular, the exponential characteristic of the BJT causes a large reduction in small-signal gain as the quiescent bias setting is reduced, and the lower quiescent settings display a disconcerting amount of gain expansion. It will be shown in Chapter 4, following [2], that if some allowance is made for the interaction of the input termination and the nonlinear base-emitter junction impedance, much more linear responses can be obtained. This confirms the general observation that BJTs (and especially the mod-

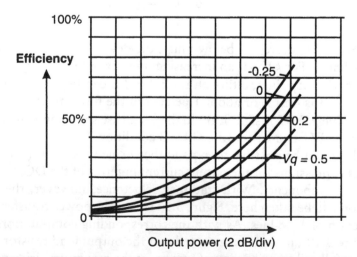

Figure 3.14 BJT Class AB efficiency versus output power (compare with equivalent FET plot, Figure 3.11).

ern GHz derivative, the HBT) can outperform FETs in terms of the tradeoff between linearity and efficiency for variable envelope signals. In any event, it can be stated that gain expansion is an easier form of linearity to correct than gain compression, and there are numerous mechanisms in a practical design which will almost inevitably cause some cancellation and nulling effects in the IM or ACP response of a Class AB BJT PA.

3.6 Effect of I-V "Knee"

The analysis in the previous sections can now be extended to consider the effects of a more realistic knee voltage on the above results. These effects will be very similar for FET and BJT devices. In transistor amplifiers, the knee voltage will always be a significant percentage of the DC supply, and in portable battery-powered PA applications the effect of the knee voltage can dominate the whole design strategy.

Figure 3.15 shows the modified ideal FET I-V characteristics, which include a realistic turn-on region. The V_k parameter is normalized to the DC supply V_{dc}, and the transistor current has the modified form:

$$I_d = v_g \cdot g_m \cdot I_{max}\left(1 - e^{-\left(\frac{v_{ds}}{V_k}\right)}\right) \tag{3.16}$$

where v_{ds} is the drain-source (output) voltage, also normalized to V_{dc}.
So when $v_{ds} = V_k$,

$$I_d = v_g \cdot g_m \cdot I_{max} \cdot \left(1 - e^{-1}\right)$$

and the current at the knee voltage will be reduced to about 63% of its saturated value at higher voltages. If the value of V_k is normalized to the DC supply voltage, a

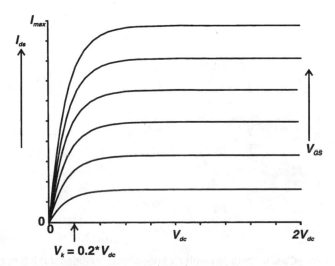

Figure 3.15   Knee region modification to ideal I-V characteristics.

value of 0.1 will represent a realistic value for many practical cases. Such cases would be a MESFET or BJT operating at 5v supply, or an LDMOS device having a 4v knee operating at 24v supply.

The device current is now a function of both the input and output voltages, and prediction of the output voltage becomes a recursive problem. This difficulty can be circumvented by specifying an output voltage swing. The current can then be computed using the basic transconductance relationship multiplied by the drain voltage dependency, as given in (3.16). The output voltage amplitude parameter, V_{out}, is normalized to the DC supply voltage V_{dc}.

For a Class B amplifier, $V_q = 0$; $V_s = 1.0$; $V_k = 0.1$; $V_{out} = 1.0$.

The waveforms for the above set of conditions are shown in Figure 3.16(a). This should be compared directly with the ideal (zero-knee) Class B condition shown in

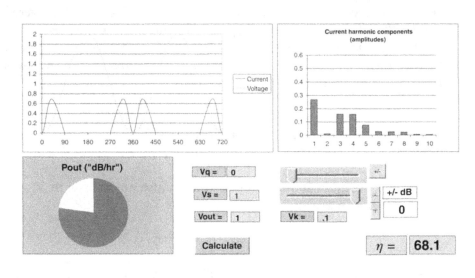

(a)

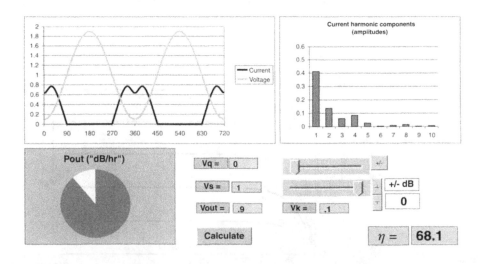

(b)

Figure 3.16   Class B operation with I-V knee effects included: (a) full output RF voltage swing, and (b) swing reduced by 10%.

Figure 3.7. The striking feature is the bifurcated current waveform; this is a logical result of having the drain voltage drop down to zero. Although the bifurcated current pulse actually has a relatively low DC component, the RF output power is more than 2 dB lower than the ideal case. The efficiency, however, is still holding up at 70%.

This represents a major loss of RF power and PUF. The obvious first step toward retrieving some lost performance is to reduce the output voltage swing, so that the current will not drop down to zero. Figure 3.16(b) shows the result for the following set of input parameters:

$$V_q = 0; V_s = 1.0; V_k = 0.1; V_{out} = 0.9$$

The RF output power has recovered to within 1 dB of its ideal value, and the efficiency stays, fortuitously, at the same value. Further experimentation with the V_{out} parameter shows that this is close to an optimum condition: about 1 dB less than ideal power, and less than 10% down in ideal efficiency. (Further reduction V_{out} causes the efficiency to drop for only marginal inrease in RF power.) Ironically, or perhaps serendipitously, the resulting value for the output load resistor comes out to be 0.98, essentially the same value predicted using the most ideal model. This kind of result, where different idealities in simple models cancel each other's detrimental effects, is by no means uncommon and is a strong argument in favor of this kind of analysis.

3.7   Input Drive Requirements

Throughout this treatment of conventional high efficiency modes, it has been stressed several times that they have a significant disadvantage of requiring much higher drive levels, as much as 6 dB higher, than a linear Class A amplifier using the same device. This is particularly troublesome at higher RF frequencies, where linear gains of 10 dB are often a luxury, and anything better would involve much higher cost in terms of device technology or packaging. Indeed, at frequencies higher than 10 GHz, almost all power amplifiers are of the Class A type, for this reason.

As mentioned in Section 3.3, it seems a matter of experimental fact that bipolar transistors, and to a lesser extent FETs as well, do not display such a big difference between Class A and Class B power gain. In some respects, the BJT is tailor-made for Class B operation, where at zero base bias the large RF drive signal naturally switches the forward diode on and off, and the collector current follows suit. The key point is that the input impedance changes substantially during the RF cycle, so that the actual increase in RF drive power required for Class B operation cannot be simply predicted. The FET, on the other hand, is a more naturally linear amplifier, and with a sinusoidal drive requires the cutoff characteristic to do most of the required wave-shaping. The gate impedance will remain constant over the RF cycle, other than some smaller effects arising from the voltage dependent depletion capacitance, so the extra 6 dB of drive would appear to be inescapable and possibly represents a significant disadvantage of FETs over BJTs in these applications.

Taking the idealized results for efficiency and power gain, Figure 3.17 quantifies the problem of high efficiency PA design at higher frequencies, where the avail-

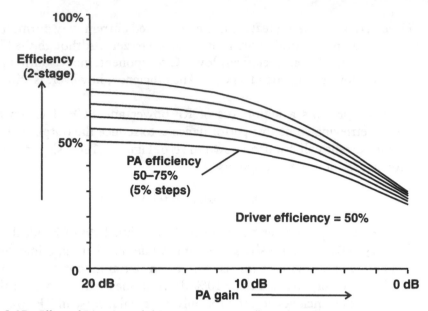

Figure 3.17 Effect of PA gain and drive power on overall system efficiency.

able gain can easily drop well below 10 dB. For example, a Class B PA stage giving an output efficiency of 75% and 6 dB power gain (corresponding to a typical linear gain of 12 dB) will display an overall efficiency of 55% when teamed up with a driver which itself has 50% efficiency. Under these conditions, there is clearly a tradeoff between the efficiency and gain of the PA stage. It may well be more efficient, from an overall system viewpoint, to run the PA stage in a mid-AB condition where the power gain may be 3 dB higher than in Class B; taking some typical numbers, a PA efficiency of 65% could be expected to have 9 dB of power gain, resulting in a two-stage efficiency of 57%. This is hardly a significant improvement in overall efficiency, but it demonstrates that there is little point in striving to achieve highly efficient PA operation if the resulting power gain is significantly lower.

In a multistage amplifier chain, the drive signal to the PA stage may not be sinusoidal. In fact, there are some possibilities for shaping the drive signal waveform, especially its second harmonic content, to alleviate the drive requirements for Class AB or B operation. Figure 3.18 shows a typical, and nearly optimal, situation. If it were possible to contrive a driver which caused a Class-B type wave shape at the PA input, all of the conventional advantages of Class B operation could be obtained without the disadvantage of heavy drive power. In fact, the fundamental component of the shaped voltage drive waveform in Figure 3.18 is the same as a sinewave which swings between the same limits as in Class A operation. So taken in isolation, this PA stage with wave-shaped drive will display the same fundamental power gain as a Class A stage but the efficiency of Class B. There would be additional advantages of such an arrangement, in that the PA transistor is only operating over its linear region, and the large negative swings of voltage have been eliminated. This, potentially, would mean that the device could be operated at higher supply voltage than in a conventional over-driven configuration.

Techniques for creating such a shaped drive signal require the generation of sufficient second harmonic content, either through nonlinear operation of the driver or

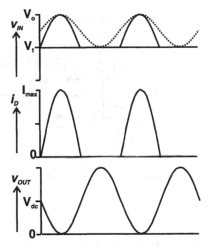

Figure 3.18 Class B operation with a wave-shaped RF drive signal. Fundamental component of drive signal (dotted) is half of the amplitude required if drive signal were sinusoidal.

by feeding back some second harmonic from the output of the PA stage. It is even possible that the nonlinear C-V characteristic of the input junction could be harnessed to do some beneficial wave shaping [3]; this will be considered further in Chapter 4. It should be noted that an inappropriate phasing of the second harmonic will conversely cause efficiency degradation. The desired mix of harmonics can be further tailored by the frequency response of the interstage match. Figure 3.19 suggests one such implementation. The driver stage can be operated in an "inverted" Class B mode, where the device is biased close to its saturation region, and the negative-going RF drive swings the current down to zero for less than half of the cycle. If the driver output load looks like a sufficiently broadband resistor, the corresponding RF output voltage will take the desired form for driving a Class B PA—see Figure 3.19(c).

Such an amplifier will have only moderate efficiency, due to the nonoptimum current waveform. A simple estimate for the efficiency of the driver amplifier shown in Figure 3.19 can be made as follows:

- Current waveform: Inverted half wave rectified cosinewave, peak I_{max}, Fundamental component amplitude, $I_1 = I_{max}/2$, DC component, $I_{dc} = (1 - 1/\pi) \cdot I_{max}$;
- Voltage waveform: half wave rectified cosinewave, peak I_{max} (Class B waveform)
  Peak voltage, $V_{pk} = \pi \cdot V_{dc}$,
  Fundamental component amplitude, $V_1 = \pi \cdot V_{dc}/2$.

So RF power at fundamental $= (V_1 \cdot I_1)/2$
$$= (I_{max} \cdot \pi \cdot V_{dc})/8$$
DC power $= V_{dc} \cdot (1 - 1/\pi) \cdot I_{max}$
Efficiency $= (\pi/8)/(1 - (1/\pi))$
$$= 57.6\%.$$

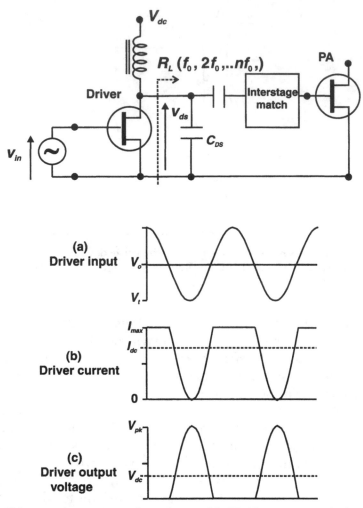

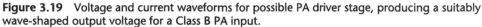

Figure 3.19   Voltage and current waveforms for possible PA driver stage, producing a suitably wave-shaped output voltage for a Class B PA input.

This is an acceptable efficiency for a driver stage. Unfortunately, the input of the PA transistor will not look like a broadband resistor, and a broadband interstage matching network will need to be designed, which transforms the input impedance of the PA stage to the necessary resistive load, which has been assumed in the waveforms of Figure 3.19. Such a network may be quite feasible, especially when the match is only required at harmonic frequencies, rather than over a continuous bandwidth.

As stated at the beginning of this section, this is an advanced topic worthy of further research, and what has been presented does not constitute a complete solution. It can however be speculated that this condition, or some diluted version of it, is one which has been utilized a good deal more than it has been intentionally designed. Judicious tuning of bias and interstage matching in a multistage PA often reveals a complex behavior, seemingly with as much tradeoff between power, linearity, and efficiency as that which is obtained through tuning of the output match. Essentially, almost any amount of second harmonic in the drive signal to a PA stage, with the appropriate phasing, will reduce the apparent overdrive required to realize higher efficiency modes, and this will undoubtedly be a factor in many practical situations.

3.8 Conclusions

This chapter has presented conventional reduced conduction angle modes with an emphasis on quantitative analysis using idealized mathematical models. The classical A, AB, B, and C modes have been analyzed in detail, and the effects of finite knee voltage and weakly nonlinear effects in the linear region have been quantified, both for FET and BJT models. The effect of PA gain on overall efficiency has been examined, along with the possible tradeoffs. Finally, some possibilities for improving the performance of multistage PAs using drive signal wave shaping have been proposed.

Most of the analysis in this chapter has adopted the conventional assumption of shorted harmonics in the output load. This is an important issue in the practical realization of high efficiency RF amplifiers and will be considered in more detail in the next chapter.

References

[1] Doherty, W. H., "A New High Efficiency Power Amplifier for Modulated Waves," *Proc. IRE*, Vol. 24, No. 9, September 1936, pp. 1163–1182.
[2] Cripps, S. C., *Advanced Techniques in RF Power Amplifier Design*, Norwood, MA: Artech House, 2002.
[3] White, P. M., "Effect of Input Harmonic Termination on High Efficiency Class B and Class F Operation of PHEMT Devices," *Proc. IEEE Intl. Microw. Symp., MTT-S 1998*, THIF-05.

CHAPTER 4

Class AB PAs at GHz Frequencies

4.1 Introduction

The classical PA design theory described in the last chapter is based on a number of idealizing assumptions, both in the active and passive element models employed. Although the nature of these idealizations was clearly defined, it is a curious fact that expectations of a practical PA design or product are still widely based on the classical results, regardless of deleterious effects that must come into play at GHz frequencies. Such effects as parasitic capacitance, especially on the input of RF power transistors, and the turn-on characteristic of the I-V device response will surely modify the classical results in a substantial manner. Yet there has been an endemic culture, especially pervasive in the PA customer and user community, which has nurtured a blind spot for such matters of detail. If it's a Class B amplifier, the expectation for efficiency is 78.5%, right on the button, or something is wrong with the design. So in this chapter we have rather more at stake than merely digging deeper and working out the more detailed aspects of PA design in the real world.

The problem is complicated by the fact that Class B PAs *can* sometimes deliver 78% efficiency, and even a bit more, if various performance compromises are made with power and linearity. But the manner in which this is achieved seems to have rarely been seen in print, largely due to the difficulty for the GHz designer to observe current and voltage waveforms, along with the widespread use of empirical design methods. We will see that when dealing with imperfect devices and imperfect circuits, the bread can sometimes fall butter side up; the goal here is to understand how and why. As a starting point, we will pick up on the analysis of a practical alternative to the classical short-circuit Class AB harmonic termination. This will lead into some surprisingly new theoretical territory, where the classical results for power and efficiency can still be obtained using a harmonic impedance environment that is far from classical but at the same time highly practical.

Nonlinearities in the device I-V characteristics, and especially the turn-on region, are not strictly speaking effects that are limited to GHz frequency operation, but it is frequently observed that these effects can themselves be frequency dependent. The effects of the IV knee were quantified in Chapter 3. In this chapter the effects of a nonlinear transfer characteristic will be considered, with special emphasis on the impact this can have on the classical linearity results. Time and again, PA designers find a particular device which exhibits "sweet spots" in Class AB operation, where spectral distortion products drop into nulls over restricted power drive levels. These effects can usually be attributed to the fortuitous cancellation of the nonlinearity of the classical Class AB current truncation mechanism and the inherent nonlinearity of the device transfer characteristic.

The effects of the nonlinearity of the input impedance of the RF power transistor, especially the *Cgs* varactor in junction FETs, is a vastly neglected area which will be shown to have a major impact on PA design, and especially the detailed design considerations for interstage matching between driver and PA stages.

4.2 Class AB Using a Capacitive Harmonic Termination—The Class J PA

4.2.1 Theory

For reasons which will become apparent in due course, the circuit shown in Figure 4.1 will be defined as a Class J configuration. It is the very simplest of circuit configurations, consisting of a device and a fundamental matching network. Although the detailed properties of matching networks have yet to be discussed (in Chapter 5), it is assumed that the reader is at least familiar with the basic low-pass matching network which consists of a shunt capacitance and a length of transmission line. Such a circuit has no specific harmonic termination network; the harmonic impedances presented to the device will be a function of the matching elements (C_p, Z_0, θ) and also, most significantly, the output capacitance of the device (C_{ds}). Such a configuration can be considered as an approximation to an ideal classical Class AB circuit, so long as the device capacitance is large enough to approximate to a short circuit at the frequencies of interest. As discussed in Chapter 5, a convenient criterion for the requisite "largeness" of the device output capacitance can be estimated in terms of the ratio of its capacitive reactance to the loadline resistance of the device at the fundamental frequency. It will be seen that as long as this ratio is unity or lower, near to classical results can be expected, in terms of efficiency, power, and nearly sinusoidal voltage waveforms.

Depending on the frequency and the device technology in use, this ratio can be much higher than unity. Table 4.1 shows how the ratio varies widely depending on the device technology and the frequency. Especially problematic are devices run at low supply voltages, which includes most GaAs types. In such cases it could be assumed that a more complicated circuit topology will be required, including for example such elements as short-circuited stubs at harmonic frequencies. Practical experience however seems to indicate that the situation is less well defined. In this section, we will see that a combination of a lower capacitance value ($X_{Cds}/R_L > 1$) and a substantial *reactive component* in the fundamental load can work together to restore the power and efficiency to values close to, or even greater than, the corre-

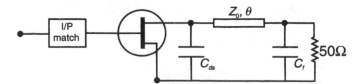

Figure 4.1 "Class J" PA circuit (biasing components omitted).

Table 4.1   Technology and Frequency Dependence of the X_{cds}/R_L Parameter

	C_{ds} (pF/mm)	V_{dc} (V)	I_{max} (mA/mm)	R_L (Ω–mm)	XC_{ds}/R_L
GaAs PHEMT, 2 GHz	0.25	10	350	57	5.6
GaAs PHEMT, 10 GHz	0.25	10	350	57	1.1
Si LDMOS, 850 MHz	1	28	200	280	0.7
Si LDMOS, 2 GHz	1	28	200	280	0.3
GaAs HBT, 2 GHz	0.5	3	250	24	6.7
GaN FET, 2 GHz	0.2	40	600	133Ω	3
GaN FET, 10 GHz	0.2	40	600	133Ω	0.6

sponding classical values. The corresponding voltage and current waveforms will however be far from classical. This will have some important impact on conventional thinking about the requirements of PA matching networks, and goes some way to quantifying why such a simple circuit topology can sometimes give excellent results in terms of power and efficiency. Such "good" results are frequently observed, but not well explained, when using commercial load-pull systems for PA device evaluation.

For the purposes of analysis, the circuit of Figure 4.1 can be redrawn schematically as shown in Figure 4.2. The transistor is now assumed to be a current generator, representing a Class B condition where the RF current waveform is a halfwave rectified sinewave. Initially it will be assumed that the fundamental matching circuit branch is a low-pass structure which blocks all harmonic components. So the transistor current I_T can be defined as a halfwave rectified sinewave,

$$I_T = I_{max} \sin\theta \quad 0 < \theta < \pi$$
$$= 0 \qquad\qquad \pi < \theta < 2\pi$$

and the fundamental current I_F flowing in the matching branch has the form

$$I_F = I_1 \sin(\theta + \phi)$$

where the magnitude I_1 and the phase ϕ are at this point considered as *independent* variables, whose values will ultimately define the various circuit elements. So the current flowing into the shunt capacitor, Ic, is given by

$$I_C = I_{DC} - I_T - I_F$$

where

$$I_{DC} = \frac{I_{max}}{\pi}$$

so that the voltage developed across the capacitor, and hence also the device output voltage V_o, is given by

$$V_o(\theta) = \frac{1}{\omega C}\left[\int_0^\theta \left(I_{max}\sin\theta - \frac{I_{max}}{\pi} - I_1\sin(\theta+\phi)\right)d\theta\right] + V_{off}, \quad 0 < \theta < \pi$$

$$= \frac{1}{\omega C}\left[\int_0^\pi \left(I_{max}\sin\theta - \frac{I_{max}}{\pi} - I_1\sin(\theta+\phi)\right)d\theta\right.$$

$$\left. + \int_\pi^\theta \left(-\frac{I_{max}}{\pi} - I_1\sin(\theta+\phi)\right)d\theta\right] + V_{off}, \quad \pi < \theta < 2\pi$$

$$(4.1)$$

which can be further evaluated to give

$$\omega C V_o(\theta) = I_{max}\left[1 - \frac{\theta}{\pi} - \cos\theta\right] + I_1\left[\cos(\theta+\phi) - \cos\phi\right] + V_{off}, \quad 0 < \theta < \pi,$$

$$= I_{max}\left[2 - \frac{\theta}{\pi}\right] + I_1\left[\cos(\theta+\phi) - \cos\phi\right], \quad \pi < \theta < 2\pi$$

$$(4.2)$$

where V_{off} is an arbitrary DC offset which will be assigned a value such that

$$V_o(\theta) \geq 0$$

In order to make a rational comparison with the classical (harmonic-shorted) results, the zero-knee assumption will continue to be used, and the value of V_{off} of most interest will therefore be the case where the minimum value of V_o grazes zero at the minimum value(s) of V_o.

To obtain a value of V_{off}, it is necessary to determine the global minimum value of V_o for given values of the independent variables I_1 and ϕ. This is a cumbersome process to complete analytically, and the analysis is more conveniently continued using direct computation. For any given choice of the independent variables I_1 and ϕ, the V_o function can be computed for a whole RF cycle, and the minimum point determined. This point is then offset to zero by suitable choice of V_{off}, and the whole voltage function is scaled such that the mean value is unity; this can be achieved by suitable choice of the capacitive reactance scaling factor ($1/\omega C$).

Figure 4.3 shows voltage waveforms, for three specific choices of I_1 and ϕ. The first case, representing a higher choice for I_1 ($I_1 = 2$), resembles a sinewave, although some low harmonic content is still evident. The other two cases ($I_1 = 0.7, 0.6$) show an increasingly nonsinusoidal voltage waveform, which by inspection appears to have a higher fundamental component but also has a significant phase shift from the current. Essentially, the reactive voltage components being generated by harmonics flowing into the capacitor allow a higher fundamental component. By performing the appropriate Fourier integrals on (4.2), it is possible to determine the in-phase and quadrature components of V_o, and consequently the normalized values of fundamental impedance in each of the three cases (all impedances in Figure 4.3 are normalized to the loadline resistance). The normalized capacitive reactance, used as a scaling factor as described above, is also a function of the choice of I_1 and ϕ, and all of these values, with the corresponding computations of RF output power and efficiency, are summarized in the table in Figure 4.3.

These three computations show some surprises. In particular, the second case, which displays a strikingly nonsinusoidal waveform, gives the same power and effi-

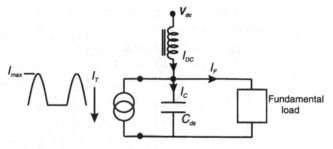

Figure 4.2 Class J PA schematic.

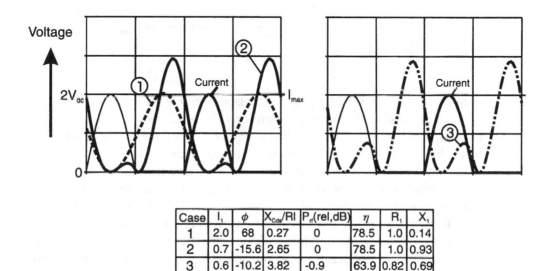

Case	I_1	ϕ	X_{Cds}/Rl	P_n(rel,dB)	η	R_1	X_1
1	2.0	68	0.27	0	78.5	1.0	0.14
2	0.7	-15.6	2.65	0	78.5	1.0	0.93
3	0.6	-10.2	3.82	-0.9	63.9	0.82	0.69

Figure 4.3 Analysis of Figure 4.2 circuit, three cases (see text).

ciency as a reference Class B design using an ideal harmonic short. This despite also showing a fundamental load which has a substantial overall reactive component (i.e., allowing also for the reactance of C_{ds}). Moving back to case 1, the value of the normalized C_{ds} reactance, X_{Cds}/R_L, has moved well below unity, and as already surmised this indicates operation which is getting closer to classical Class B performance. But even in this case, it can be seen that some reactive tuning of the fundamental has restored the efficiency to the full 78.5% value. Case 3 shows a strongly nonsinusoidal voltage waveform, with high second harmonic content, and a much higher value of X_{Cds}/R_L.

Assuming that the C_{ds} value in this analysis represents the device output capacitance, the X_{Cds}/R_L parameter clearly has important physical significance. It is independent of device peripheral scaling, and is in essence a performance index for a particular device technology. It is therefore useful to consider X_{Cds}/R_L as a dependent variable, by plotting it as the x-axis value, as shown in Figure 4.4. As the values of the input variables, I_1 and ϕ, are varied over a wide range, the corresponding values of efficiency and X_{Cds}/R_L are plotted out. Each point on the swept curves represents a particular solution of (4.2), using a value of V_{off} such that the voltage

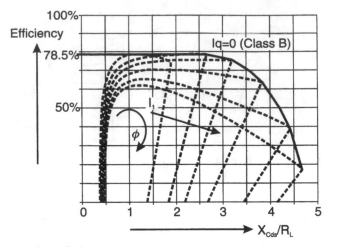

Figure 4.4 Class J mode analysis.

minimum grazes zero. The points of special interest are those which represent the maximum possible efficiency solution for a given value of X_{Cds}/R_L; these points have been joined up to give the heavy line in Figure 4.4. A remarkable conclusion emerges from this plot: the efficiency and power can be maintained at their classical Class B values for values of X_{Cds}/R_L up to about 2.5. Above this value, the power and efficiency which can be obtained without voltage clipping drops sharply. The key to obtaining the maximum efficiency lies in the reactive tuning of the fundamental; the waveform at the critical point is in fact case 2 in Figure 4.3. There is however a price that has to be paid for obtaining maximum efficiency in cases where the X_{Cds}/R_L value is substantially greater than unity, in the form of an increased peak voltage compared to the sinusoidal case.

Figure 4.4 refers specifically to the zero bias case, where the quiescent current parameter $I_q = 0$. Figure 4.5 shows a set of solutions for (4.2) for a deep Class AB case, $I_q = 0.1$. Clearly, the useable range for X_{Cds}/R_L is substantially increased, as the second harmonic current component is reduced. Further computation enables an

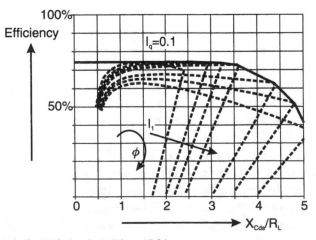

Figure 4.5 Class J mode analysis, deep Class AB bias.

approximate optimum condition to be identified, where a setting of $I_q = 0.2$ still enables 70% efficiency to be obtained at an X_{Cds}/R_L value of 5.

All of the possible solutions to (4.1), which are summarized in Figures 4.4 and 4.5, appear to represent modes of PA operation that are not covered by standard mode classifications. The key features are a fundamental load with a substantial reactive component and reactive harmonic terminations that can be physically realized using the device output capacitance. The generic term "Class J" is proposed to categorize such PA operation.[1]

4.2.2 Practicalities

The above analysis opens up some interesting new possibilities for high efficiency Class AB PA design. The traditional view that harmonic short-circuits are a necessity to obtain the classical results would appear to be usefully compromised. In particular, the possibilities for broader band design are worth further investigation. It is however necessary to revisit some idealizations which were still in place. As with any PA analysis, the use of a realistic device IV characteristic, including the knee turn-on region, will have a substantial impact on the results for power and efficiency. Additionally, and as has already been noted, a typical single section low-pass matching network may not completely block second harmonic current. The impact of the IV knee can be considered in much the same way as for the classical Class AB, and indeed in most PA textbook treatments, the DC supply voltage has to be increased so that the voltage minimum is pulled safely out of the turn-on region. This strategy keeps the conclusions intact, the only change being an increase in the DC supply, which has an impact on output efficiency represented by the expression

$$\eta_{knee} = \frac{V_{dc} - V_{knee}}{V_{dc}} \eta_{ideal}$$

so that if V_{knee} is about 10% of V_{dc}, the impact on all efficiency results will be a factor of 0.9, or a reduction from 75% to 67.5%. But this is the same impact that the IV knee will have on any configuration. The matching issue is however more specific to the present analysis and merits quantitative treatment.

We return therefore to the physical schematic shown in Figure 4.1, which shows the device matched by a length of transmission line and a shunt capacitor. One of the key aspects of this whole section is to recognize the widespread use of such a physical topology, both in final PA board layouts and load-pull characterization systems. Unfortunately, despite the apparent simplicity of this circuit, analytical expressions for impedance at the device plane are cumbersome, and direct computation will be used. Figure 4.6 shows a typical case, where the fundamental matching elements have been designed to match a device having a loadline resistance of 5Ω. Following the design strategy outlined in Section 4.2.1, the fundamental load has an almost equal reactive component, in order to obtain maximum efficiency for a relatively high value of X_{Cds}/R_L, approximately 4 in this case (note that the ratio is

---

1. The Class J mode, as defined here, differs significantly from the Class E, F, xx modes defined by other authors (see [1], Chapter 8), in that a physical circuit, rather than a specific harmonic environment, is used.

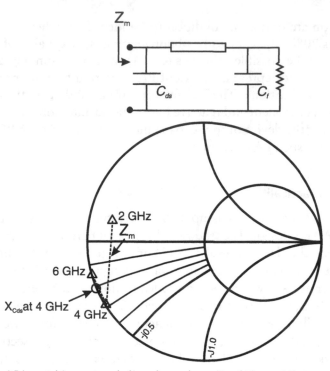

Figure 4.6   Class J PA matching network, based on schematic of Figure 4.2.

defined using the X_{Cds} value at the fundamental). It can be seen that the fundamental matching network places an inductive component in parallel with the C_{ds} capacitor at the second harmonic frequency of 4 GHz, effectively increasing its shunt reactance value from the unmodified reactive value which was assumed in the analysis; the discrepancy in this particular case is a factor of about 1.5.

So the assumption that the fundamental matching network blocks higher harmonic currents is somewhat approximate in this case, although such an assumption is frequently made in analyzing problems of this nature.[2] In defense of the approximation, it should be noted that the present example uses a high value of X_{Cds}/R_L; designs using values closer to unity will show less discrepancy. The discrepancy can be further reduced if the practical implementation of the circuit is considered in more detail. At higher frequencies, and even at 2 GHz for higher power devices, the shunt tuning capacitor C_f will probably be realized using an open circuit stub. This would apply especially if the circuit is considered to be a representation of a typical load-pull test bench, where stubs, or low impedance tuning "slugs," are commonly used to implement the shunt tuning element. Figure 4.7 shows this practical realization, using a 25Ω SCSS to implement the tuning capacitor, and the circuit now shows a harmonic impedance environment much closer to the idealized analysis. It is important to note that the use of a tuning stub is not just fortuitously providing the required second harmonic termination; the fundamental matching network branch is actually presenting a high second harmonic impedance at the device plane, so that the harmonic impedance is dominated by the C_{ds} capacitor.

---

2.   See, for example, the switch mode Class E analysis in Chapter 7.

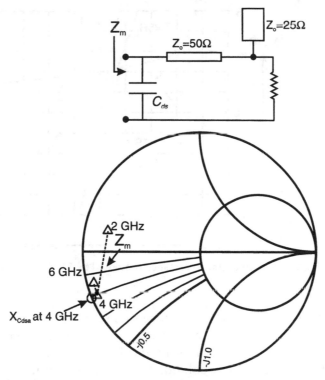

Figure 4.7   Modified circuit topology to give improved second harmonic termination.

The comparison between the matching circuit topologies shown in Figures 4.6 and 4.7 demonstrates an important general principle which has much wider implications for PA design than the specific example under consideration. Basically, the choice of matching topology can have an important impact on the PA performance, even if various alternatives achieve the same matching function at the fundamental.

It is at this point worth revisiting the purpose of the analysis in this section. The question should now be asked, is it not the case that most of the above analysis becomes superfluous if a classical design approach is used, and a physical harmonic short circuit is included in the matching network? The real issue here is what exact form does such a network have? The simple quarter-wave SCSS is in itself an approximation to an ideal solution; it has a bandwidth limitation which can approach severe proportions in applications requiring more than a few percent bandwidth, and in any case does not perform its short-circuit function at odd harmonics. Even harmonics above the second will also likely see a reactive termination due to dispersion effects in the SCSS and parasitic reactances associated with its short circuit termination. The SCSS bandwidth issue is illustrated in Figure 4.8, where it can be seen that even moderate bandwidths require the use of a stub whose characteristic impedance is inconveniently low. The values of SCSS characteristic impedance have been normalized to the loadline resistance R_L and in most practical RFPA designs R_L will have a value less than 10Ω, and in many cases will be closer to 1Ω. The use of a shunt capacitor of moderate value (Figure 4.8 shows a value corresponding to $X_{Cds}/R_L = 2$) starts to look like a better all-around solution for Class AB

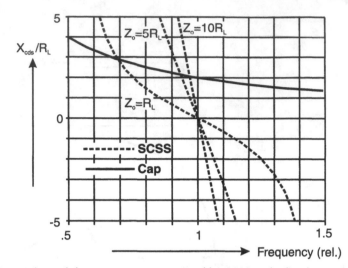

Figure 4.8 Comparison of shunt reactance presented by SCSS and a fixed capacitor.

harmonic termination. The reactance sign reversal of the SCSS above resonance is especially troublesome, given the analysis in the last section.

So we head further towards a radical conclusion; there may be many practical applications where the best solution for harmonic termination requirements in Class AB PA design is to persuade the transistor manufacturers to emplace additional output capacitance on their devices. This would apply particularly in applications where a device is being used at a frequency much lower than its upper frequency limit, for example most GaAs devices below 4 GHz, and Silicon LDMOS devices below 1 GHz. The MMIC designer, of course, has freedom to include such a capacitor in the design, but this option is not available when using higher power packaged discrete power transistors. The immediate objection to such a strategy is typically stated as an increase in output matching Q-factor and a corresponding reduction in the bandwidth of the fundamental match. Figure 4.9 however shows that the Q-factor impact will typically be quite small for X_{Cds}/R_L values greater than about 2.

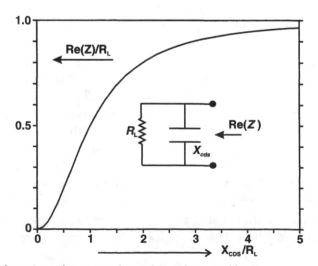

Figure 4.9 Step-down impedance transformation ratio caused by X_{cds}.

The widespread use of load-pull measurements as the basis for PA design seems to have defined a pragmatic approach to matching network design that has no place for a physically distinct harmonic termination. It is therefore suspected that a high proportion of such designs do in fact use the theoretical "Class J" modes of PA operation described in this section. The inability to observe current and voltage waveforms creates something of a smokescreen, which in turn engenders a false sense of confidence that classical Class AB operation is being obtained, based on the power and efficiency numbers being measured. Class J may have been an unrecognized workhorse for the RFPA designer. In Chapter 5, a CAD design example will further illustrate Class J operation. Chapter 8 will pick up the theme of Class J, extending the concept to qualify it as a potentially high efficiency mode which crosses paths with widely publicized assertions about the realization of Class E operation at GHz frequencies.

4.3   Nonlinear Device Characteristics

The classical PA theory in Chapter 3 assumes that the device transfer characteristic is perfectly linear beyond the threshold point. This is an intuitive assumption for idealizing a FET characteristic but, as further shown in Chapter 3, is fundamentally unacceptable for a bipolar device. In considering the necessary nonlinearity of a BJT transfer characteristic, it was shown that very satisfactory Class AB PAs can be designed, even from the viewpoint of linearity. This section therefore considers the impact of nonlinearities in the device characteristic, with particular emphasis on FET devices. This subject has been considered in more detail in another book by this author [1] and the present section will only summarize the results of this previous treatment.

As a starting point, we consider the ideal transconductive device characteristic, shown in Figure 4.10. The key elements of such a characteristic are an abrupt cutoff and a linear active region, both of which will represent a substantial idealization for any practical device. The ideality of a completely linear active region is intuitively obvious; however, it is important to recognize that an ideally abrupt device turn-on is equally and importantly a very idealized concept. If the quiescent bias point is set close to the turn-on, or threshold, point, it will always be possible to reduce the

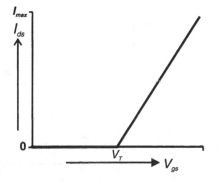

Figure 4.10   Ideal transconductance characteristic.

signal drive level such that the range of deviation around this point will no longer display abrupt behavior. In other words, there will always be a magnified scale of voltage and current where the threshold point becomes continuous rather than abrupt. If the device is being used to amplify a communications signal with zero-envelope amplitude crossing points, there will be a range of low-level envelope amplitudes where this nonideal behavior becomes an important source of nonlinearity. Interestingly, however, it is a form of nonlinearity which can in some circumstances act in a way that neutralizes the fundamental nonlinearity of a classical Class AB amplifier. Thus nonideal transfer characteristics are not necessarily detrimental to the ideal Class AB behavior described in Chapter 3, and just like the Class J amplifier have been widely used without being widely recognized. It is this aspect of the device transfer characteristics that forms the focus of this section.

A simple example of the possibilities that nonabrupt, nonlinear device transfer characteristic can offer is a square-law device, shown in Figure 4.11. It is a straightforward matter [1] to show that for any selection of quiescent bias point, other than the actual threshold point itself, such a device will exhibit higher efficiency than a Class A linear device, provided the operation is kept inside the square-law region. Furthermore, such an amplifier will not have the undesirable odd degree distortion processes that cause intermodulation distortion. The optimum point is the quiescent level of 25%, where the device will give the same output power as a Class A device having an ideal linear transconductance, but with an efficiency of 66.7%. Figure 4.11 shows the two transfer characteristics so far considered in this section. It is tempting to speculate whether the abrupt characteristic represents some form of limiting case to a set of increasingly sharp characteristics which share the same properties of linear operation and increasing efficiency. Some more extensive analysis [1] shows that this is indeed the case, and a set of such characteristics is shown in Figure 4.12.

The characteristics of Figure 4.12 are generated using truncated polynomial series which contain only even degree terms, yet retain overall features which represent realizable characteristics of real devices. The two extreme cases are basically undesirable goals for the device fabrication process definition; the square-law device has lower efficiency, and the abrupt device is fundamentally unrealizable. But the intermediate cases would appear to be the kind of targets for RF power transistor characteristics of which device technologists should be aware. The general folklore

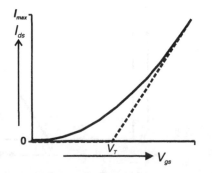

Figure 4.11   Square law transconductance characteristic.

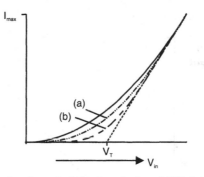

Figure 4.12 Even degree transfer characteristics (see text and [1]). Intermediate cases: (a) uses second and fourth degree nonlinearity and can potentially deliver 70% efficiency, and (b) uses even degrees up to the eighth and can deliver 74% efficiency.

that "sloppy" turn-on characteristics should be avoided, and that transfer characteristics beyond the threshold region should be as linear as possible, may be intuitive but appears to be seriously flawed. Figure 4.13 shows how the nonlinearity in the transfer characteristic shown in Figure 4.12(b) is able to "engineer" a current waveform that has most of the advantages of an abruptly truncated deep Class AB waveform, but without the troublesome odd degree nonlinearities that conventional Class AB produces. As noted in Chapter 3 and [1], such a characteristic can in principle be obtained more easily using a bipolar device than a FET.

Unfortunately, until such time as these concepts are tested out, the PA designer has to deal with devices that have sharper turn-on characteristics which display "weakly expansive" behavior, as shown in Figure 4.14. Such a device shows a soft threshold, a region of small signal gain expansion where the transconductance increases with output current, then a saturation region. It is instructive to consider the impact of such device characteristics on the classical results for Class AB PA efficiency and linearity, in particular the conduction angle plots shown in Chapter 3. The device characteristic of Figure 4.14 was introduced in Chapter 1 when defining strong and weak nonlinearities. The transfer function was generated using the expression

$$i_o = 3v_{in}^2 - 2v_{in}^3 \ldots\ldots 0 < v_{in} < 1$$
$$i_o = 0 \ldots\ldots\ldots\ldots v_{in} < 0$$
$$i_o = 1 \ldots\ldots\ldots\ldots v_{in} > 1$$

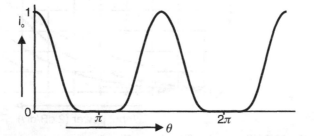

Figure 4.13 Current waveform resulting from nonlinear transconductance characteristic shown in Figure 4.12, case (b).

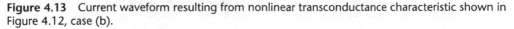

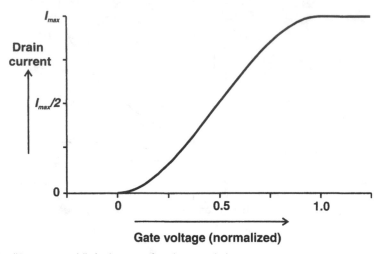

Figure 4.14 "Strong-weak" device transfer characteristic.

and the resulting modified efficiency and linearity curves as a function of conduction angle are shown in Figures 4.15 and 4.16. Figure 4.15 shows that the impact on peak efficiency is quite small, whereas the gain characteristics shown in Figure 4.16 show significant changes from the classical curves (also shown). Most notably, the modified device characteristics cause the Class A mode to become significantly non-linear, and low quiescent bias settings, in the region of 10%, now show quite linear behavior over a wide range of input drive levels. This is due primarily to the expansive square-law characteristic in the turn-on region compensating for the compressive effect of Class AB truncation of the current waveform.

Such fortuitous linearization of deep Class AB modes can be observed over certain critical bias ranges in many RF power transistor types. These effects should always be treated with caution, inasmuch as their repeatability over different process runs, sometimes over many projected years of production schedules, must be

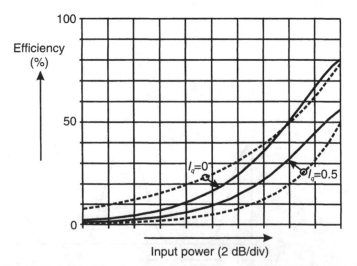

Figure 4.15 PBO efficiency for Class A and Class B modes using strong-weak device model (corresponding results for linear transconductance shown dotted.

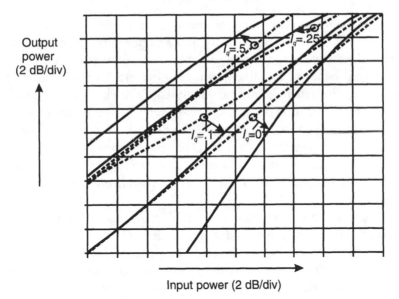

Figure 4.16 Linearity characteristics for Class AB modes using strong-weak device model (corresponding ideal linear transconductance shown dotted).

maintained. There are also other mechanisms which can affect the linearity and general performance characteristics of an RFPA and are associated with the voltage dependency of the capacitive parasitics in the device. This is the next topic for consideration, especially given its increasing impact at higher frequencies.

4.4 Nonlinear Capacitance Effects in RF Power Devices

4.4.1 Introduction

One of the principal issues that differentiates low frequency, or analog, circuit design from RF, or microwave, design, is the greatly increased intrusion of parasitic reactance effects, both in the device and the circuit elements. In this section, we will consider the three main capacitive parasitics in any RF power transistor: input, output, and feedback. Particular emphasis will be given to the effects of a voltage-dependent input capacitance; in the case of a FET device, the gate to source capacitance, C_{gs}. It will be seen that for any PA design above 1 GHz, and even at lower frequencies in higher power designs, the distorting effects of the input varactor can have a major impact on the conclusions and expectations which come from classical analysis. This will frequently mandate that the input matching network needs just the same careful attention to harmonic terminations as is widely observed for the output. This is a much neglected area in modern PA design and has clearly important implications for the design of interstage matching networks.

4.4.2 Nonlinear Capacitors—Characterization and Analysis

The analysis of the nonlinear properties of voltage-dependent capacitors has a long history, reaching something of a zenith in the 1960s when varactor multipliers were the only means of generating significant power levels at GHz frequencies using solid state components [2]. Both theory and practice support the generalization that non-linear capacitors, or varactors, are a good deal more difficult to deal with than the more well-behaved resistive nonlinear elements to which many textbook authors safely confine their nonlinear analysis. "Quirky" is an appropriate term for varactors, given that a simple L-C-R series combination, which incorporates a varactor as the capacitive element, can display chaotic behavior under some conditions [3]. Cozy theoretical axioms, such as the stipulation that only higher harmonics can be generated in a nonlinear device, are typically blown away with varactors, which can be observed to show "period multiplication," or frequency division, effects.

The first important issue is to establish a basis for characterizing and simulating varactor effects in a microwave device. It is important to understand these basics when using CAD simulation tools to predict the effects of varactors in RFPA designs. In particular, the very definition of capacitance is frequently, and fiendishly, changed by device physicists from the fundamental high-school physics definition

$$C = \frac{Q}{V}$$

to a differential form,

$$C = \frac{dQ}{dV}$$

Most semiconductor physics books adopt the differential form without even as much as commenting about the deviation from fundamental electrostatics that it represents. The actual reason for the change would appear to be the likely measurement method, which will almost certainly be some form of AC impedance characterization. In the RF range, the appropriate small signal impedance of the device in question will be measured over a wide frequency range for a variety of bias voltages. The assumption is then made that the "smallness" of the excitation allows the extraction of an equivalent circuit which contains fixed element values, whose voltage dependency can then be determined by repeating the model extraction process for the various bias settings. Small signal AC measurements of this kind will yield capacitor values that represent the *differential capacitance*, which is the slope of the voltage-charge characteristic at the respective bias setting. The validity of this process depends critically on the assumption that the impedance measurements are indeed performed at suitably low RF excitation levels.

The realization that bias-dependent s-parameter nonlinear models characterize the differential capacitance has important implications in the actual circuit analysis formulation. The instantaneous current in any capacitor, $i(t)$, is given by

$$i(t) = \frac{\partial q(t)}{\partial t}$$

and since (at least in our youth) $q = Cv$, it would seem reasonable to write

$$i(t) = \frac{\partial}{\partial t}(Cv)$$

where now C is itself a function of v and cannot be taken out of the derivative bracket. Not so, however, if C is defined to be the differential form; then

$$C(v) = \frac{\partial\, q(v)}{\partial v}$$

and

$$
\begin{aligned}
i(t) &= \frac{\partial q}{\partial t} = \frac{\partial q}{\partial v}\frac{\partial v}{\partial t} \\
&= C(v)\frac{\partial v}{\partial t}
\end{aligned}
$$

(4.2)

a result which would appear to ignore the fact that C is a variable and not a constant. It is the change of definition of capacitance to the derivative form which makes (4.2) correct, despite still looking shaky to the well-drilled calculus student.

Unfortunately, not all circuit simulators are in agreement about this change of definition, and some care needs to be exercised. In the RF domain, it is unlikely that capacitance could be measured in absolute terms, and some form of impedance measurement is the only practical path. But a more general application could be imagined, where the capacitance change is caused by physical movement, and would be characterized using the changes in mechanical dimensions. For this reason a general purpose CAD engine such as SPICE has to be told how to deal with a non-linear capacitor of the differential type. This can be done using a subcircuit, as shown in Figure 4.17. The nonlinear function which describes the (differential) capacitance variation with voltage,

$$C = f(V)$$

is used to define a voltage generator in a subcircuit, whose output voltage is equal to the voltage integral of the capacitance,

$$F(V) = \int f(V)dV$$

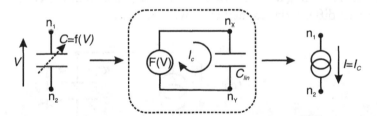

Figure 4.17   Procedure for simulating nonlinear capacitance using dependent voltage and current generators.

and whose dependent nodes are the nodes to which the nonlinear capacitor is connected. An ammeter measures the current flowing into a fixed linear capacitor, which is connected across the subcircuit voltage generator. This current is proportional to the required nonlinear capacitor current, since

$$I_C = C_{lin}\frac{dV}{dt} = C_{lin}\frac{dF}{dV}\frac{dV}{dt} = C_{lin}C(V)\frac{dV}{dt}$$

which can be forced to flow between the two nodes in the actual circuit using a current dependent current source.

The above procedure is recommended for any nonlinear capacitor model in SPICE, since the resident element library models either use absolute capacitance or create convergence problems when used for high fractional capacitance changes [4]. Harmonic balance programs which have been specifically aimed at microwave applications appear to handle the analysis correctly, but some caution is still advised. Unfortunately, experience seems to show that the inclusion of one or several nonlinear capacitors in a circuit greatly increases convergence problems.

4.4.3 Input Varactor Effects on Class AB PAs

A quick taste of what can happen to an intended Class AB PA design when the device input is a varactor, rather than a fixed capacitor, can be obtained using a circuit and device model which have been carefully chosen to be representative, but analytically tractable. Such tractability, unfortunately, is a short-lived luxury in this zone. Figure 4.18 shows a considerable simplification of the device input, where the input generator is assumed to have a sufficiently high impedance that it can be modeled as a sinusoidal current source, and the input varactor has a hyperbolic C-V characteristic, given by

$$C(v) = \frac{C_o}{\left(1-\left(\frac{v}{V_0}\right)\right)^m}$$

Such a characteristic has some general features that are quite typical of reverse-biased microwave diodes; most notably a "punch-through" region where the capacitance is asymptotic to its lowest value, C_o, and a fairly steep region of increasing capacitance, which will ultimately be limited by the diode becoming forward biased and other physical factors. The value of m will vary substantially from one device type to another, and is usually quoted somewhere in the $0.5 < m < 1$ range. Assuming, as will now always be the case, that the capacitance characteristic refers to differential capacitance, we can write

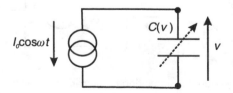

Figure 4.18 Schematic for basic varactor analysis.

$$I_o \cos \omega t = \frac{C_o}{1 - \left(\dfrac{v}{V_0} \right)} \frac{dv}{dt}$$

assuming for analytical convenience that $m = 1$.

Integrating both sides gives

$$\frac{v}{V_0} = 1 - e^{\frac{I_o}{V_0 m C_o} \sin \omega t} \tag{4.3}$$

This voltage is plotted in Figure 4.19 for various values of the parameter $\dfrac{I_o}{\omega V_0 C_o}$.

Although simplifications have been made in order to reach these results for the input varactor voltage, the results are quite instructive. The important observation is that the device input voltage is significantly distorted from its original sinusoidal form. If this varactor was the C_{gs} of a device, biased to operate in a deep Class AB mode, the varactor distortion of the voltage would oppose the intended reduction in conduction angle that would occur if the voltage remained sinusoidal, as is always assumed in classical analysis. Furthermore, it is clear that the voltage distortion appearing across the C_{gs} capacitor will cause some significant changes in the harmonic components of the resulting transconductive output current. This merits some further investigation, in order to obtain a more quantitative picture of how serious these potential effects are in specific, or typical, cases of RFPAs.[3]

The first step is to use a more familiar-looking voltage generator, along with a generator internal resistance. It is also helpful to find a C-V function which does not have a discontinuity. C-V profiles will change substantially from one device type to another, even within the same general technology; this is the problem with simula-

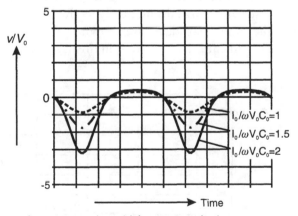

Figure 4.19   Response of varactor to sinusoidal current excitation.

---

3.   It should be noted that some RF power device types do not use reverse biased junctions, for example, all forms of MOS technology, including LDMOS.

tions of specific cases, and the major advantage of symbolic analysis wherever it is possible. A suitable function [5] is

$$C(v) = C_o + C_v \left(\frac{v'}{\sqrt{1 + v'^2}} \right)$$

(4.4)

where $v' = k_1 v + k_2$

The basic function is asymptotic to −1 for negative v, and to +1 for positive v, so C_o is an offset which must be greater than C_v. The scaling and offset parameters C_v, k_1 and k_2 can be used to place the steepest part of the C-V characteristic at a suitable voltage, and to adjust the steepness itself. Figure 4.20 shows this function plotted with $C_o = 0.5$ pF, $C_v = 1$ pF, $k_1 = 2$, $k_2 = 1$V, and can be considered representative up to the junction turn-on in the region of $v = 1$; the key difference from the hyperbolic function is the removal of the physically impossible discontinuity at $v = V_o$, which will cause simulation problems.

Figure 4.21 shows a representative circuit for analyzing the effects of an RF power transistor having an input varactor characteristic. It is now necessary to be more specific about the input model of the device, in particular the scaling factor on the capacitance and the associated series resistance. The plateau capacitance has been set at 2.5 pF, and the series resistance to 5Ω. These values would correspond, for example, to a GaAs FET device having an I_{max} in the 200–400 mA range. The matching elements have initially been set to give a conjugate match at 2 GHz, assuming the varactor is at its plateau value.

Figure 4.22 shows the varactor voltage waveforms using the capacitor C_{fund} to tune the fundamental matching, with the open circuit stub inactive. This response shows broadly similar characteristics to the current source response shown in Figure 4.19, and will clearly have an important effect on attempts to use the quiescent bias setting of the device to control the output conduction angle. The simulation does however assume that the input generator presents a broadband 50Ω source at all frequencies, an assumption likely to be invalid in practical circuits. Figure 4.23 shows how the response can change dramatically as the impedance environment is moved around from the somewhat ideal case shown in Figure 4.22. Although the element values were adjusted empirically to obtain this response, it should be noted that this

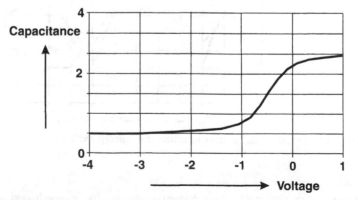

Figure 4.20   Varactor characteristic for simulation.

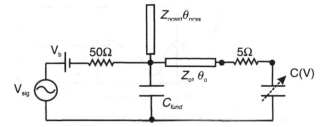

Figure 4.21   FET Cgs varactor simulation circuit.

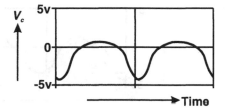

Figure 4.22   Varactor voltage for Figure 4.21 circuit, and varactor characteristic of Figure 4.20; element values: $V_{sig} = 3V$, $V_b = -1$, $\theta_{ocss} = 0°$, $\theta_o = 47°$, $C_{fund} = 5.4$ pF, $Z_o = 50\Omega$.

procedure closely replicates the kind of tuning which is frequently done on PA circuits to optimize their performance, either using a load-pull system or by physical modification of board level components. The asymmetrical nature of this response has some potential use, and will be revisited in Chapter 8.

It is clear from a brief inspection of these simulations that the input varactor can have a major impact on the current wave shaping that is fundamental to Class AB PA operation. Unfortunately, with so many assumptions about the varactor characteristic itself, the scaling levels, and the frequency, it is not worthwhile to engage in a lengthy simulation exercise on a particular set of choices for these values. The key point is to recognize that the input tuning of a PA device can have as much of an impact on power, efficiency, and linearity as the actual quiescent bias point setting, and can easily override the intended effects of a classical Class AB design.

It has been suggested that a harmonic short circuit, placed as close to the device input as possible, can restore the input voltage to the intended classical sinewave. Although this may appear to be an analogous strategy for removing harmonics to that which is mandatory at the output of a Class AB PA, the actual requirements at the input are quite different. The harmonics generated by the varactor can in principle be manipulated to give better performance than reverting to sinusoidal excitation. There is an additional issue, which is that a simple quarter-wave SCSS will only short even harmonics. Figure 4.24 shows the effect of such an SCSS on the waveforms of Figure 4.22, where the SCSS is placed at the closest accessible point external to the device, which is the input of the gate resistance. Clearly, in this particular case, it would seem that the distortion was mainly even harmonic in nature, and the varactor voltage has reverted to an approximately sinusoidal form, whose amplitude and phase vary with the matching element values.

It would be reasonable to regard the transformation from Figure 4.23 to Figure 4.24 as a strong recommendation that input harmonic shorts should always be used in order to nullify the nonlinear effects on the device input. Experience, however,

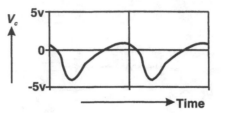

Figure 4.23 Varactor voltage for Figure 4.21 circuit, and varactor characteristic of Figure 4.20; element values: $V_{sig} = 3V$, $V_b = -1$, $\theta_{ocss} = 68°$, $\theta_o = 83°$, $C_{fund} = 0$ pF, $Z_o = Z_{ocss}$ 50Ω.

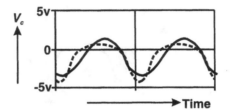

Figure 4.24 Varactor voltage from Figure 4.22; effect of $\lambda/4$ SCSS placement at device input.

leads us up an enticing but hazardous alternative path. Tuning both the fundamental and the harmonic impedance environment on the input of many PA designs can have beneficial effects on power, efficiency, and linearity. As demonstrated in Figure 4.23, even if there is no specific harmonic tuning network, the act of tuning the fundamental will almost certainly change the harmonic environment as well. Not only will the fundamental tuning change the harmonic impedances, the choice of matching topology will also have a strong effect.

All of these effects are at work whenever a PA device is tuned on a test bench. In principle, and as we have just seen, the effects should be predictable using CAD models and simulators. The problem of analysis is however complicated and challenging, but possibly worth pursuing. For example, it may be possible to define a varactor characteristic which more readily enhances the efficiency performance, in much the same way some nonlinear transconductance characteristics were developed in Section 4.3. What seems to be lacking in this area at present is any form of design strategy, other than essentially trial-and-error approaches on the computer screen or the test bench. The above simulations still represent significant simplifications to the simulation of a full PA circuit. In particular, there is the effect of the feedback capacitance, which will have a similar varactor characteristic to the much larger input varactor. There has been a long tradition among PA theorists to assume their active devices can be approximated to be unilateral, but in practice the feedback effects will provide another source for wave shaping on the input. Indeed it can be argued that both input and feedback capacitance have to be considered together in any realistic model.

Although, it can be fairly assumed that a CAD simulator will take account of all these nonlinearities as long as they are included in the device model, PA designers are frequently frustrated by convergence problems and resort to more empirical techniques such as load-pull. Output load-pull techniques are familiar enough for evaluating output matching in PAs, but the addition of input harmonic load-pull tuning is an additional expense and inconvenience that makes heavier demands on the operator, be it human or computer.

4.5 Conclusions

This chapter has identified various effects which come into play when designing PAs in the GHz region and which not only modify the results, but should also change expectations about the results. In particular, the classical implementation of a harmonic short circuit at the output of a PA in the GHz region requires some radical rethinking from the ubiquitous but now impractical SCSS. Nonlinear effects in the device characteristics, both the transfer characteristic and the nonlinear parasitic capacitances, become important considerations at GHz frequencies, but their effects can in some cases be harnessed to good purpose.

References

[1] Cripps, S. C., *Advanced Techniques in RF Power Amplifier Design*, Norwood, MA: Artech House, 2002.
[2] Maas, S. A., *Nonlinear Microwave Circuits,* Norwood, MA: Artech House, 1988.
[3] Jefferies, D. J., G. G. Johnstone, and J. H. B. Deane, "An Experimental Search for the Conditions for the Existence of Chaotic States in Class C Bipolar Transistor Amplifiers," *Int. J. Electronics*, 1991, Vol. 71, No. 4, pp. 661–673.
[4] Kundert, K., "Modelling Varactors," available at http://www.designers-guide.com.
[5] Personal communication with D. Root.

CHAPTER 5

Practical Design of Linear RF Power Amplifiers

This chapter will take the theoretical, and somewhat idealized, results and design equations of the last chapter and develop these into practical designs. The main focus, therefore, will be on the derivation and analysis of matching circuit networks and configurations which are suitable for matching the fundamental and also provide the necessary harmonic terminations. Having established a suitable topology and network element values, only then will designs be put to the test using a nonlinear simulator.

A logical starting point is a survey of matching techniques for higher frequency applications. The wireless communications bands represent a transition region between the lower frequencies, where the matching process is largely dominated by magnetically coupled devices, and the microwave bands, where almost all matching is done using transmission lines. With the availability of good quality miniature passive components, the wireless communication bands allow the use of hybrid matching techniques, incorporating both lumped elements and transmission lines. Although the bandwidths required are generally much lower than in ECM and Satcom applications, it will be shown that awareness of some of the broader band techniques can pay dividends, since generally speaking a broader band match will be more robust in a high volume production environment.

In order to put designs to the test using a simulator, it is necessary to develop a suitable model for the device. Following up on comments made in Chapter 1 (Section 1.3) concerning the lack of good models for commercially available devices, the process by which a useable model can be derived using simple data-sheet parameters will be illustrated in Section 5.5. Such models cannot necessarily be expected to predict the final performance accurately, in absolute terms, but can be invaluable in evaluating circuit topologies and their sensitivities to component tolerances. This will be demonstrated in Section 5.6, where several configurations of a Class B circuit are designed and analyzed.

The key requirement, stressed in the last chapter, for a short circuit termination at all harmonics, poses the main design challenge when designing these kinds of amplifiers. Although simple solutions can be found in the form of resonant transmission line stubs, it will be seen that in many cases this is either inconvenient, in terms of board or chip space, or works optimally over too small of a bandwidth. Alternatives must therefore be considered, and this has often resulted in make-do solutions where simple "pi" networks, which primarily serve to match the fundamental, are also utilized in a harmonic termination role. Empirical design methods, coupled with results from load-pull measurements, often demonstrate that quite acceptable results can be achieved without the additional complexity and band-limiting effects of harmonic resonators.

It is therefore appropriate to consider, in a quantitative fashion, the detrimental effects of imperfect, but more convenient, matching topologies. This forms the main subject for Sections 5.7 and 5.8, and connects at this point with the Class J discussion and analysis in Chapter 4. It will be shown that the use of a simple capacitive harmonic termination, with a load impedance readjusted to be somewhat reactive, can give very satisfactory results without the use of narrowband resonators. The Class J approach therefore lends itself to designs requiring broader bandwidth, and this will be illustrated with CAD design examples for both a GaAs FET and an HBT.

5.1 Low-Pass Matching Networks

Figure 5.1 illustrates the basic matching problem. A terminating resistance R_O (typically, but not necessarily, a 50Ω interface) has to be transformed to a different, usually resistive impedance R_T. Since we are dealing with RF power transistors, it will be assumed that usually $R_T < R_O$. This problem represents a fairly simple specific case in the very generalized theory of matching filter synthesis, a classical example of *a priori* RF design. Such methods are still widely used when the required matching bandwidth extends to an octave or more, however for wireless applications a simplified approach is usually more appropriate.

Indeed, the most widely used solution is shown in Figure 5.2(a). It is usually described as a single section low-pass network, and consists of a capacitive reactance shunting the (higher) load resistance, along with a series resonant inductive reactance. At higher frequencies, the series section would be realized using a short length of moderate to high impedance (often just a 50Ω) transmission line, and the shunt capacitance would be a discrete capacitor, or an open or short-circuited stub, of (preferably) lower characteristic impedance and suitable length. This network could be fairly described as the "Occam's Razor" of high frequency design. Notwithstanding the extensive technical literature on matching network design, it is still surprising what can be achieved with this simplest of solutions.

The design of such a network in distributed form [Figure 5.2(b)] is standard fare in any introductory RF design class, although the basic lumped element form is equally important in lower frequency RF applications. The bandwidth of such a matching network is a strong function of the transforming ratio $m = R_O/R_T$, becoming quite narrow when this reaches into double digits which is quite common in RFPA design. The bandwidth itself should be subject to careful definition when defining PA matching networks; Figure 2.11 shows that a simple definition based,

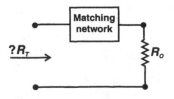

Figure 5.1 Basic RF matching problem.

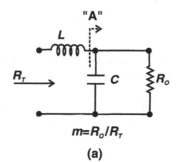

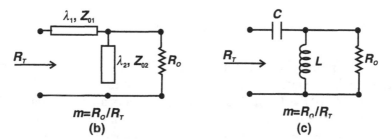

Figure 5.2 Basic RF matching networks: (a) lumped element low-pass, (b) distributed low-pass, and (c) lumped element high-pass.

say, on a 2:1 VSWR mismatch could be quite misleading for maintaining power performance within acceptable limits. In assessing the bandwidth of a particular matching network for power matching applications, the 0.5 dB load-pull contour boundary, using the equations derived in Chapter 2, should be used rather than the corresponding 0.5 dB mismatch, or 2:1 VSWR circle.

Simple lowpass networks are so widely used in narrowband RF design that it is worth deriving the design equations. These equations are a rather poor relation of a much grander general theory [1], but have important and frequent practical uses.

At plane A, the impedance is

$$Z_A = \frac{R_o}{1 + j \cdot R\omega C} = \frac{R_o(1 - jR_o\omega C)}{1 + (R_o\omega C)^2}$$

so the transformed real component at plane A is

$$\mathrm{Re}(Z_A) = \frac{R_O}{1 + (R_O\omega C)^2} = \frac{R_O}{1 + \left(\dfrac{R_O}{X_C}\right)^2}$$

So for a required transformation ratio $m = R_o/R_T$ [where $R_T = \mathrm{Re}(Z_A)$],

$$X_C = \frac{R_O}{\sqrt{m - 1}} \tag{5.1}$$

and

$$X_L = R_T \cdot \sqrt{m-1} \qquad\qquad (5.2)$$

For example, a network to match 50Ω down to 1Ω, at a center frequency of 1.9 GHz, would require the following element values:

Shunt capacitance: $m = 50$

$\qquad\qquad X_c = 50/7\,\Omega$ (from 5.1)

$\qquad\qquad C = 11.7\ \text{pF}\,(1.9\ \text{GHz})$

Series inductance: $X_L = 7\Omega$ (from 5.2)

$\qquad\qquad L = 0.59\ \text{nH}\,(1.9\ \text{GHz})$

A few points are worth noting here. The value of series inductance (0.6 nH) is quite small, by any standards, and may exceed the typical inductance of a transistor package. Also, the 12 pF capacitance represents an additional realization problem, in that a standard surface mount component will have sufficient series inductance to make a substantial impact on the nominal capacitor value. Not only that, but the Q-factor of the capacitor will need to be carefully considered if significant power losses are to be avoided. In this case, for example, if the capacitor has a specified Q-factor of 50, it will have an equivalent parallel resistance of $Q/\omega C$, or 2500/7 (taking $X_c = 50/7$), about 360Ω. It might be thought at first that such a high value could be neglected in comparison to the 50Ω termination, but in fact a simple calculation shows that it represents a power loss of about 1 dB. This not only represents a substantial loss of expected performance, but also a possible fire hazard in higher power applications.

Figure 5.3 shows a plot of the frequency response of this matching circuit, for a set of values of transforming ratio m. This plot uses a standard mismatch transmission loss format, where the assumption is that the useful bandwidth is defined by the 2:1 VSWR points, which correspond to a transmission loss of about 0.5 dB (shown dotted). On this basis, the network designed for a 50:1 transformation shows a 9% bandwidth. Figure 5.4 shows a plot based on a load-pull model, as described in Chapter 2. In order to generate this plot, the general formulation for load pull contours derived in Chapter 2 is used to convert a load impedance into an RF power measured relative to the optimum power with a termination of R_{opt} (=1Ω in this case). It can be seen that using this criterion, the 50:1 transformation network bandwidth is reduced to about 5% if the 0.5 dB contour is used as defining the useable range. (Note that in Figure 5.4, the mismatch curves have break points, due to the discontinuous nature of the load-pull contours.) Such a bandwidth would appear to be marginal, even for wireless communication bands, and this can, in principle, be resolved by using more matching sections.

The design of multisection matching networks has been well covered in the literature, and is now available in more user-friendly (and, unfortunately, user-ignorant) CAD formats. It is, however, quite easy to determine the approximate values of more complex networks by multiple application of (5.1) and (5.2). The matching problem is broken into two smaller steps by matching first to an intermediate value of resistance, and then using a second network to transform to the final value. A useful rule of thumb is that the terminating and intermediate impedances should be in a

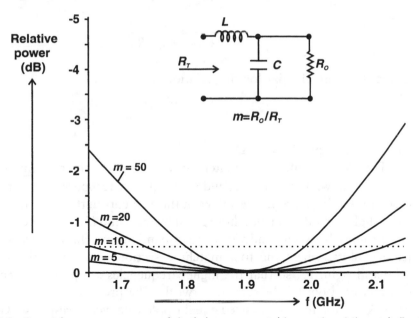

Figure 5.3   Swept frequency response of single low-pass matching section. Mismatch (in dB) is computed using conventional linear circuit theory for a mismatched source of resistance R_T.

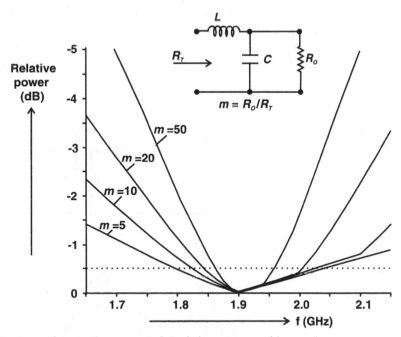

Figure 5.4   Swept frequency response of single low-pass matching section.

harmonic sequence. In this example, if the intermediate resistance is R_m, then the matching impedance R_t and the termination R_o will be related according to:

$$\frac{1}{R_O} \cdot \frac{1}{R_M} = \frac{1}{R_M} \cdot \frac{1}{R_T} \tag{5.3}$$

giving the more familiar looking relationship,

$$R_M^2 = \sqrt{R_M \cdot R_O}$$

so in the example, $R_M = 7.1\Omega$.

The element values and matching trajectories of the resulting two-section network are shown in Figure 5.5, and the frequency response is shown (for the $m = 50$ case only) in Figure 5.6. The values of the elements in the individual networks have to be slightly modified from those given by (5.1) and (5.2) in order to obtain the best compromise between bandwidth and in-band mismatch; this can be seen on the trajectory plots, where the first matching section overshoots the R_m target at the midband point F_M, so that the higher frequencies are past resonance; this is then compensated out by the second network. The process, as described, results in an approximation to a synthesized equal ripple Chebychev response. Generally speaking, the greater the overshoot of the R_m target, the wider the overall bandwidth, but the greater the in-band mismatch.

Such networks can be synthesized exactly using low-pass prototypes and appropriate transformations [1], although the process of transforming a Chebychev low-pass prototype into a low-pass, rather than a band-pass, final topology is not such a straightforward process. For the narrower bandwidths encountered in wireless communcations, the above approximate technique, coupled with some final optimization on a linear simulator, will usually yield the same result in a fraction of

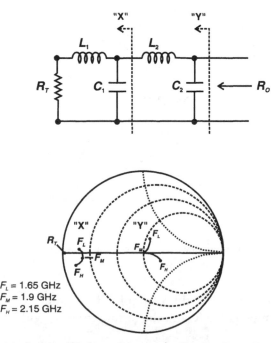

Figure 5.5 Smith chart trajectory of two-section match. Circuit parameters: $R_T = 1\Omega$, $R_o= 50\Omega$ (m = 50), $L_1 = 0.214$ nH, $C_1 = 32.3$ pF, $L_2 = 1.6$ nH, $C_2 = 4.3$ pF.

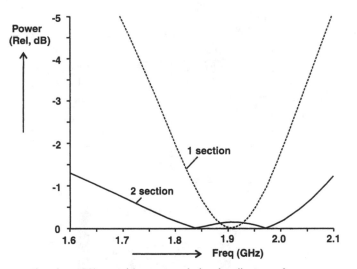

Figure 5.6   Two-section (*m* = 50) matching network; load-pull power-frequency response compared to single section response (dotted).

the time. Needless to say, the same answers can also be obtained by typing the required matching parameters into a CAD synthesis program. Each approach will give element values close to those shown in Figure 5.5. The two-section frequency response is clearly much broader in bandwidth, as shown in Figure 5.6. However the element values pose even greater realization problems than in the single section case. One has to conclude that matching down to 1 ohm at these frequencies is a marginal affair, especially in a PC board environment. If impedances get this low, chip and wire hybrid techniques probably have to be used to realize the necessary element values.

A typical practical application for this matching problem is illustrated in Figure 5.7. The design parameters would be typical for a high power LDMOS type device which is specified as giving 25W RF output in a Class A or AB mode, with a DC supply voltage of 28V. The corresponding DC current would be about 2A, corresponding to a drain efficiency of 45%. Using the loadline approach for Class A design developed in Chapter 2, and allowing a knee voltage of 3V (for a typical LDMOS device), the loadline resistance is (28–3)/2, or 12.5Ω. Note that this is hardly the low, unmatchable impedance with which these kinds of transistor are frequently discredited. Unfortunately, this kind of device will have a typical output capacitance of around 25 pF. Taking a convenient value of 23 pF, a simple application of (5.1) reveals that the real part of the loadline impedance measured outside the 23 pF capacitor is just about 1Ω at 1.9 GHz. (R_o = 12.5, X_c = 3.64, R_t = 0.98Ω.) So the "unmatchability" of RF power transistors is really more a function of their output capacitance than the loadline impedance itself. This causes some consternation when applying the loadline technique to large RF power transistors such as this; designers often have a hard time reconciling the simple loadline values with measured load-pull parameters measured outside the package. It is, however, a simple consequence of the increasing effect of the output capacitance at higher frequencies, and the measurement plane issue discussed in Section 2.5.

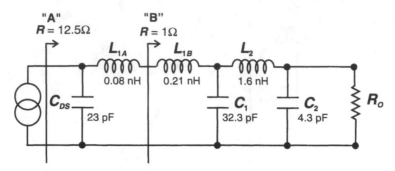

$$R_{OPT} = (28 - 3)/2 = 12.5\Omega$$

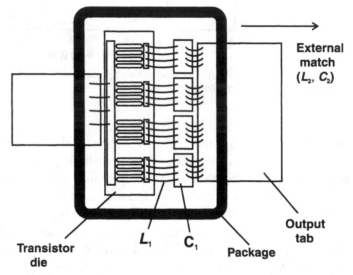

Figure 5.7 Typical high power 2 GHz transistor; two-section matching network and physical configuration.

This is of little comfort to the circuit designer, who still has to match to 1 ohm, although we will see in due course that the output capacitance can take on a more positive role in harmonic termination effects. The almost impossible-looking element values shown for the two-section matching network in Figure 5.7 can actually be realized quite accurately in chip-and-wire form, shown also in Figure 5.7. The two-section matching network designed above can be used directly, but an additional series inductor ($L_{1A} = 0.08$ nH) is required to complete the step-down lowpass transforming section mandated by the transistor output capacitance. The 0.21 nH series inductor is made up of several equal parallel bondwires running symmetrically from different parallel cells of the transistor, and the large shunt capacitor (C_1) will be a custom-made piece of metalized ceramic, whose dimensions match up with the transistor footprint. The spacing of the transistor and the capacitor, and the exact profile of the wire bonds have to be carefully controlled, preferably with automated equipment. The ceramic material has to be carefully chosen, in order to have an acceptable Q factor and practical physical dimensions. In more extreme cases, not only do the capacitor dimensions have to be customized, but even the ceramic material itself has to be specially mixed.

As shown in Figure 5.7, the first chip-and-wire matching section needs to be placed inside the transistor package, and so the designer only has to supply the second section, which will appear to have a more moderate matching requirement (about 7Ω in this case). In this manner, the designer, or user, is largely oblivious of the manufacturing nightmare which lies safely concealed inside a typical high power RF transistor package. The increasing volume demands of the wireless communications industry has however addressed this issue impressively, and such assembly can now be performed with great precision and repeatability using automated machines. This in turn has changed the climate for the designer who can now look to using multiple section networks without the traditional fear of manufacturing and tolerance problems.

In many cases, the RF designer will be dealing with less extreme matching problems. The simple matching networks so far considered can have both fully and partially distributed forms. Usually, the shunt capacitor will be realized as a lumped element component, and the series inductor as a microstrip transmission line. High-pass versions can also be implemented; the same formulae can be used, interchanging the capacitive and inductive reactances. Figure 5.2(c) shows the configuration, and the values for a 1Ω to 50Ω transformation will be:

$$L = 0.59 \text{ nH}$$
$$C = 11.7 \text{ pF}$$

These are the same values as in the low-pass circuit but with a different configuration. In practice, the additional matching compensation required due to the output capacitance of the transistor will cause there to be a significant difference between low-pass and high-pass element values. This high-pass network has the same properties in terms of in-band performance as the low-pass counterpart, but is not generally as favored in RF design. One problem is that the orientation of the inductor and capacitor in the high-pass configuration cannot be realized physically without additional series elements, although the real reason may be the relative ease with which a low-pass network can be adjusted or "tweaked" by moving a shunt capacitor along a length of microstrip line.

An interesting variant of the basic two-element low-pass matching network is the "Pi" form, shown in Figure 5.8. In its generalized asymmetrical form this network is quite relevant to PA stage matching, because there will always be some output capacitance associated with a transistor output. As we have already seen, this can have a major impact on the matching network design. The design equation for the symmetrical form is also of interest, due to its similarity to a transmission line transformer:

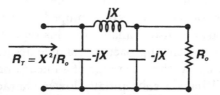

Figure 5.8 Symmetrical Pi form of low-pass matching network.

$$R_T \cdot R_o = X^2 \text{ where } X_L = X_C = X$$

so in the above example, $R_T = 1$, $R_o = 50$, $X = 7.07\Omega$ giving at 1.9 GHz, $L = 0.59$ nH, $C = 11.85$ pF.

These values are very similar to the single two-element section; the additional placement of a 7Ω capacitive reactance in shunt with the 1Ω impedance level at this point only requires a minor retuning of the existing elements. The symmetrical network replicates the action of a quarter-wave transmission line more closely, over an extended bandwidth, than the asymmetrical two-section network, and it also forms a basis for making hybrid structures on high dielectric material, which can simulate the effect of very low impedance transmission lines.

5.2 Transmission Line Matching

An important alternative matching technique is the quarter-wave matching transformer. In its simplest form, illustrated in Figure 5.9(a), a quarter wavelength transmission line of characteristic impedance Z_T is placed between the terminating impedance R_0, and the required match R_T. Transmission line theory gives the well-known result:

$$Z_T^2 = R_O \cdot R_T \tag{5.4}$$

As with the lumped element match, the bandwidth will be a strong function of transformation ratio, and for high ratios multiple sections can be used. For the 50:1 transformation example in the last section, the quarter-wave has slightly higher bandwidth (12%) but requires an uncomfortably low characteristic impedance of 7.1Ω. Such a low impedance would require very thin dielectric board, or a very wide line width, which would no longer necessarily behave as a simple one-dimensional TEM transmission line. Up to 2 GHz, the larger physical size of a quarter-wave section usually means that a lumped element approach is preferred. There are some important caveats, however, to consider in making this judgment. The transmission line can be etched into the board pattern, and would require no additional component placement; the Q-factor requirements of an equivalent lumped element network may incur substantial extra component cost; larger RF power transistors require a wide physical interface with the circuit board, which is conveniently provided by a low impedance transmission line. The main disadvantage, which is physical size, can be alleviated by using higher dielectric constant materials and transmission line configurations, although this may impact the cost.

As with the lumped element matching networks, broader bandwidth can be obtained by the use of multiple sections. The intermediate impedances are selected in the same way as for the multisection lumped element networks (5.3). More precise synthesis methods are also available for this type of matching network [2], but as with the lumped element approach, the simple design equations given here are sufficient to put a design within the range of a linear optimizer. The problem is that the first section will have a substantially lower characteristic impedance [see Figure 5.9(b)] than the single section transformer and thus presents even greater realization

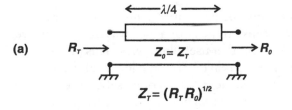

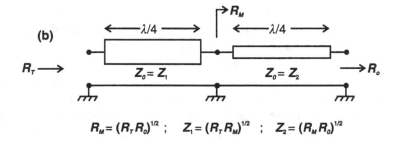

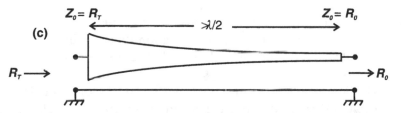

Figure 5.9   Transformers: (a) single quarter-wave transformer, (b) two-section quarter-wave transformer, and (c) tapered transmission line transformer.

problems. Such networks consume a fair amount of space and are not used much in active RF matching design, especially with the availability of higher quality capacitors.

There is a further logical extension to the multiple quarter-wave transformer, which is the tapered transmission line [see Figure 5.9(c)]. If several quarter-wave sections are cascaded, the steps in line width at the junctions become quite small, and can be smoothed out to give a line whose impedance transitions gradually from one termination impedance to the other. This tends to have the general effect of removing the band-pass characteristic at the high frequency end, while increasing the frequency of the low end roll-off. The long, slow, impedance taper many wavelengths long represents a physically inconvenient but quite admirable method of achieving very wide bandwidth impedance transformation. There has been much published analysis on this class of transmission line [3], but their applications have been largely limited to very specialized applications, such as fast pulse transmission. It falls into a general class of older microwave engineering tricks which are worth an occasional revisit.

Transmission lines have certain properties that cannot be replicated using lumped element networks. The ability to match two complex impedances with a single transmission line of prescribed length and characteristic impedance is one example [4]. The "transparency" of a half-wavelength line, regardless of its characteristic impedance, is another useful property. This property will be utilized in Chapter 12 in formulating a passive harmonic load-pull tuner.

Transmission line elements can, conversely, be used to simulate the effects of lumped element networks. At higher microwave frequencies, it is common practice to design a matching network using a prototype lumped element structure, and then translate it into distributed form. Shunt capacitors become open circuit shunt stubs (OCSS), shunt inductors become short circuit shunt stubs (SCSS), and series inductors, more restrictively, can be replaced by short lengths of series high impedance line. Series capacitors pose a problem at higher microwave frequencies in that they do not have a simple distributed equivalent. They are generally avoided as matching elements altogether due to parasitic and tolerancing issues. Chip capacitors are used for the larger values required for DC blocks, but have to be carefully modeled, since they are quite often being used well above their self-resonant frequencies.

In portable wireless communications applications, transmission line structures are in general avoided, mainly due to board space (e.g., in handsets) and the extra cost of board material with suitable dielectric properties. There is an additional problem concerning the need to maintain a free area around an RF distributed element to avoid coupling effects; this also can lead to layout problems in densely packed boards. Although these considerations may be less critical in high power, higher "ticket" products such as basestation PAs, it will be assumed in this chapter that nondistributed, or partially distributed, networks will be required in some applications.

The Smith chart was, and still is, the universal tool of the RF designer for matching network design. Although the quantitative part has been largely taken over by the computer, it still remains a good tool for visualizing the action of a multiple element matching network, with or without transmission lines. The use of a polar reflection diagram to perform transmission line calculations actually predates Smith's chart; the unique contribution of Smith was to lay a grid of resistance and reactance (and/or conductance and admittance) on top of the polar reflection plot, giving the world a universal impedance calculation and visualization device. Due to the fact that the chart is still widely regarded as a transmission line calculator, Figure 5.10 is included here to show the basic impedance trajectories of all four lumped element types (series/shunt L,C) and a general (nonnormalized) transmission line from any point.

The transmission line trajectories are probably less familiar, other than the $Z_o = R_o$ case. The lumped element trajectories define a boundary, inside which any impedance can be matched to the original starting impedance using a prescribed length of transmission line of suitable characteristic impedance. This is worth noting in higher power applications, where lumped element Q-factors can seriously degrade the performance of a conventional matching network.

5.3 Shorting the Harmonics

The matching networks discussed in the previous sections will be used primarily to match the fundamental load impedance, R_{OPT}, as defined in Chapter 3, to the 50Ω

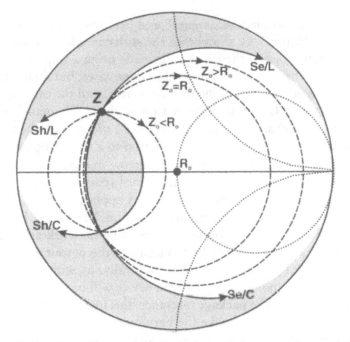

Figure 5.10 Smith chart impedance trajectories for lumped element and transmission line elements from a generalized impedance Z. Any impedance in the nonshaded area can be matched using a single length of transmission line of suitable characteristic impedance.

termination, or to the input of the next stage in the case of a driver. The need for a short circuit at harmonic frequencies is an additional requirement that will initially be realized using a separate network. Ideally, this network will not interact with the fundamental matching circuit; in practice some readjustment may be necessary. It is important to note that, as shown in Figure 3.2, the most important harmonic to deal with is the second; in fact in an ideal Class B device the odd harmonics will be zero. Even when operating in deep Class AB mode, it will be seen in Chapter 6 that some odd harmonic voltage components can be quite beneficial and probably need not be completely shorted.

The classical method, as described in Chapter 3, was to use a high-Q parallel resonant circuit on the output of a tube amplifier. The capacitor value was chosen such that its reactance at harmonic frequencies was many times lower than the fundamental loadline resistance. This was not a difficult task when the loadline was maybe 10 kΩ. In the case of a transistor, the capacitor value quickly reaches unrealizable levels. For example, at 1.9 GHz, the 25W LDMOS device shown in Figure 5.7, a shunt reactance 10 times lower than the 12.5Ω loadline resistance would correspond to a shunt capacitor value of 67 pF, which would require to be shunt resonated at the fundamental by an inductance of 0.1 nH. These values really are across the dividing line of realizability at this frequency, although it will be shown later (Section 5.8) that a more modest shunt capacitance at the transistor output can still give useful performance.

The simplest solution for a second harmonic short is a quarter-wave short-circuited stub (SCSS) at the fundamental frequency. This has the advantage of being an open circuit at the fundamental, and presents a short circuit at successive even harmonics. Figure 5.11 shows that the stub can additionally provide a convenient bias insertion point, if the RF short is made using a suitable bypass capacitor. Such an arrangement is certainly practical in some cases, but there are two problems. One concerns the bandwidth of the stub resonator, and the other is a more general issue concerning the placement of the harmonic short. The bandwidth is a fairly strong function of the characteristic impedance, Z_o, of the stub; the bandwidth over which it appears as an acceptable approximation to a short improves as Z_o is reduced, as shown in Figure 5.12. Conveniently narrow microstrip widths, such as 70–100Ω can be meandered around to reduce board space, but exhibit narrow bandwidths which are too small for some practical PA applications.

Placement of the resonator causes additional problems. Figure 5.13 shows a typical schematic for a transistor with a matching network and a harmonic resonator. The harmonic short needs to be connected to the output of the device, defined by its internal current generator terminals; typically, as with the matching network, the closest accessible point for a packaged device will be the tab of the package. Depending on the value of the package reactance, this may have a serious de-tuning effect on the resonator, and may make the harmonic short difficult to realize at the correct point. Unless the package reactance is quite low compared to the series-inductive matching element, the detuning effect cannot be simply eliminated by adjusting the stub length. One simple remedy would be for the manufacturer to provide a separate, isolated, pinout from the main drain (or collector) tab for the purposes of providing the necessary short circuit. Most of the lower cost devices available for handset PA applications have multiple output pinouts which are adapted for this use.

Unfortunately, the situation is less promising for users of higher power RF transistors, which usually have one large tab. If there is internal matching as well, as described in the previous section, the transistor output becomes a truly distant and inaccessible point for the user. In these applications, the final network is often arrived at by making load-pull measurements, with the general intent of letting the harmonics take care of themselves. This scenario will be considered in a more quan-

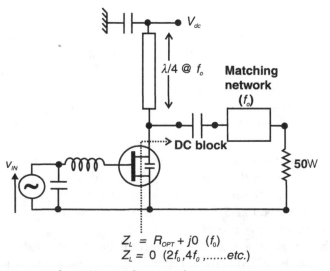

$$Z_L = R_{OPT} + j0 \quad (f_o)$$
$$Z_L = 0 \quad (2f_o, 4f_o, \ldots\ldots etc.)$$

Figure 5.11   Quarter-wave short-circuit stub as even harmonic trap.

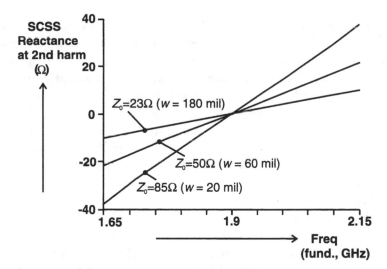

Figure 5.12   Reactance of short-circuit quarter-wave stub at second harmonic frequency. Dimensions are widths and lengths for 0.031 FR4 microstrip implementation.

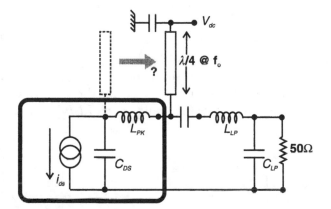

Figure 5.13   Quarter-wave SCSS placement problem for packaged device.

titative fashion in Section 5.8; it turns out that in many cases, the output capacitance can provide a built-in solution to the harmonic termination problem. This has an ironic twist, in that a transistor technology with a much higher MAG (and hence, usually, lower parasitic capacitance) may not perform in PA circuits at lower frequencies as well as a device using a lower gain process, unless more specific harmonic termination is provided.

5.4   A Generic MESFET

We are now in a position to design and simulate a high efficiency amplifier. The first step is to select a suitable transistor, and for most of the design examples in this book a generic MESFET is used. The term "MESFET" is chosen deliberately to be generic, and is intended to be representative of the various RF and microwave FET processes that are currently used. This model can be used to simulate Si and GaAs discrete devices available from several manufacturers in low cost packages; with some minor changes to the parameter values it can also be representative of numerous GaAs MMIC foundry processes. These devices would be used primarily in lower voltage applications, but the general design principles are equally applicable to higher voltage and higher power technologies such as LDMOS and newer high bandgap devices such as gallium nitride (GaN) and silicon carbide (SiC). In general, the same simulation files can be used with these alternative processes by simple scaling of voltages, currents, and suitable adjustments to the parasitic capacitor values. A simulation representing BJT applications will be presented in Section 5.9.

Figure 5.14 shows the PSpice generated I-V characteristics, and the list of model parameters. For illustration purposes, and for maximum compatibility with different implementations, the Statz model ("B" device, Level=2, [5]) is used. In this particular case, the device in question does not come supplied with Spice model. Such is frequently the case in RFPA design. There is usually enough basic information on a data sheet, however, to build up a useful model. The following procedure was followed in this case:

1. Set the V_T parameter to the device pinchoff.
2. With the *alpha* and *lambda* parameters at default values (2, 0.3) adjust the *beta* parameter to give the specified *Idss* (for V_{ds} and $V_g = 0$). (Data sheets normally specify *Idss*, not I_{max}.)
3. Sweep the I-V charactersitics and adjust *b, alpha, lambda* to get the best fit. The *alpha* parameter mainly controls the knee voltage, *lambda* controls the slope, and *b* controls the "tail" of the *Ids/Vgs* characteristic.

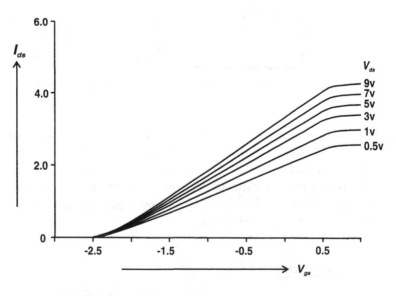

```
b1 n1 n2 n3    genfet 10.0
.model genfet gasfet level=2 beta=.310 vto=-2.5
+ alpha=2.0 b=3.0 cds=.001 pf cgs=.001 pf cgd=.06 pf
+ rg=.001
```

Figure 5.14 Generic MESFET Spice model. Values for Cds, Cgs are small values to eliminate voltage dependency (see text).

4. Some readjustment of *beta* may be necessary, especially after setting *lambda* to a nondefault value.

5. The parasitic elements *CDS, CGD, CDS, RG, RD, RS* can be identified from a linear model, which is usually available either in the data sheet or from the manufacturer. Spice assumes these to be zero bias values, so these may need some final adjustment in the next step (6). Most implementations of Spice have built-in equations which characterize the voltage dependency of the junction capacitances, and these can be unrepresentative when dealing with the large voltage swings encountered in power amplifiers. A simple and recommended initial remedy is to neutralize the voltage variations, either by using external linear elements or by setting internal program defaults.

6. Run a simulation of the device, with 50Ω terminations, over a broadband linear sweep, and compare the forward gain (Mag s_{21}) and the reverse gain (Mag s_{12}) with the data sheet s-parameters. Final adjustments can be made to match up s_{12} (primarily by varying *CDG*) and s_{21} (primarily by varying *lambda* and *CGS*).

This procedure gives a useful but approximate model, and can be applied equally easily to any kind of FET device type. There is usually a problem getting a close fit to both the DC and RF gain parameters, especially if Spice routines are available to compare s_{11} and s_{22}. In particular, it is frequently found necessary to change *beta*, or V_p, in order to get a good agreement on s_{21}, which for an RF design is usually judged to have more importance. This problem has triggered a major expenditure of effort over the last few years, in attempts to explain the problem and to generate a grand unified model ("GUM") which works for both DC and RF accurately. But useful results can be obtained with the approximate model described here.

5.5   A 2W Class B Design for 850 MHz

A design for a Class B amplifier will now be attempted, assuming a supply voltage of 5v.

Step 1: Establish I_{max}, R_{opt} (Effective peak current, fundamental load impedance).
From the results of Chapter 3, the optimum load for Class B operation is the same as the linear loadline impedance, and can be approximated by

$$R_{opt} = (V_{dc} - V_k)/(I_{max}/2)$$

So a value for I_{max} needs to be obtained from the I-V plot shown in Figure 5.14; this is not as easy to define on a real device as was assumed in the idealized models; the key point is that we have to assume that for any realistic design, the peak RF current will coincide with low output voltage. This will tend to make the effective saturation characteristic much harder than appears on an I-V plot; 2.5A will be used for this example, this being the current at 0.5v Vds, +0.5v Vgs.

$$\text{So } R_{opt} = (5 - 0.5)/1.25 = 3.6\Omega$$

Clearly, this value is an approximation based on the fact that there will be a compromise between how far the voltage swings into the knee region, and the desire to use the maximum possible current swing for good PUF. It is therefore instructive to run a few basic simulations with a simple load resistor and harmonic "tank" circuit, in order to observe this tradeoff in a more quantitative manner. Figure 5.15 shows the simple schematic used for this simulation.[1] The parallel resonator across the load resistor has unrealisitic values for practical realization, but is a simple and convenient way of simulating a harmonic short. It should be noted in Figure 5.13 that in practice a large DC block would normally be required to prevent the shunt resonator inductance from shorting out the DC supply. When dealing with RF loads as low as a few Ohms, the value of this capacitor has to be correspondingly high. One of the features of Spice is that a DC voltage source is treated as an RF short, so that in the simulation the series DC blocking capacitor can be eliminated by returning the RF ground connections to the DC supply point, saving some computing time. Fortunately, the more practical configurations considered in due course have no need for the DC block, or can incorporate it at a higher impedance level. A simple input match has also been included in the simulation. This low-pass network transforms the 50Ω termination down to a lower value, increasing the voltage swing at the transistor gate. It should be noted that in setting values for the amplitude of the sinusoidal input generator, the voltage swing at the gate itself is of most relevance in setting the mode of operation and drive level; these two voltages will have an approximately linear relationship until strong compression occurs.

Figure 5.16(a) shows a set of current waveforms for different values of RF load resistance, for Class B bias and maximum (unsaturating) RF input drive. Higher values of R_L cause deeply bifurcated current pulses, as predicted in Chapter 3, while lower R_L values cause higher peak currents and a lower voltage swing. The resulting tradeoffs in power and efficiency are shown in Figure 5.16(b). It can be seen that the

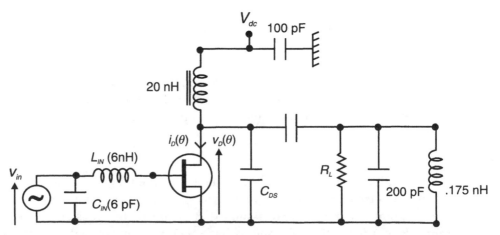

Figure 5.15   Circuit for class B amplifier simulation.

1.   The file for all of the simulations in this example (ABFD.cir) can be found on the accompanying CD-ROM.

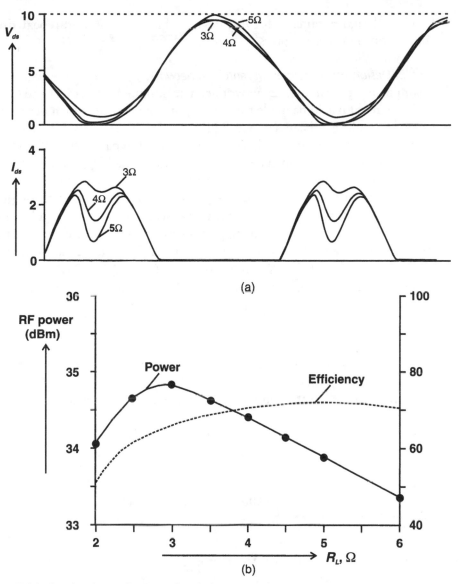

Figure 5.16 Load resistor selection, Class B design: (a) device voltage and current waveforms, and (b) power/efficiency tradeoff due to current bifurcation.

bifurcated current waveform produces higher efficiency but at relatively lower output power. This is consistent with the idealized waveforms which were analyzed in Chapter 3. The lower R_L values show higher power but lower efficiency, due to the reduced voltage swing. This general performance trait can often be observed in practice, and is largely a function of the mathematics of a bifurcated current pulse.

The lower R_L values will also display more linear input-output characteristics up to a higher power level. The final choice of R_L will depend on the application; for a constant envelope signal the higher efficiency will be more attractive but for the more usual situations where the input signal has amplitude modulation, a lower R_L value would be selected for linearity and PUF, with an inevitable loss in efficiency. In this design we will settle on a compromise value of $R_L = 4\Omega$; this is an important baseline design parameter for this device which will be used repeatedly in several simulations in other chapters.

Step 2: Design the fundamental matching network.
A simple low-pass matching network can be designed using (5.1) and (5.2); however an adjustment has to be made for the output capacitance, which in this case is taken to be the small signal capacitance C_{DS} (Figure 5.17). As explained in Section 5.1, this output capacitance effectively forms a second lowpass network, which unfortunately transforms the impedance the wrong way initially. Although in this particular example, the output capaitance of 2.5 pF has an almost negligible reactance at the design frequency of 850 MHz, the process will be followed through because in higher power designs the impact of the transistor capacitance can be much greater. We will also see, later in this chapter, that supplementing the output capacitance can be a convenient and effective way of providing an acceptable harmonic termination.

The reactance of C_{DS}, X_{CDS} is $1000/(2.\pi.).(0.85).2.5) = 75\Omega$.

From (5.1), the desired impedance of 4Ω at plane "A" is transformed down to

$$4\big/\left(1+\left(4/75\right)^{2}\right) = 3.99\Omega$$

So we can safely ignore the transforming effect of C_{DS} in this example and move on to design the matching network:

The reactance of the shunt capacitor, X_{CLP}, is given, from (5.2) by:

$$X_{CLP} = 50\big/\left(11.5\right)^{0.5} = 14.7\Omega$$

So

$$C_{LP} = 1000\big/\left(14.7 \cdot \left(0.85\right) \cdot \left(2\pi\right)\right)$$
$$= 12.7 \text{ pF}$$

The reactance of the series inductor, X_{LLP}, is given from (5.2) by:

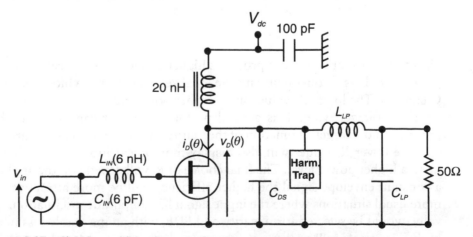

Figure 5.17   Class B amplifier schematic: low-pass match and harmonic short.

$$X_{LLP} = 4 \cdot (11.5)^{0.5} = 13.6\Omega$$

So the series inductor $L_{LP} = 13.6 / (2\pi).(0.85) = 2.5$ nH.

Step 3: Add a suitable harmonic trap; readjust fundamental match (if necessary).
The circuit will initially be evaluated using the ideal parallel resonant tank circuit used in Step 1 above. More practical harmonic traps will then be evaluated and compared. So the only adjustment initially is to absorb the C_{DS} into the calculated value for the parallel resonant capacitor.

Step 4: Run simulation.
The drain current and voltage waveforms are shown in Figure 5.18. These results should be compared directly with the $R_L = 4\Omega$ case in Figure 5.16. The results are very similar due to the near-ideal nature of the harmonic trap.

Step 5: Use a more realizable harmonic trap.
The simplest realizable harmonic trap is a quarter-wave SCSS at the fundamental frequency. In the spirit of realizability, the characteristic impedance needs to be on the higher side so that it can be meandered into a small space; 50Ω is chosen initially. It is also assumed that the stub can be physically located at the output of the transistor; this means having a package with very low series inductance, or with some form of separate pin access as already discussed. For applications such as this, below 1 GHz, some of the better package designs may have sufficiently low inductance for this simple arrangement to work satisfactorily. Figure 5.19 shows the result of rerunning the previous simulation with all other parameters unchanged. Comparing these waveforms, and power/efficiency calculations with Figure 5.18, it can be seen that there is a significant difference now, mainly due to the presence of odd harmonics in the voltage waveform. The stub, of course, is an open circuit and performs no function at odd harmonics. But the differences are surprisingly positive; the bifurcation of the current pulse has gone, due partly to a slightly lower

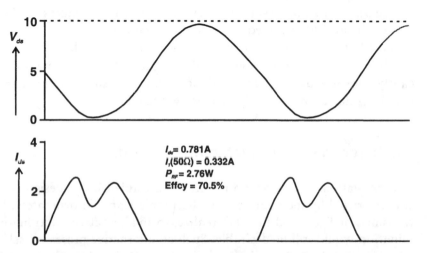

Figure 5.18 Device voltage and current waveforms: low-pass matching section and idealized harmonic trap (see Figure 5.17).

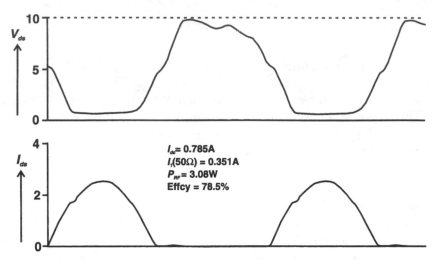

Figure 5.19 Device current and voltage waveforms: ideal harmonic trap replaced by 50Ω quarter-wave short circuit stub.

peak-to-peak voltage swing, but simultaneously the fundamental power has increased by about 0.5 dB compared to the ideal tank circuit, with a corresponding increase in efficiency to 78.5%.

This improved performance represents a fortuitous discovery of the possible beneficial effects of retaining odd harmonic components in the output voltage waveform of a power amplifier. The squaring effect of the odd harmonics gives a voltage waveform with lower peak to peak swing and a similar or slightly increased fundamental component. The matching network element values could, in fact, be now modified to present a higher value of R_L. These issues will be treated in a more formal way in Chapter 6, under the headings of Class F and Class FD modes. In principle, this technique can raise the efficiency up to about 90% if some power reduction can be tolerated.

Step 6: Frequency sweep.
Figure 5.20 shows the result of sweeping the circuit of Figures 5.18 and 5.19 over a modest bandwidth centered on the design frequency. The satisfactory performance obtained at the design frequency can be seen to roll off quite fast, which can be attributed to the narrow resonance of the 50Ω stub. The bandwidth could be significantly improved by lowering the Z_0 of the stub, but such lower values present layout problems due to their rapidly increasing line width.

5.6 The Pi Section Power Matching Network

The simulations in Section 5.5 resulted in a viable Class B design whose bandwidth was restricted by the quarter-wave harmonic trap. Unfortunately, low impedance transmission lines are difficult to realize, so an alternative solution for harmonic termination would still be desirable. Such an alternative presents itself in a serendipitous manner when attempting to design high efficiency amplifiers using devices with high output capacitance; a large capacitance will provide a low reactance to higher

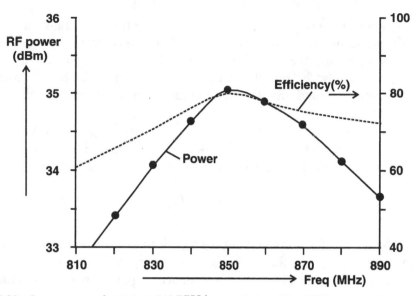

Figure 5.20   Frequency performance, λ/4 SCSS harmonic trap ($Z_0 = 50\Omega$).

harmonics while being resonated out at the fundamental. This turns out to be an irony of the highest order, in that more exotic, high GHz device technologies used at lower frequencies in the UHF or low microwave range may give disappointing results in PA circuits unless special provisions are made for harmonic termination. Lower frequency devices, with higher parasitic capacitances will often deliver higher efficiency in circuits which lack such provision. This has some important implications for load-pull characterization as well as amplifier circuit design.

So we return to reconsider the pi-section match, shown in Figure 5.21. In principle, it will always be possible to make the input (drain or collector) capacitance indefinitely large so that the harmonics will be terminated with a very low reactance. The remaining low-pass section can then resonate out the input capacitance and provide the necessary transformation at the fundamental. The central issue becomes how large does C_{DS} have to be in order to be realizable and at the same time have enough effect as a quasi-harmonic short. There is, of course, the additional issue of the extra impedance transformation required due to the larger capacitance.

These issues will be treated in a quantitative manner shortly, but as an initial attempt the network in Section 5.5 will be reconfigured with a much larger input capacitance; a value corresponding to a reactance of $-j.R_L$ at the fundamental will be initially selected. The rationale for this selection is as follows: the capacitive

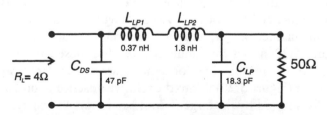

Figure 5.21   Pi-section power match and harmonic trap (850 MHz).

reactance at the second harmonic is half of the fundamental value; in a Class B circuit, the second harmonic component of current is just under half of the fundamental component; the resulting (undesirable) second harmonic voltage will be (approximately) 90° out of phase with the fundamental. These factors all contribute to reducing the undesirable voltage component to about one fifth of the fundamental amplitude.

Once again, (5.1) and (5.2) can be used to design the network (Figure 5.21). First, the desired final R_L value of 4Ω is transformed downwards, now much more substantially, by the larger shunt capacitor. The transformation, m, is given by:

$$m = 4 \Big/ \left(1 + (1)^2\right) = 2$$

So here is the downside of the large drain capacitance; instead of matching the fundamental to 4Ω, the target has now dropped by a factor of 2, to 2Ω. The 2Ω impedance has to be brought to resonance by part of the series inductor which follows it; this value is given by (5.2):

$$X_L = 2 \cdot (m - 1)^{0.5} = 2\Omega$$

giving a series inductance L_{LP1} of 2/(2.π.0.85) = 0.37 nH.

This is a very low value, lower than any package inductance. Fortunately, it can be absorbed into the inductance L_{LP} which forms the lowpass transforming network; this can now be designed using a value of $m = 25$:

$$X_{LLP} = 2 \cdot (24)^{0.5} = 9.8\Omega$$

$$\text{So } L_{LP2} = 1.8\text{nH}$$

This value can now be added to the 0.37 nH required to resonate the C_{DS} capacitor, giving a final value of 2.2 nH for the series inductance. The matching shunt capacitor, C_{LP} can be determined using $m = 25$ in (5.1):

$$X_C = 50 \Big/ (24)^{0.5} = 102\Omega$$

$$\text{So } C_{LP} = 18.3\text{pF}$$

Finally, the actual value of C_{DS} will be 1000/(4.2π.0.85) = 47 pF. This is still an uncomfortably large value for this frequency, which could not be realized using a simple surface mount component. It could, however, be integrated onto the chip itself or a high frequency capacitor chip could placed inside the package. (Note that a lower frequency MOS or an LDMOS device would provide a substantial amount of this capacitance *in situ*, with no additional elements.)

Figure 5.22 shows the simulation results; these can be compared directly with Figure 5.20. The midband performance is somewhat lackluster in comparison to the SCSS circuit (Figure 5.20) due to the relative absence of odd harmonics which will be essentially shorted by the large output capacitor. But there is some improvement in performance variation versus frequency, for a configuration that will be much more

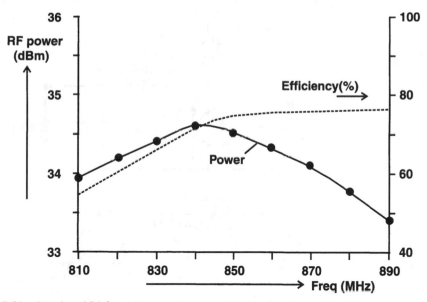

Figure 5.22 Simulated PA frequency performance, Pi network ($X_{cds} = -jR_L = 4\Omega$).

convenient to realize in practice. The question now arises, for the pi network, as to whether the intuitive value of C_{DS} used in this example ($-jR_L$) represents the best compromise between realizability and performance. This question is now addressed in a more quantitative manner.

5.7 Pi Section Analysis for PA Design

The schematic diagram of Figure 5.23 shows an ideal reduced conduction angle mode amplifier, with pi-section matching network. The intent is to select a value of C_{DS} that satisfies the following requirements, as nearly as possible:

1. The value shall be larger than the intrinsic transistor output capacitance.
2. The capacitive reactance shall be low enough to reduce the second harmonic of voltage at the transistor output to levels where negligible degradation of power and efficiency (compared to an ideal short) are obtained.
3. The value should be realizable, as a minimum using on-chip or off-chip (hybrid style) capacitors.
4. The reactance shall be high enough such that the downward impedance transformation of R_L, as measured at the transistor current generator (plane "A"), still results in an impedance level that can be matched over the required bandwidth using realizable element values.

This list clearly contains several contradictory requirements, and a quantitative analysis is needed in order to examine the tradeoff carefully in specific cases. The general analysis of the nonlinear problem represented in Figure 5.23 is quite cumbersome, and in the usual spirit of *a priori* design methodology, some simplifying assumptions will be made:

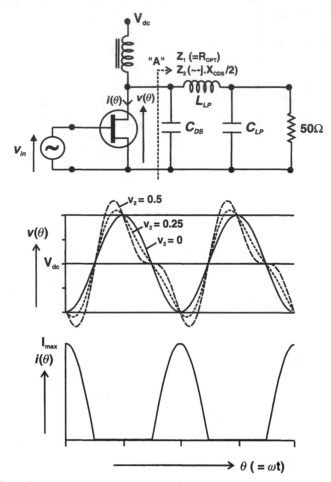

Figure 5.23 Effect of reactive second harmonic voltage component (resistive fundamental termination).

1. The pi section will always be designed such that the fundamental impedance presented at plane A will be resistive and equal to the appropriate value of R_L for the selected mode of operation.

2. The only additional harmonic considered will be the second. This is justified for the following reasons:

   (a) All odd harmonics are much lower level than the second in AB to B operation (see Figure 3.2) and the capacitive reactance is proportionately lower at higher harmonic frequencies.

   (b) Higher even harmonics are much lower than the second, and the capacitive reactance is proportionately lower.

3. The low-pass matching section will be assumed to have a high Q factor, and thus presents a relatively high reactive impedance in shunt with the C_{DS} capacitor. The second harmonic impedance will therefore be well approximated by the reactance of C_{DS} alone.

With these assumptions, the analysis becomes quite straightforward. The fundamental component of current I_1, given by (3.3) sees a termination of R_L only, and so a fundamental sinusoidal voltage component of amplitude

$$V_1 = I_1 \cdot R_L \qquad (5.5)$$

will be developed at the transistor output. The second harmonic component will have an amplitude given, based on the above assumptions, by

$$V_2 = I_2 \cdot X_{CDS} / 2 \qquad (5.6)$$

where X_{CDS} is the reactance of the C_{DS} capacitor at the fundamental; the factor of 2 accounts for the frequency of the second harmonic. The second harmonic voltage will, of course, be 90° out of phase with the fundamental.

Initially, it is clear that as I_2 is reduced by moving to lighter AB modes of operation, the overall effect of the second harmonic will, in any case, be reduced. Assuming that Class B operation then represents the worst practical condition, Figure 5.23 shows the effect on the voltage waveform as X_{CDS} is increased from the ideal short circuit to take on values which are normalized to the fundamental load resistance R_L (note that the normalization of X_{CDS} is performed at the fundamental).

The modified voltage waveform appears to increase the original (ideal) sinewave peak value in a symmetrical fashion. So the RF voltage swing is trying to drop below zero, which will cause a collapse of the current at a point close (but no longer coincident with) the current peaks. Clearly, the current can no longer be sustained and any one of three actions[2] can be taken to accommodate the increased voltage swing:

1. Reduce the value of R_L.
2. Reduce the drive level to reduce I_1.
3. Increase the supply voltage.

All three actions will have a detrimental effect on efficiency. (3) will maintain the RF power output, but is excluded here because in PA applications the DC supply is often an important specified design goal, and may in any case be limited by transistor breakdown considerations. Reducing the drive level, (2), will also increase the conduction angle, which complicates the conclusions on efficiency. The analysis will proceed, therefore, on the assumption (1) that the fundamental voltage swing must be reduced, through adjustment to R_L, to accommodate the increase in peak-to-peak swing caused by the second harmonic.

To determine the necessary reduction in the value of R_L for a known value of V_2, it is necessary to find the peak value of the voltage function

---

2.   It turns out that there is another possible action, which is to introduce a reactive component in the fundamental match. This was discussed in Chapter 4 (Class J) and will be further considered in a later section of this chapter. The Class J approach does, however, still involve an increase in voltage swing which differentiates it from the three choices considered here.

$$v(\theta) = \cos(\theta) + v_2 \cdot \sin(2\theta) \qquad (5.7)$$

where v_2 is the second harmonic voltage amplitude normalized to the fundamental. The functional dependency of the peak-to-peak value of $v(\theta)$ on v_2 is surprisingly cumbersome if derived analytically and is more easily obtained by direct numerical computation, as shown in Figure 5.24. Once a representative value for this peak voltage, v_{pk}, is obtained, it is a straightforward matter to determine a revised value of R_L, R_{L2}, which will be required to remove the peak negative voltage swing:

$$R_{L2} = \left(\frac{1}{v_{pk}}\right) \cdot R_L \qquad (5.8)$$

So the maximum unsaturated power, P_2, will be reduced from the "ideal" power (i.e., with $v_2 = 0$, corresponding to ideal harmonic short) P_1, according to:

$$\frac{P_2}{P_1} = \left(\frac{1}{v_{pk}}\right) \qquad (5.9)$$

since the assumption has been made that the drive level, and current waveform, will be maintained the same as in the ideal case.[3] Also, by the same assumption, the DC components of current and voltage will be the same as the ideal case, so (5.9) represents a corresponding reduction in efficiency due to the presence of second harmonic in the voltage waveform.

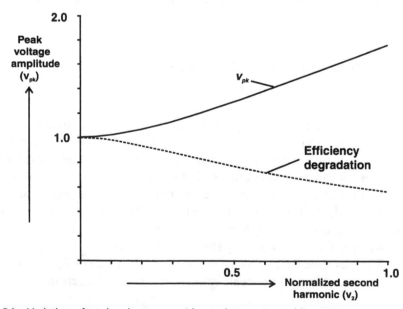

Figure 5.24   Variation of peak voltage, v_{pk}, with quadrature second harmonic component, v_2. Corresponding reduction in power (and efficiency) is shown dotted.

---

3.   Note that the reactive component of voltage, v_2, is also scaled by the same factor as v_1 in this formulation. This has been accounted for in computing the values of XCDS in Figure 5.25.

The results of (5.5) through (5.9) can now be combined into an overall design guide chart, shown in Figure 5.25. The key issues addressed in this chart are as follows:

1. For a chosen class of operation (AB-B), how much degradation in power and efficiency will be observed for a given value of device output capacitance (drain-source, collector-emitter), if voltage clipping is to be avoided.
2. If the degradation is unacceptable (e.g., for high F_t transistors with negligibly small output capacitances), how much larger can this capacitance be made (using on-chip or off-chip-in-package supplementary capacitors) without an unacceptable downward transformation of the external load, R_L.

So for each class of operation (for clarity, six conditions are shown; others can be easily interpolated), the two sets of curves in Figure 5.25 show power and corresponding efficiency degradation. The fundamental load transformation is also plotted against the capacitive reactance, which is normalized to the ideal load resistance at the *fundamental* frequency.

Two examples, which have already been used in this text, will serve as useful indications of the significance of this chart. First, returning to the current design example in Sections 5.5 and 5.6, the device was determined to have the following parameters:

$$R_L = 4\Omega$$

$$X_{CDS} = 75\Omega$$

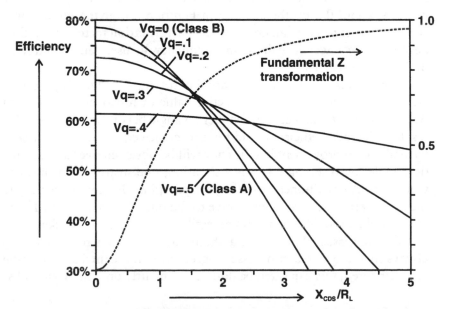

Figure 5.25 Degradation in efficiency and RF power output due to imperfect second harmonic short provided by a simple shunt capacitive reactance, X_{CDS}. The step-down impedance transformation ratio caused by X_{CDS} is also shown (dotted).

noting that the X_{CDS} reactance is defined to be the fundamental frequency reactance.

It can be seen that the normalized value of $(75/4) = 18.75$ is actually beyond the range of the plot of Figure 5.25, indicating a major degradation in performance. So the intuitive value for X_{CDS} which was taken in Section 5.6, corresponding to a normalized value of unity on the chart, can be seen to give an efficiency of 74%, as compared to an ideal case of 78.5% (note that the Spice simulation gave 71% in this case; see Figure 5.22). But as shown on the impedance curve, this corresponds to a factor of 0.5 step-down transformation of the required fundamental match, as viewed outside the device. This is probably around the best compromise for this particular case, but there are few designers around who are eager to place a 47 pF capacitor across the output of a "spendy" microwave device. Indeed, there is an additional consideration of bandwidth, which may drive the designer to a less aggressive point on the chart. A normalized value of 2, for example, for the capacitive reactance would result in an expected efficiency of 64%, corresponding to a power reduction of 0.9 dB. The corresponding capacitor value (23 pF) would be more palatable at this frequency, and would have much less impact on the bandwidth available from a simple single section matching network.

Different applications draw different conclusions. A typical set of parameters was considered in Section 5.1 for a higher power device, of the LDMOS type, also considered earlier:

$$R_L = 12.5\Omega \text{ (based on 28v supply at 2A)}$$

$$C_{DS} = 23 \text{ pF}$$

$$X_{CDS} (1.9 \text{ GHz}) = 3.6\Omega$$

so in this case, the intrinsic capacitance of the device itself puts us at a normalized value of about 0.3 on the abscissa in Figure 5.25, giving almost undegraded efficiency at the expense of an impedance transformation factor of 0.1. So some devices, despite being lower in gain at the frequency in use, can sometimes come with a built-in harmonic trap.[4]

The curves in Figure 5.25 are interesting in that they seem to display a well-defined pivotal point at a reactive value of just over twice the load resistance. Above this point, reducing the conduction angle beyond a light class AB condition can result in degraded efficiency, and for a reactance equal, approximately, to this value, virtually no improvement in efficiency will be observed over a wide range of conduction angle (bias) settings. This is a situation, one suspects, that has frustrated many would-be PA practitioners over the years; despite what the elementary textbooks say, the device efficiency seems insensitive to the quiescent bias point. The remedy is simple enough, although not always accessible in the case of a packaged device.

The key issue about using a shunt capacitor as a harmonic trap is that it is always there, to a greater or lesser degree. Even if the intrinsic capacitance has to be supplemented, this can often be done conveniently on-chip. External stubs have the

---

4. The output capacitance of LDMOS devices seems to be decreasing with time; current process generations may show a factor of at least 2 lower values of C_{DS}/watt.

advantage that they do not perturb the fundamental impedance at their resonant frequency, but can be awkward to realize when the device is packaged, especially if there are internal matching elements. The assumption that the Q-factor of the low-pass matching section is high enough that the second harmonic impedance at plane "A" can be approximated by just the shunt reactance of C_{DS}, becomes questionable when the impedance transformation ratio has lower values. Figure 5.26 shows a plot of the actual impedance at the second harmonic, versus this idealized assumption, for a range of m values. It can be seen that, in general, for higher values of transformation ratios the assumption is justified. Additionally, lower m values represent situations where much lower relative values of X_{CDS} are more realizable. So these guidelines form a very useful starting point for optimizing a design using a nonlinear simulator.

The assumptions used in this section to obtain design guidelines are approximate. The performance chart shown in Figure 5.25 is intended only as a guideline, and an illustration of some important but not well-known issues in higher efficiency PA design. The performance degradations shown by the curves in Figure 5.25 could be reduced by various means, in particular by tolerating a small amount of voltage clipping at the high end of the input drive power range, corresponding to the peak envelope power (PEP) of the input signal. Another approach, possibly widely used but not recognized, is the Class J concept described in Chapter 4, whereby the extra peak-to-peak voltage swing caused by a second harmonic voltage component (Figure 5.23) is phase-shifted to remove the baseline zero-crossings, at the expense of

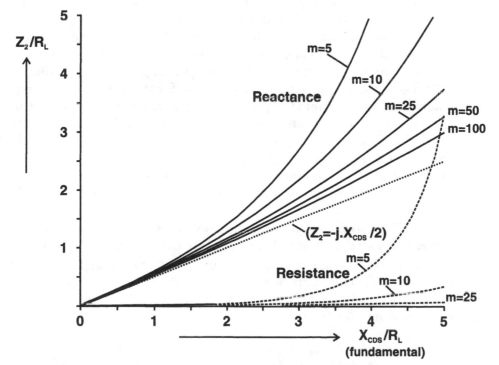

Figure 5.26 Second harmonic impedance (Z_2) at plane "A" (see Figure 5.23). For higher values of m (fundamental impedance transformation) and lower values of X_{CDS} (shunt capacitive reactance), the assumption that the second harmonic impedance is a capacitive reactance of $X_{CDS}/2$ can be seen as a fair approximation for design purposes.

higher voltage peaks. This chapter includes a Class J design example (Section 5.8) using the same generic MESFET device, to illustrate this particular tradeoff. The Class J approach in no way invalidates the underlying assumptions in the derivation of the curves in Figure 5.25, which represent a strict adherence to a predefined allowable peak voltage limit.

5.8   Class J Design Example

The results and conclusions in the previous section were obtained using some simplifying assumptions in order to come up with some general guidelines as to the range of capacitance values that are required to take on the role of an approximate harmonic short. It was assumed all along that the device would still be operated in classical Class AB mode, with a resistive fundamental termination. Chapter 4 introduced the concept of Class J design, where the output capacitance is used to provide a specific reactive termination at the second harmonic, in order to generate a nonsinusoidal waveform with a substantial second harmonic component. It would seem appropriate therefore to use the Class AB simulation and device model developed earlier in this chapter to explore the possibilities of Class J operation.

In order to set some objectives, and obtain useful approximate component values, the **PA_Waves** "postulator" will be used.[5] Figure 5.27 shows the **PA_Waves** screen set up for a typical Class J design. The knee voltage parameter **Vk** has been set to a median value of 0.1, approximately representing a device with a 1V knee voltage running off a 5V supply (note that the normalized **Vk** parameter represents the voltage at which the current reaches 67% of its saturated value). An in-phase component of second harmonic has been added to the fundamental voltage, giving it a triangular shape and raising the minimum level. This in turn enables the voltage waveform to be scaled up, relative to the DC supply voltage, as shown in Figure 5.27. The key Class J design maneuver is to shift the phasing of the voltage and current waveforms, without changing their shape, such that the second harmonic voltage and current components are in phase quadrature. This has been achieved in Figure 5.27 by empirical adjustment to the phase angle settings of **V1** and **V2**, where the **V2** angle is always changed by twice the amount of the **V1** angle. The result shown in Figure 5.27 represents a realizable design with power and efficiency higher than that which could be obtained using conventional Class B (shorted harmonics) operation.

The **PA_Waves** computations for the various impedances can now be un-normalized for a specific device and operating frequency. It has already been established that the loadline design value of this device is around 4Ω, and since the **PA_Waves** impedance calculations are already normalized to the loadline resistance, they can be directly scaled from the displayed values for a specific design. So in this case, the key design impedances are:

- Fundamental: $4 \cdot (1.2 + j1.1)$;
- Second harmonic: $4 \cdot (-j1.25)$.

---

5.   See Appendix A for a description of loading and using this program.

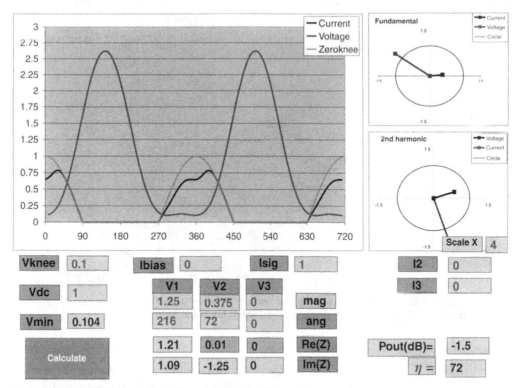

Figure 5.27   PA_Waves postulation of a Class J design.

The second harmonic reactance leads directly to a shunt capacitance value, 18.7 pF. This uses the approximation in Section 5.7, that the low-pass topology of the fundamental match will cause the enhanced device output capacitance to dominate the second harmonic impedance. The resulting output asymmetrical pi-section network design is shown in Figure 5.29, and an impedance plot for the first four harmonics is shown in Figure 5.28.

Figure 5.29 shows the MWO simulation file which incorporates this network, and Figure 5.30 shows the computed voltage and current waveforms at the maximum selected drive level. These waveforms are remarkably similar to the PA_Waves postulation and represent a power of 34.8 dBm and an output efficiency of 79.7%. The swept power performance of this design is shown in Figure 5.31, where it can be seen that the waveforms represent a fairly well saturated condition. Nevertheless, the swept response is remarkably linear up to a 1 dB compression point of 34 dBm output, where the efficiency is still 71%.

The enhanced output capacitor value of 20 pF is very significant. On the chart shown in Figure 5.25, the normalized X_{CDS}/R_L ratio for this value at 850 MHz comes out to be 2.3. Thus for Class AB operation, with no second harmonic voltage component and no voltage clipping, the chart indicates a much lower efficiency value for linear operation, only just higher than 50%. This design example therefore further emphasizes the importance of the Class J concept, but the tradeoff issue is a higher peak voltage. In designs such as this one which use a device having a very low output capacitance, some extra capacitance needs to be added, but a device such as an

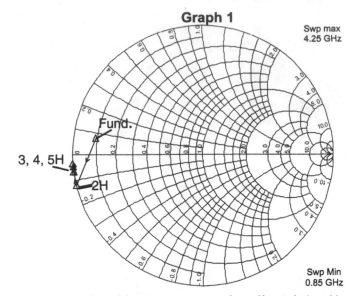

Figure 5.28 Pi section network, and frequency response, for a Class J design. Note reactive component at fundamental, and capacitive second harmonic termination.

LDMOS FET would have maybe 5–10 times higher capacitance per watt of power compared to a GaAs MESFET (or PHEMT) and could be potentially designed to work well in the Class J mode. As commented in Chapter 4, in the usual absence of any direct waveform probing at GHz frequencies, the probability of fortuitous Class J operation is quite high, especially when load-pull techniques are used (see Chapter 12).

5.9   HBT Design Example

Having been a research curiosity for more than a decade, the Heterojunction Bipolar Transistor (HBT) underwent a major promotion into front-line commercial RFIC design during the 1990s. The benefits of single supply operation, and the possibilities of using many of the circuit techniques employed in analog IC design helped to propel the technology into mainstream use. From a PA viewpoint, the advantage of a single supply was paramount, as was the gradual realization that smaller physical geometries could be used for a given RF power output than comparable GaAs FET or Si MOSFET implementations. What seems to have been largely missed is that the bipolar transistor quite probably makes a better, more linear Class AB PA. The basis for this speculation has been presented in a previous book [6], and summarized in the present Chapter 4.

This previous analysis will not be repeated here, but the key conclusion is that unlike a FET PA design, the gain linearity and efficiency will both be a sensitive function of the input termination. It is additionally clear, and well established, that in order to make a BJT work at all in a Class AB mode, it will be necessary to provide a suitable path for the desired harmonic current components flowing in the base-

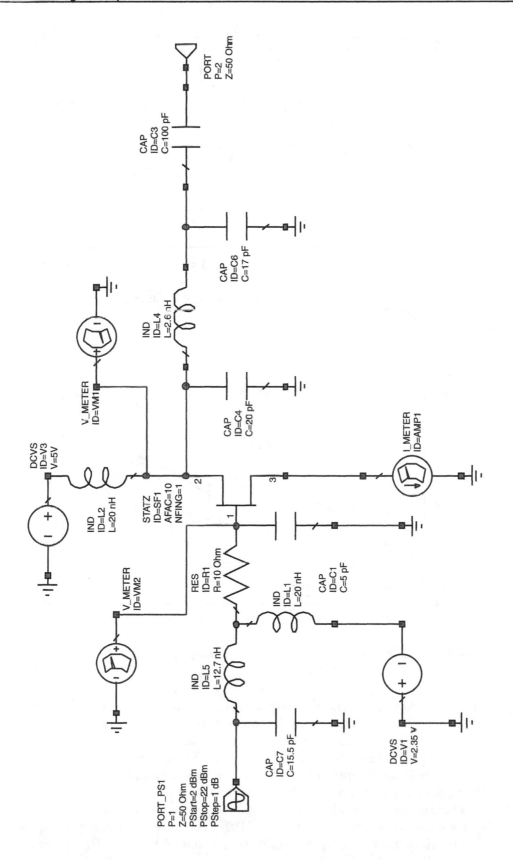

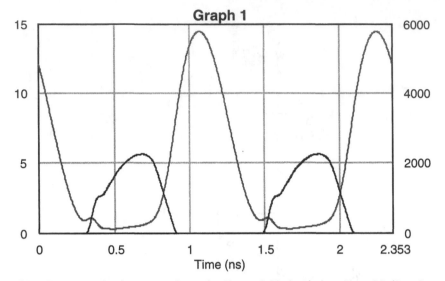

Figure 5.30 Current and voltage waveforms for Figure 5.29 simulation; Pin = 22 dBm, Pout = 34.8 dBm, output efficiency = 79.7%.

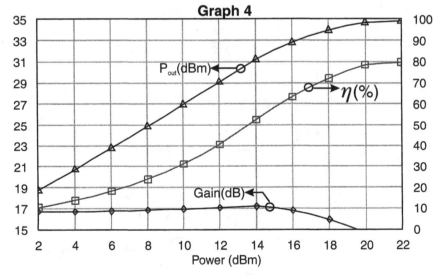

Figure 5.31 Swept power performance of Class J PA design.

emitter junction. Just as with the output harmonic path, the resident parasitic capacitance may, or may not, do part of the necessary job. Additionally, the source impedance presented to the base-emitter junction, primarily but not exclusively at the fundamental, will have a very sensitive effect on the gain linearity in Class AB mode. This will be in stark contrast to a FET device, where excluding nonlinear varactor effects, the input match mainly determines the PA gain.

These design issues, along with the underlying analysis, will be illustrated here by a specific simulation of a typical HBT PA. Since the main application for HBT PAs is in the mobile handset industry, a Gummel-Poon model has been scaled to represent a typical GaAs or InGaP HBT with a collector current of 800 mA, giving a device capable of just under 1 Watt peak RF output at 2 GHz using a 5V supply.[6]

In order to illustrate some of the input matching effects, the circuit shown in Figure 5.32 will be simulated first. Given the success of the previous Class J design, the output match has been developed along much the same lines, using an estimated loadline resistance of 10Ω. The input match initially consists of a broadband resistive source, whose resistance can be varied (note that Microwave Office has the useful ability to vary the characteristic impedance of a port termination). This does not represent a realizable, or practical, configuration at microwave frequencies. It does, however, represent an empirical adjustment that was much used in the days of discrete, highly linear BJT audio amplifiers. Following this ancient trail, Figure 5.33 shows several power sweeps for the amplifier for three different values of this source resistance. The interesting aspect of these curves is that careful inspection shows that best linearity does not trade inversely with efficiency, but rather with gain.

Clearly, the source impedance plays a big part in determining the final tradeoff between gain, gain linearity, and efficiency. This appears to be such a sensitive dependency that many PA designers may not fully trust the CAD analysis, and wish to perform some empirical characterization for a specific device, using an input load-pull tuner. Here we select the $R_s = 1.5Ω$ case as representing a satisfactory compromise. The task of transforming this input circuit into a practical configuration is not, at the outset, quite as simple as designing a transforming network from R_s up to 50Ω. Such a network, if designed using the conventional and convenient low-pass section, will not present the same impedance at the harmonic frequencies as does the simulated broadband termination. On the other hand, the value of R_s is sufficiently low in this case that a simple even harmonic short using a quarter-wave SCSS seems worth a try. This modified input match is shown in the final full simulation file shown in Figure 5.34, and the swept power amplifier performance is summarized in Figure 5.35. The gain characteristic shows some gain expansion at backed-off drive levels, which crosses into gain compression at higher levels. This particular gain signature can be very beneficial when dealing with modulated signals, and will be discussed further in Chapter 9.

It should be noted that the base bias has been defined using a DC voltage source. This may appear to be a departure from conventional small-signal BJT biasing techniques, which use current bias, derived using a potential divider on the base and a source resistor. Such biasing is only suitable for small signal, Class A designs, and it can be readily observed using the simulation that the PA performance is highly sensitive to the series impedance of the bias supply. This sensitivity somewhat mirrors the same sensitivity on the output, and needs careful consideration in modulated signal environments. Just like the output bias, this impedance should be as low as possible, but this may conflict with low frequency stability

---

6. Most handset PAs are required to operate from even lower supply rails, maybe as low as 2.5V. This requires some extra large geometry scaling and involves some deviations from standard design techniques due to extensive operation in the device knee region. For clarity in the demonstration of BJT PA design for wider applications, a somewhat higher voltage is used here.

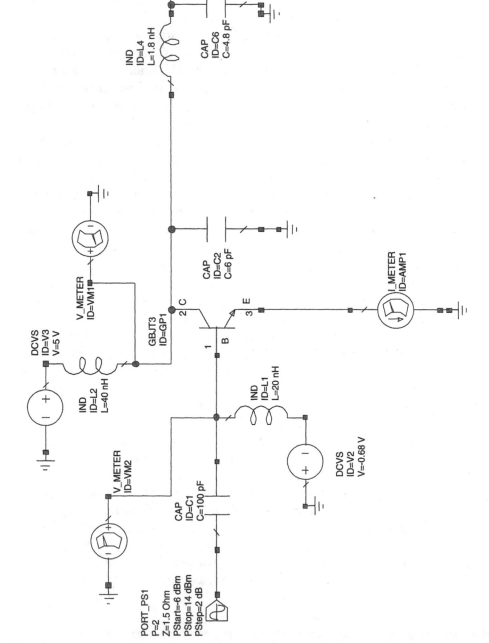

Figure 5.32   Simulation schematic for HBT PA design (broadband input; see file HBT_FIG_5_32 on CD-ROM).

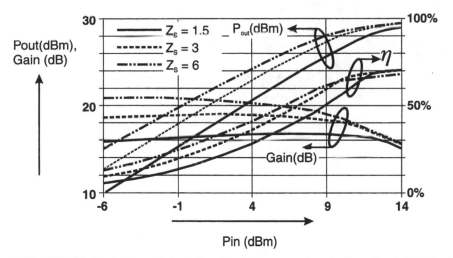

Figure 5.33 HBT PA simulation with broadband resistive source termination; $Z_s = 1.5\Omega$, 3Ω, 6Ω.

requirements (see Chapter 11). Great care has to be taken due to the sensitivity to small changes in voltage and temperature, and the more substantial current drain from the base bias supply. In practice, some form of control circuitry will be required in order to keep the bias setting stable with temperature, and to handle device-to-device parameter variations. It is common practice to use a form of current mirror (see Chapter 11), both in RFIC designs and in larger discrete BJT power devices, to define a stable bias point.

5.10 Conclusions

This chapter has taken the basic, somewhat idealized concepts of Chapter 3 to develop a practical, *a priori* method for designing reduced conduction angle high efficiency amplifiers. Techniques for impedance matching and low reactance harmonic terminations have been analyzed quantitatively. Some surprising results have been obtained in regards to the role played by RF transistor output capacitance. Notwithstanding the inconvenience of the low impedance transformation this can cause, the beneficial effect on harmonic termination may lead designers to add more output capacitance in some cases, especially when dealing with a "hot" technology at the frequency of interest.

It has also been observed that odd order harmonics can enhance power and efficiency performance, an issue that will be taken up again in more specific terms in Chapter 6 under the "Class F" heading. The Class J concepts developed theoretically in Chapter 4 have been shown to pan out very well in two CAD design examples, using both FET and HBT device models.

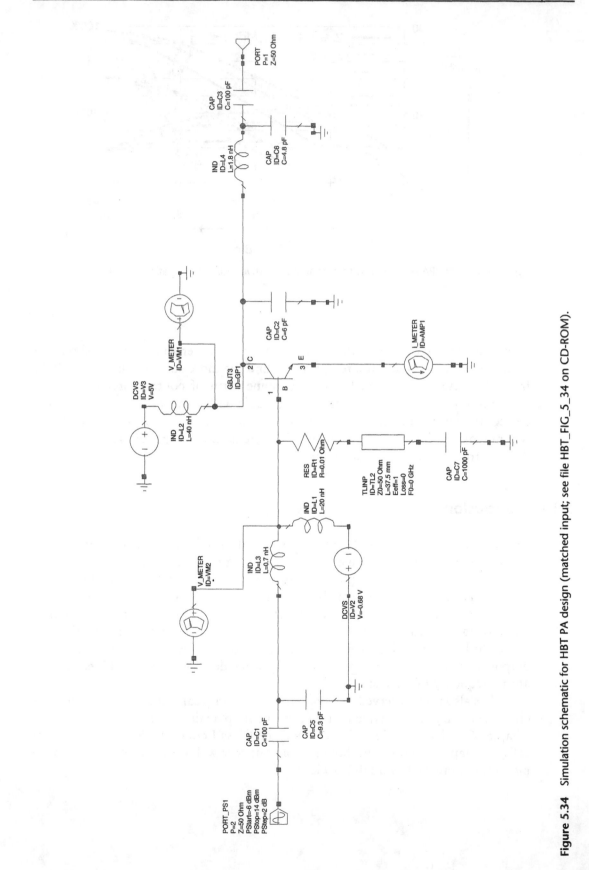

Figure 5.34 Simulation schematic for HBT PA design (matched input; see file HBT_FIG_5_34 on CD-ROM).

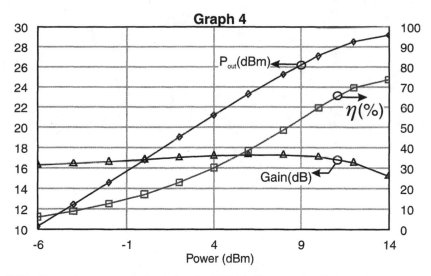

Figure 5.35   Swept input power simulation for HBT PA (Figure 5.34 schematic).

References

[1]   Cristal, E. G., "Impedance Transforming Networks of Low-Pass-Filter Form," *IEEE Trans. Microwave Theory & Tech.*, MTT 13, No. 5, September 1965, pp. 693–695.

[2]   Cohn, S. B.,"Stepped Transmission Line Transformers; Optimum Design," *IEEE Trans. Microwave Theory & Tech.*, MTT 3. No 3, April 1955, pp. 16–21.

[3]   Ahmed, M. J.,"Exponential, Cosine Squared, and Parabolically Tapered Transmission Lines; Impedance Equations," *IEEE Trans. Microwave Theory & Tech.*, T-MTT January 1981, pp. 67–68.

[4]   Day, P. I.,"Transmission Line Transformation Between Arbitrary Impedances Using the Smith Chart," *IEEE Trans. Microwave Theory & Tech.*, T-MTT 23, No. 9, September 1975, pp. 772–773.

[5]   Statz, H., et al., "GaAsFET Device and Circuit Simulation in Spice," *IEEE Trans. Electron Devices,* ED-34, 1987, pp. 160–169.

[6]   Cripps, S. C., *Advanced Techniques in RF Power Amplifier Design*, Norwood, MA: Artech House, 2002.

CHAPTER 6

Overdriven PAs and the Class F Mode

6.1 Introduction

In the power amplifier modes so far considered, the RF current and voltage swings have been carefully constrained to keep inside the strongly nonlinear boundaries; the device currents have not been allowed to exceed their saturation points, and the voltages have not been allowed to clip or distort in any way. This has been quite deliberate, and design rules have been based upon the presumption that either form of limiting action would be undesirable. This is a reasonable assumption when the application in question is a modern wireless communications system; linearity is usually a critical specification which has to be met, regardless of the possibilities for higher power and efficiency which may incur more strongly nonlinear behavior. Depending on the modulation system, and the specific application for the PA in the system, there are actually many situations in which linearity, especially amplitude linearity, can be traded for efficiency and RF power output. These applications range from single channel, constant envelope systems such as those using FSK and GMSK which can tolerate high levels of amplitude distortion, through to intermediate cases such as QPSK and DQPSK systems which can tolerate significant amounts of amplitude distortion, within the constraints of spectral mask regulations.

This chapter considers the behavior of conventional Class A, AB, and B modes in "overdrive" condition. It will be seen that two distinct effects of overdrive require separate consideration; the saturation of device current when heavily driven at the input can result in higher RF power but typically produces little improvement in efficiency other than for low conduction angles. A more interesting scenario is the deliberate use of voltage clipping, which can give significant efficiency improvement. Although voltage clipping inevitably introduces undesirable odd degree distortion effects, it will be shown that there can be a useful tradeoff between efficiency enhancement and linearity which can be utilized in low or intermediate envelope amplitude applications.

In the course of this discussion and analysis, two new modes, Class F and Class D, will make their appearance. Although each of these has a concise definition, in terms of cognizant RF voltage and current waveforms, they both represent specific and idealized conditions which can only ever be approximated in practice. Indeed, there exists a substantial literature describing circuit configurations [1, 2] which strive to realize the benefits that the basic mathematical definitions appear to offer. This has lead to some confusion about what really constitutes a Class F or a Class D amplifier, and the policy adopted in this chapter, and the rest of the book, is to use the terms generically rather than specifically. By "Class F," therefore, we refer to a continuum of possibilities where the voltage waveform at the output of the active device is gradually transformed, by some means or other, from a sinewave to a squarewave.

Initially, a simple Class A amplifier will be analyzed under conditions of current and voltage overdrive. This will illustrate the differences between the two kinds of overdrive and establish some methods for simplified analysis. Then the Class F and Class D conditions will be defined and analyzed, followed by some more specific analysis on reduced conduction angle modes under voltage clipped conditions. These kinds of amplifiers will be seen to approximate closely to Class F operation and can approach the desirable Class D mode under extreme conditions of overdrive and clipping.

One ongoing theme which will emerge in this chapter is that the benefits of "rail-to-rail" voltage swings are less attractive for real devices than ideal switches; driving a device into a voltage clipping regime usually brings the benefit of higher efficiency but almost always at the expense of lower peak current swing, and consequently lower RF power. This means that the ideal, zero-voltage knee assumption which has been used extensively so far to analyze amplifiers in their linear, or weakly linear, regions, has to be used with much greater caution in predicting the power-efficiency performance of overdriven, or limiting amplifiers. Indeed, in most cases it will be necessary to include a more realistic model for the knee region, such as that introduced in Chapter 3, in order to obtain meaningful information about the efficiency/linearity/power tradeoffs which are of paramount concern in wireless communication applications.

Some of the methods used in this chapter to analyze overdriven power amplifiers are novel and involve some innovative approximation. This is done within the ongoing principles of *a priori* design; clearly, with modern CAD tools any individual case can be analyzed in detail and with greater precision than that given by simplified theoretical analysis based on conspicuously approximate methods. The key issue is that the analysis techniques yield a broader insight into the circuit and device interaction effects, enabling design tradeoffs and circuit topologies to be selected as a preliminary step in the full design process.

6.2 Overdriven Class A Amplifier

The starting point for this discussion of overdrive and clipping effects is the basic, loadline matched, Class A amplifier. Initially, all of the idealizations used to analyze this circuit as a linear amplifier will be used; the device will have an ideal strongly nonlinear transconductance characteristic (as shown by the solid line in Figure 1.3), and a zero knee voltage. It should be noted straightaway that this analysis may appear to be deviating along a path which separates FETs from BJTs; a bipolar device would not be expected to have such a sharply defined current-limiting characteristic. In fact, the analysis still has relevance for BJTs since the forward biasing of the base junction can cause apparent current limiting due to the fixed impedance level of the input RF source. In any event, this divergence will not apply for most of the subsequent analyses in this chapter, which consider voltage, rather than current, limiting.

Figure 6.1 shows a set of waveforms for input voltage drive, output current, and voltage. As before, we assume the device to be ideally transconductive so that the current i_D is, conveniently, still a directly dependent function of input voltage, v_{GS}. As the drive level exceeds the maximum linear swing, a symmetrically clipped current waveform is generated; the lower cutoff is familiar and similar to a Class AB situation, but the upper limiting action is due to the idealization of the saturation characteristic. Note however that the bias point is still at the Class A position, midway between saturation and cutoff; this is not a Class AB amplifier.

The voltage waveform poses an immediate problem; just as with reduced conduction angle modes, the strongly nonlinear action of the device has generated a waveform with strong harmonic content, and the voltage waveform will be a function of the harmonic impedances, as well as the current waveform itself. Not only that, but the resulting voltage will have a recursive effect back on the original current waveform, if it drops below the knee voltage. This is a highly interactive and surprisingly complicated problem to solve analytically, and throughout this chapter it will be necessary to accept some idealizations and approximations. The first such case we consider is that of a broadband resistive termination, and that is the termination assumed in Figure 6.1, for the purposes of drawing the voltage waveform. The resistor is assumed to have the same loadline value that would be designed for optimum Class A operation. Such a termination is, in principle, realizable at higher RF frequencies. At lower frequencies, where broadband transformers are frequently used, it would also be a realistic impedance model.

Assuming an ideal, zero voltage knee characteristic, it is clear that the voltage will clip at twice the rail (supply) level if the load resistor has the ideal linear loadline value, as discussed in Chapter 2. So the voltage waveform is a simple Ohm's Law scaling of the current waveform, the negative sense being due to the direction of cur-

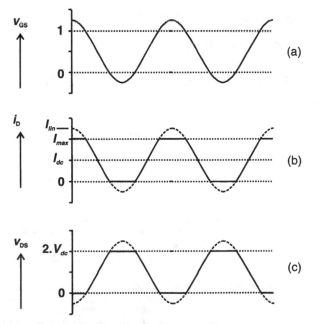

Figure 6.1   Overdriven Class A waveforms.

rent flow specified for the active device; the load behaves as a sink and the device behaves as a generator.

In order to determine the power and efficiency characteristics of this amplifier, it is necessary to obtain expressions for the fundamental component of current flowing in the resistor. It is of course assumed that the only RF energy of interest is the fundamental; the harmonic energy being dissipated in the load resistor will be filtered out by subsequent circuitry. Figure 6.2 shows the mathematical formulation used to perform the necessary Fourier analysis of a symmetrically clipped cosine wave. It is important in this analysis to keep track of the linear current amplitude, I_{lin}, which represents the amplitude of the undistorted current cosine wave which would result if the device remained completely linear. This will be used as a reference level for calculating the gain compression. The first part of the analysis is to determine the fundamental component of a cosine wave, amplitude I_{lin}, clipped at the prescribed level of I_{max}, which is a characteristic of the device. We define a clipping angle, α, such that clipping occurs for a total conduction angle of 2α, and α is given by

$$I_{dc} + I_{lin} \cdot \cos(\alpha) = I_{max}$$

so that

$$\cos(\alpha) = \frac{I_{max}}{2 \cdot I_{lin}} \tag{6.1}$$

(assuming that $I_{dc} = I_{max}/2$ for Class A bias).

The clipped cosine wave of current is therefore defined by

$$i(\theta) = I_{max}, \, 0 < \theta < \alpha, \, 2\pi - \alpha < \theta < 2\pi$$

$$i(\theta) = \frac{1}{2} \cdot I_{max} + I_{lin} \cdot \cos\theta, \, \alpha < \theta < \pi - \alpha, \, \pi + \alpha < \theta < 2\pi - \alpha \tag{6.2}$$

$$i(\theta) = 0, \, \pi - \alpha < \theta < \pi + \alpha$$

This is clearly an even function and so the fundamental RF component will be given by

$$I_1 = \frac{2}{\pi} \int_0^\pi i(\theta) \cdot \cos\theta \cdot d\theta$$

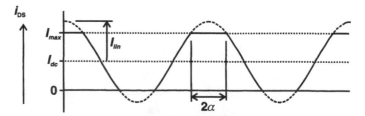

Figure 6.2 Symmetrically clipped sinusoidal current waveform.

so that

$$I_1 \cdot \frac{2}{\pi} = \int_0^\alpha I_{max} \cdot \cos\theta \cdot d\theta + \int_\alpha^{\pi-\alpha} \left(\frac{I_{max}}{2} + I_{lin} \cdot \cos\theta \right) \cdot \cos\theta \cdot d\theta$$

which, after some simple integration and rearrangement, gives

$$I_1 = I_{max} \cdot \frac{2}{\pi} \cdot \left(\frac{2\sin\alpha\cos\alpha + \pi - 2\alpha}{4 \cdot \cos\alpha} \right) \qquad (6.3)$$

This expression can be checked by confirming that when $\alpha = 0$, $I_1 = (1/2) \cdot I_{max}$, corresponding to the maximum linear power condition. The extreme case of $\alpha = \pi/2$, where the clipping extends over a half cycle, producing a squarewave, needs some more subtle reasoning to reach the expected value for I_1 of $I_{max} \cdot (2/\pi)$. [One has to avoid dividing through by zero, in the form of $\cos\alpha$. In fact, the bracketed expression has a limiting value of unity when $\alpha = \pi/2$; this can be proved by setting $\alpha = (\pi/2) - \delta$ and letting δ tend to zero.]

So for a given value of I_{lin}, where $I_{lin} > I_{max}/2$, a value for α can be determined from (6.1), and a value for the fundamental component of current can then be determined from (6.3). The ratio I_{lin}/I_1 corresponds to the current amplitude compression, which can be simply squared to obtain the power gain compression. So a power gain plot can be generated, centered around the maximum linear condition, where $I_{lin} = I_{max}/2$. For lower levels of I_{lin}, the device will follow a linear characteristic, and for higher levels, the power, gain compression, and efficiency can be examined. If the maximum linear power, P_{lin}, is defined to be

$$P_{lin} = \frac{I_{max}^2}{8 \cdot R_L}$$

then the fundamental RF power is given by

$$P_{rf} = \frac{I_1^2}{2 \cdot R_L}$$

so the RF output power can be plotted for a range of I_{lin} values which can be related to the drive power, and the resulting values of I_1 can be related to the maximum linear output power, as shown in Figure 6.3. The efficiency can also be plotted, assuming that the value is 50% at the maximum linear power point, and subsequently the DC power remains the same, due to the symmetry of the waveforms in Figure 6.1. So any increase in RF output above the clipping level will result in a corresponding increase in efficiency.

The power curve in Figure 6.3 will look familiar to anyone who has measured the compression characteristic of an RF amplifier. The key point is that the RF power does not clip as sharply as the voltage and current waveforms; the fundamental component of a clipped sinewave is significantly higher than the amplitude represented by the clipping level. For example, in the power transfer curve shown in Figure 6.3, 1 dB more RF power can be obtained than the linear power, at the

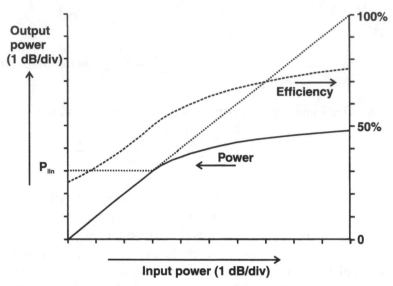

Figure 6.3 Class A amplifier Pin-Pout characteristic (solid) and efficiency (dashed), showing power and efficiency increase in overdrive conditions.

expense of 1 dB compression; in other words, the amplifier has to be driven 2 dB harder in order to extract 1 dB more output power. This so-called "1 dB compression point" has become almost synonymous with the maximum linear power in some technical circles, and this can be misleading. In particular, it can be seen that at the 1 dB compression point more RF output power is being obtained for no increase in DC consumption, representing an efficiency of about 63%. So Class A amplifiers, driven to the 1 dB compression point, can show similar efficiency to a Class AB amplifier. As the device is driven still harder, 71% efficiency can be obtained at the 3 dB compression point, however the efficiency curve is asymptotic to a value of $(8/\pi^2)$, or about 81%. This represents the limiting case of squarewaves for both current and voltage, which although converting DC energy to RF energy at 100% efficiency generates substantial amounts of energy at odd harmonics and is therefore of limited value as an RF device (see Chapter 7).

Returning to more modest levels of overdrive, it is clear that both power gain and linearity are being seriously compromised to obtain higher efficiencies. It should be noted however that the power gain is also compromised in reduced conduction angle modes, and in applications which have low sensitivity to amplitude distortion, an overdriven Class A amplifier may represent a better compromise between power, power gain, and efficiency. It should also be fairly pointed out that the assumption of a broadband resistive termination will not represent the harmonic terminations of a typical narrowband matching network. In practice the power transfer and compression characteristics shown in Figure 6.3 are nevertheless quite representative of tuned Class A amplifiers (compare, for example, with a typical measured characteristic using a tuned load shown in Figure 2.2).

The above analysis has assumed that the primary overdriving mechanism is the input drive; the current waveform has a symmetrically clipped characteristic due to the strongly nonlinear features in the transfer characteristic. The RF load resistor was selected such that the RF voltage still remains just equal to the available swing set by the DC supply rail. In practice, such an overdrive technique can cause significant additional nonlinear behavior, in particular AM-PM effects, whereby the input elements of the transistor, both resistive and reactive, start to show drive dependency and alter the phase of the amplifier gain from its linear value. Nevertheless, overdrive has been shown to increase RF power and efficiency, and takes the PUF well above unity for a Class A device. It is therefore logical to ask whether equivalent improvements can be obtained by overdriving amplifiers operating in reduced conduction angles modes, from Class AB through to Class B. This will be considered in the next section.

6.3   Overdriven Class AB Amplifier

The same approximations and idealizations will be assumed, both for the reduced conduction angle modes as discussed in Chapter 3, and the overdrive conditions in the previous section. This means that the knee will still be ideal, and the transfer characteristic of the transistor will have a sharp saturation and cutoff characteristic. In all cases, the full harmonic short will be assumed as for classical reduced conduction angle analysis, and the load resistor will always be selected so that the maximum voltage swing does not exceed the limits set by the rail voltage. The possibilities offered by voltage clipping are substantial, and will be considered separately in later sections of this chapter.

Figure 6.4 shows the formulation for analyzing a reduced conduction angle mode current waveform with provision included for overdrive into the saturated region. The key aspect is that the quiescent bias point remains the same as for "linear," or unclipped, operation (shown dashed in Figure 6.4) and is used to define the nominal conduction angle. As the input drive level is increased, the current pulse starts to flatten, and the effective conduction angle in the overdrive condition changes from the nominal value. For this analysis, it is assumed that the load resis-

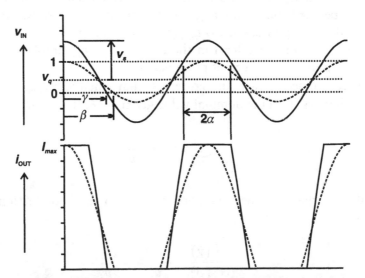

Figure 6.4   Class AB amplifier, with nominal conduction angle α (dashed), showing overdrive condition (solid) having modified conduction angle γ. Ideal, hard limiting saturation is assumed.

tor will be set to give an output sinusoidal voltage swing of amplitude V_{dc}, at the highest level of drive; this is consistent with allowing only the current, and not the voltage, to clip. The transfer characteristic of the device is defined as

$$i_D(q) = v_G(\theta), \, 0 < v_G(\theta) < 1;$$
$$= 0, \, v_G(\theta) < 0$$
$$= 1, \, v_G(\theta) > 1$$

so that the input voltage is normalized such that $v_G(\theta) = 0$ at the cutoff point, and $v_G(\theta) = 1$ at the saturation point. The output current, $i_D(\theta)$, is normalized as a ratio to the device saturation current I_{max}, so that $i_D(\theta) = 1$ at the saturation point.

The input voltage $V_G(\theta) = v_Q + v_s . \cos\theta$, where v_Q is the quiescent bias setting which determines the mode of operation and the nominal conduction angle, and v_s is the amplitude of the sinusoidal input signal; the input drive power is assumed to be proportional to $(v_s)^2$.

The nominal conduction angle, β, is given by

$$\cos\beta = \frac{v_Q}{v_Q - 1} \tag{6.4}$$

noting, as in Chapter 3, that conduction physically occurs for a duration of 2α, so that, for example, $\beta = \pi$ for the Class A condition, where $v_Q = 1/2$, and $\beta = \pi/2$ for the Class B condition, where $v_Q = 0$.

The signal amplitude for maximum "linear," or unclipped operation is

$$v_S = v_{LIN} = 1 - v_Q$$

so that v_{LIN} represents the usual level of drive considered for "classical" operation with a specified conduction angle β. (Unfortunately, as shown in Chapter 3, operation at drive amplitudes below v_{LIN} may not be linear, due to the dependence of conduction angle on drive level in the AB regime; v_{LIN} simply represents the level at which clipping of the current pulse begins.)

It can be seen clearly in Figure 6.4, that when $v_s > v_{LIN}$, the actual conduction angle changes from the nominal value given in (6.4). This new value, γ, is given by

$$\cos(\gamma) = -\frac{v_Q}{v_S} \tag{6.5}$$

combining this with the expression for the nominal conduction angle, β, in (6.4), gives

$$\frac{\cos(\gamma)}{\cos(\beta)} = \frac{1 - v_Q}{v_S} = \frac{v_{LIN}}{v_S} \tag{6.6}$$

so that in the overdrive condition, where $v_s > v_{LIN}$, γ will be a lower angle than β in the Class AB region ($\pi < \beta < \pi/2$), but in the Class C region, the overdriven conduction angle will be higher.

The clipping angle, α, is given by

$$\cos(\alpha) = \frac{1 - v_Q}{v_S} \qquad (6.7)$$

So the device output current is defined by

$$
\begin{aligned}
&i_D(\theta) = 1, 0 < \theta < \alpha, 2\pi - \alpha < \theta < 2\pi \\
&i_D(\theta) = v_Q + v_S \cdot \cos(\theta), \alpha < \theta < \gamma, 2\pi - \beta < \theta < 2\pi - \alpha, \qquad (6.8)\\
&i_D(\theta) = 0, \gamma < \theta < 2\pi - \gamma
\end{aligned}
$$

Expressions for the DC and fundamental component of $i_D(\theta)$ are therefore

$$I_{dc} = \frac{I_{max}}{\pi} \cdot \left[\int_0^\alpha \cdot d\theta + \int_\alpha^\gamma \left(v_Q + v_S \cdot \cos(\theta) \right) \cdot d\theta \right] \qquad (6.9)$$

(recognizing that $i_D(\theta)$ is an even function, so the integral need only be evaluated for a half cycle and the result doubled).

$$I_1 = 2 \cdot \frac{I_{max}}{\pi} \cdot \left[\int_0^\alpha 1 \cdot \cos\theta \cdot d\theta + \int_\alpha^\gamma \left(v_Q + v_S \cdot \cos(\theta) \right) \cos\theta \, d\theta \right] \qquad (6.10)$$

As before, since the final results will be in the form of data plots, these integrals will be left in their present form for direct computation.

All the time, it has been assumed that the voltage swing for the selected overdrive condition will be a cosinewave, peak amplitude V_{dc}. So the RF power can be written as

$$P_{RF} = \frac{V_{dc} \cdot I_1}{2}$$

and the DC supply power as

$$P_{DC} = V_{dc} \cdot I_{dc}$$

so the output efficiency is

$$\eta - \frac{1}{2} \frac{I_1}{I_{dc}}$$

These results are plotted out in Figure 6.5. In order to compare the results for overdriven modes with the conventional results shown in Chapter 3 (Figure 3.4), a similar format has been used, although the scale for relative power has been magnified. The efficiency, and power relative to the Class A condition, are shown as a function of nominal conduction angle. Plots are shown for successive amounts of

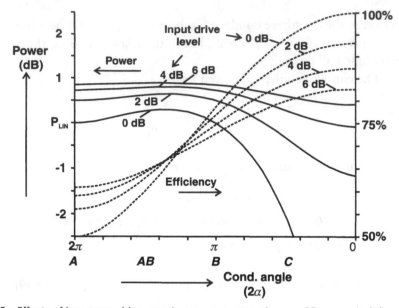

Figure 6.5 Effects of input overdrive, causing current saturation, on RF power (solid) and efficiency (dashed) of reduced conduction angle modes. Input drive levels of 0, 2, 4, 6 dB are measured relative to the power that causes maximum unsaturated current swing at the specified conduction angle.

overdrive (2, 4, 6 dB), where it should be noted that the actual amount of drive power still varies according to the nominal conduction angle; the overdrive is measured relative to the normal drive power for unclipped current at each point.

Several general observations can be made looking at the plots in Figure 6.5. First, it shows that significant improvements in power and efficiency only occur simultaneously for the lightest Class AB and Class A cases; once the conduction angle reaches Class B conditions, a modest increase in power is offset by a reduction in efficiency from the nominal 78.5% value. Shorter conduction angles, into the Class C region, show a quite dramatic increase in power from the nominal values, but closer scrutiny shows that this is really a manifestation of the increase in effective conduction angle, γ, as shown in (6.6) above; the power increase is accompanied by a corresponding decrease in efficiency from the unsaturated (0 dB drive) case.

These modest power increases, coupled with low or even negative efficiency improvement, can be attributed to the mathematical properties of the flat-topped current pulses. Such a waveform has a higher fundamental component than its equal amplitude, unclipped counterpart, but has a higher DC component. So when coupled with a sinusoidal voltage waveform, the flat-topped current pulse dissipates more power as heat in the transistor. It might perhaps be thought that a more optimum situation would be to allow clipping both on the current and voltage peaks, as was the case for the overdriven Class A example discussed in the previous section. This would ensure that less heat would be dissipated during the clipped current peaks, but unfortunately causes substantial RF harmonic power to be generated. This case is not considered further here, beyond the Class A analysis in Section 6.2, due to its additional practical limitations in the harmonic matching requirements. It should be noted, however, that for the Class A case (see Figure 6.3), the power and efficiency increases are slightly higher than for the shorted harmonic analysis represented in Figure 6.5.

It has to be concluded that the classical reduced conduction angle amplifier, with shorted harmonics, is not well suited for efficient exploitation of the possibilities offered by overdrive. As we discovered in Chapter 4, it so happens that this is not in any case an optimum arrangement for a low impedance solid state device. The use of an even harmonic short, such as a short-circuited quarter-wave stub, significantly changes the way an overdriven device behaves, and this is the next subject for consideration.

6.4   Class F: Introduction and Theory

This section will serve as an introduction to an important aspect of all RF power amplifier engineering; the replacement of a sinusoidal output voltage at the device output with a flatter, squarer periodic waveform. This can provide important benefits in both power and efficiency, both of which can be effectively traded for linearity. Initially we will consider the problem from a purely mathematical viewpoint; practical methods for realizing the benefits will follow in later sections.

Figure 6.6 shows a sinewave, with prescribed amounts of third harmonic added.[1] It is clear that the effect of adding some in-phase third harmonic is basically to reduce the peak-to-peak swing of the composite waveform; in other words, the peak-to-peak swing is lower than the amplitude of the fundamental component. This can be expressed mathematically, assuming already that we are discussing RF voltages, as

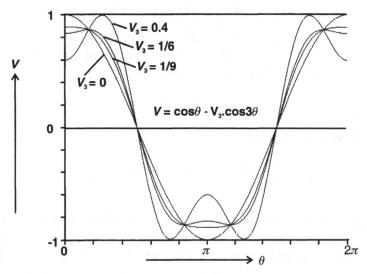

Figure 6.6   Third harmonic "squaring" effect on sinusoidal waveform.

---

1.   In order to comply with analyses in several references quoted in this section, sinusoidal, rather than cosinusoidal, waves will be used.

$$v(\theta) = V_1 \cdot \cos(\theta) - V_3 \cdot \cos(3\theta) \qquad (6.11)$$

It is fairly easy to prove that for values of V_3 such that

$$\frac{V_3}{V_1} < \frac{1}{9}$$

there is still a single peak, whose magnitude is given by

$$V_{pk} = (V_1 - V_3)$$

It is well known that for values of V_3 higher than $(V_1/9)$, the waveform starts to overshoot and has a double peak. It might be assumed (and it is a common misconception) that the onset of overshoot would represent a global minimum in the peak amplitude of the waveform, but in fact the twin-peak value, for yet higher values of V_3, continues to decrease up to the point where

$$\frac{V_3}{V_1} = \frac{1}{6}$$

at which point the peak value reaches its global minimum, given by

$$V_{pk} = \frac{\sqrt{3}}{2} \cdot V_1$$

In general, any value of V_3 between zero and $V_1/2.5$ (approx) will give a peak voltage which is lower than the original V_1 amplitude by a factor of κ, where κ has a possible minimum value of $\frac{\sqrt{3}}{2}$. This is shown in Figure 6.7.

The key issue is that if the peak voltage has a maximum permissible amplitude swing of V_{max}, then the presence of a third harmonic component allows the fundamental component amplitude to be increased from

$$V_1 = V_{max}$$

to a higher value given by

$$V_1 = \frac{V_{max}}{\kappa} \qquad (6.12)$$

This corresponds directly to an increase in fundamental power, given by the factor κ^{-1}; since the fundamental current, and the DC supply, are assumed to be unaffected by the generation of a small amount of third harmonic voltage. So in the optimum case, the fundamental power increases by a factor of $\frac{2}{\sqrt{3}}$, or about 0.6 dB.

But if the DC supply is unaffected, this translates directly into an efficiency increase; in the case of a Class B amplifier, the efficiency would increase to $\frac{\pi}{4} \cdot \frac{2}{\sqrt{3}}$, or 90.7%.

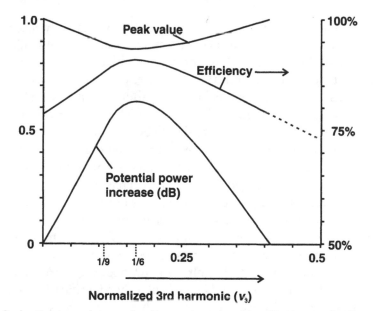

Figure 6.7   Reduction in peak-to-peak voltage swing caused by third harmonic effects; potential power and efficiency increase for Class B amplifier.

It is relevant to consider also the maximally flat case, where $V_3/V_1 = 1/9$.

In this case, $\kappa = 8/9$. So the RF power increase is 9/8, or about 0.5 dB. The corresponding efficiency, assuming Class B input conditions and current waveform, would be 88.4%. Such an amplifier, having a half-wave rectified sinewave for its RF current, and a maximally flat third harmonically enhanced sinewave for its RF voltage, has been termed a *Class F* amplifier, and is illustrated in Figure 6.8. It is fair to ask whether the Class F label should more specifically be applied to the optimum minimum κ case, giving a double-peaked voltage and 91% efficiency, although this could be regarded as an academic detail. It may appear that the maximally flat case is more practically realizable [1–3], but this impression will change when the device knee effects are fully included, as will be seen in later sections. As stressed in the

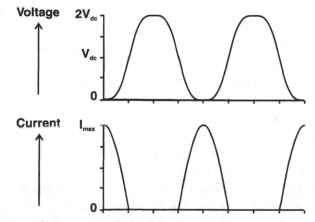

Figure 6.8   Class F waveforms, maximally flat case.

introduction to this chapter, the Class F label is of relatively recent origin and its precise definition is not as universally accepted as the venerable Class A, AB, B, and C terminology. But the underlying principles which the Class F term addresses are of wide significance and practical importance.

At this point, nothing has been said about the manner in which the third harmonic voltage component might be generated. It is clear that if the current is a precise, mathematical, halfwave rectified sinewave, that it contains no third harmonics. This causes some consternation regarding how the third harmonic voltage can be generated. But the waveforms as they stand present no mathematical contradictions, provided it is acknowledged that the RF load has to present an infinitely high impedance to the third harmonic; the mathematics has no problem with the concept that a transistor can be a harmonic voltage generator. This issue has become something of a *cause célèbre* in RFPA design, and will be considered in more depth in Section 6.6.

Clearly, the squaring-up process can be continued, by adding higher degree odd harmonic components. With multiple additional odd harmonic components, exact analysis of the waveforms, and in particular the identification of the optimum power and efficiency conditions, becomes rapidly more cumbersome. For example, taking the most important extension for practical purposes, the addition of a fifth harmonic component to the voltage waveform, the voltage can be expressed in the form

$$\frac{V(\theta)}{V_{dc}} = 1 + V_1 \sin\theta + V_3 \sin 3\theta + V_5 \sin 5\theta + \dots \text{ etc.} \qquad (6.13)$$

which can be conveniently normalized to the DC supply, giving

$$v(\theta) = 1 - v_1 \sin\theta - v_3 \sin 3\theta - v_5 \sin 5\theta + \dots \text{ etc.} \qquad (6.14)$$

where the negative signs have been introduced to indicate the required inverted voltage wave when a positive half-wave rectified current wave is assumed..

Some empirical number-crunching with the v_n coefficients yields the middle waveform shown in Figure 6.9, which has a higher fundamental component, v_1, than the equivalent case using only third harmonic (also shown). Once again, similar experimentation reveals that the optimum (maximum v_1 component) case is not the maximally flat waveform, but one which shows an overshooting ripple characteristic, with symmetrically placed maxima and minima. On this empirical basis, it might be surmised that the optimum case will be the condition where the upper maxima and lower minima are symmetrically spaced, $\pi/4$ apart. Knowledge of the location of the maxima and minima then allows a straightforward determination of the harmonic voltage coefficients v_n, using the relations $v(\theta) = 0$ and $v'(\theta) = 0$ at the minimum points. Such a surmise can be extended to the next higher case, which includes a seventh harmonic component, shown also in Figure 6.9, where the minima are spaced $\pi/5$ apart. Actual mathematical proof that these intuitive results are correct is challenging, and probably intractable, if approached in a conventional manner. Although the actual difference in efficiency between the various optimum case candidates is minimal for practical purposes, this particular issue has been the root cause of an extended debate among PA theorists. Recently, a remarkable piece of

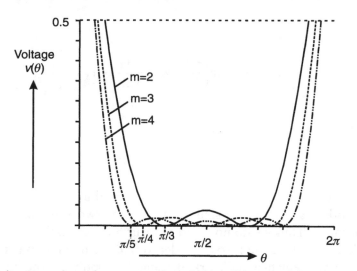

Figure 6.9   Voltage waveforms for cases $m = 2, 3, 4$ (third, fifth, seventh harmonic addition); see Table 6.1.

reasoning by Rhodes [4] appears to have solved the problem, and gives a generalized solution by recognizing the singular property of the optimum case solutions. So we may now formally state that for a voltage having the general form

$$V(\theta) = 1 - \sum_{q=1}^{q=m} V_{2q-1} \sin(2q-1)\theta \qquad (6.15)$$

and having the constraint that $V(\theta) > 0$, the set of zeros, θ_r, of $v(\theta)$ which correspond to the maximum allowable value of the fundamental coefficient v_1 are given by

$$\theta_r = \frac{r\pi}{m+1}, \, r = 1 \text{ to } m \qquad (6.16)$$

So for the case which includes third and fifth harmonics, the maxima and minima occur at conduction angles spaced symmetrically by $\pi/4$, as originally suspected. The key to understanding the singular nature of this solution is to evaluate the v_3 and v_5 coefficients. Substituting the first two (assumed) zero points, $\theta_1 = \pi/4$, $\theta_2 = \pi/2$, into $v(\theta) = 0$ (6.14), we obtain two relationships between the three voltage coefficients v_1, v_2, v_3:

$$0 = 1 \quad v_1 \sin\left(\frac{\pi}{4}\right) - v_3 \sin\left(\frac{3\pi}{4}\right) - v_5 \sin\left(\frac{5\pi}{4}\right)$$

giving

$$\sqrt{2} = v_1 - v_3 + v_5 \qquad (6.17)$$

and

$$0 = 1 - v_1 \sin\left(\frac{\pi}{2}\right) - v_3 \sin\left(\frac{3\pi}{2}\right) - v_5 \sin\left(\frac{5\pi}{2}\right)$$

giving

$$1 = v_1 + v_3 - v_5 \tag{6.18}$$

Examining (6.17) and (6.18), there is something very unexpected; we can eliminate both v_3 and v_5 in one step, by adding the two equations, leading directly to a value for v_1. Thus we do not need to incur a third relationship by consideration of the derivative function zeros in order to find the maximum permissible value of v_1. This is a mathematical hallmark of a singular condition, and justifies the selection of the $\pi/4$ minima spacing as the optimum condition. Demonstrating that this feature continues in a general case where the minimum points are assumed to be located as defined in (6.16) forms the basis for a general proof [4].

So the optimum v_1 value in the fifth harmonic case is given by

$$V_1 = \frac{1 + \sqrt{2}}{2}$$

which based on previous considerations will cause an increase in efficiency by the same factor (the normalized v_1 value) above the sinusoidal Class B case, so

$$\eta = \frac{\pi}{4} \cdot \frac{1 + \sqrt{2}}{2}$$

or about 94.8%. Table 6.1 summarizes the harmonic coefficients and efficiencies for cases up to and including the seventh harmonic.

The limiting case, as the harmonic number tends to infinity, is a voltage squarewave. Even without writing down the expressions for fundamental voltage and current, it is clear that this is an ideal RF power amplifier. The current and voltage never have simultaneous nonzero values, and the only harmonic at which power is generated is the fundamental; the current contains only even harmonics, and the voltage contains only odd harmonics.

Using the properties of a squarewave, the fundamental component of voltage, V_1, is given by

Table 6.1   Optimum Voltage Waveforms for $m = 1$ to 4

m	v_1	v_3	v_5	v_7	P (dB)	η
1	1	—	—	—	0	78.5
2	1.155	0.1925	—	—	0.625	90.7
3	1.207	0.2807	0.073	—	0.82	94.8
4	1.231	0.3265	0.123	0.0359	0.90	96.7

$$V_1 = \frac{4}{\pi} \cdot V_{dc}$$

and $\kappa = \dfrac{\pi}{4}$, so the RF power is increased from the classical Class B amplifier having a sinusoidal voltage waveform, by $4/\pi$, or about 1 dB.

This has become known as a Class D amplifier in RF applications and has its origins in lower frequency switching circuits used for DC to DC conversion (see Chapter 7). Clearly, at higher RF frequencies, finite switching speed will reduce the efficiency from the theoretical 100% value.[2]

The analysis in this section has continued to assume an ideal device model, for consistency with the previous analysis of Class AB amplifiers. In particular, the assumption of zero knee voltage needs to be reconsidered, particularly in view of the fact that the Class F voltage waveform spends a higher proportion of the RF cycle at a low value than for a Class AB type sinusoidal voltage. As will be shown in a later section, consideration of a more realistic model for the transistor knee region has a bigger impact on the power and efficiency of a Class F amplifier than for a Class AB device (as considered already in Chapter 5).

6.5 Class F in Practice

It so happens that we have already built a Class F amplifier. In Chapter 5, it was determined that the Class B amplifier design using a quarter-wave SCSS as a harmonic trap seemed to give better performance than the ideal, and unrealizable, high Q "tank" resonator. Recalling that simulation, shown in Figure 5.19(b), the voltage waveform already appears to have a distinctly square appearance. This is a simple and direct consequence of the even harmonic short, coupled with the clamping effect which takes place in the knee region of the transistor; an asymmetrical voltage clamp has been transformed into a symmetrical voltage limiter through the action of the even harmonic short. Although somewhat serendipitous in their sudden appearance, it appears that the odd degree voltage components have presented themselves without a whole lot of *a priori* design effort.

It seems, however, that explanations for the apparent spontaneous appearance of odd voltage harmonics in Class F power amplifiers do not satisfy most circuit design engineers, who do not like the idea that the required harmonic content arises in what appears to be a serendipitous manner, and more importantly appears not to respond to any direct *a priori* design procedure. This is mainly due to the deeply entrenched belief that a transistor is a transconductive, dependent current generator, so that the current waveform is entirely a function of the input signal. This is not an unreasonable viewpoint, given that so many books, including this one, frequently use this unilateral dependency as a convenient simplification for enabling

---

2. It is important to note that the Class D term is also used to describe a pulsewidth modulating amplification scheme, much used at audio frequencies; this technique has in turn been proposed at RF and is known as Class S.

analytical treatment. This approximation remains intact so long as the voltage swing keeps clear of the knee region. But the "Great Class F Puzzle" quickly evaporates when a more realistic device model is used, in particular with respect to the strong additional dependency of device current on output voltage, when the voltage enters the knee region. This is illustrated in Figure 6.10.

Figure 6.10(a) shows an ideal Class B current waveform, which results from an ideal transconductive device, biased at its threshold point. The current waveform contains no odd harmonics and remains ideally half-sinusoidal regardless of the output voltage. Such a device does pose some deeper questions about how odd harmonics of voltage could possibly be generated. It is a fact that the laws of electrical circuit analysis fall short of providing an explanation, or a solution, to the problem, and wider physical principles have to be invoked. But the device shown in Figure 6.10(b) essentially removes such problems and indicates that in the real world excursions into deeper philosophy are not required. The action of the device knee region is to clip the peaks of the current wave, *thus generating substantial amounts of third harmonic*. In practice, it is unlikely that the odd harmonic components of the device current will ever be at the vanishingly small levels that are represented by an ideal half-wave rectified sinewave.

The intractability of the ideal case using circuit theory alone, and its resolution when a more realistic device model is used, can be further illustrated by considering a comparable problem from mechanical engineering. Figure 6.11 shows a beam supported by three wires of equal length. In an ideal case, it can be stipulated that the

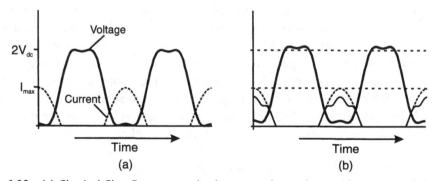

Figure 6.10   (a) Classical Class F current and voltage waveforms do not take account of device knee effects; and (b) device knee effects will usually clip the current peaks, creating substantial odd harmonic components.

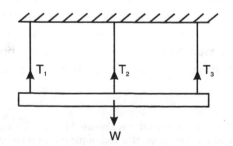

Figure 6.11   A "statically redundant" structure.

beam is rigid, and the wires are of identical length and have zero elastic modulus. If the tensions in the three wires are T_1, T_2, T_3, the laws of statics tell us only that

$$W = T_1 + T_2 + T_3$$
$$\text{and } T_1 = T_3$$

so that the values of the individual tensions cannot be resolved. Mechanical engineers call it a redundant structure for this reason.

Thus the ideal case is intractable using the laws of statics alone. But in practice the wires will never be identical. If unequal lengths, along with finite elastic properties of the wires and the beam are included in the analysis, then the system becomes fully tractable. The tension in each wire is given by

$$T_n = \lambda . x_n$$

where λ is the elastic modulus and x_n is the corresponding wire extension. So for example if the beam is assumed to be rigid and the three extensions are equal, the tensions in the wires must be equal, and given by

$$T_1 = T_1 = T_3 = \lambda x$$

So the question of what happens in the original ideal case of zero wire extension can now be deduced by extrapolation: if the beam remains completely rigid then the symmetrical geometry of the system forces the wire extensions to be equal, leading to equal tensions in the three wires. It seems acceptable to assume that this equality will be preserved as the elastic modulus of the wires tends to infinity, hence a solution to the stipulated ideal problem can be obtained. Mechanical engineers do not appear to worry themselves about the fact that in the limiting case the equations effectively state that the tensions in each wire are expressed as the product of zero extension and infinite elastic modulus;

$$0 \times \infty = \frac{W}{3}$$

They do, however, have the great advantage of devoting their lives to solving problems where the various parameters can be seen, felt, and measured. An entity such as elastic modulus is created to characterize *real* materials, not idealized artifacts such as inextensible wires. The RF designer should, perhaps, occasionally apply the same rationalization about the concept of impedance.

So in practice, as the RF voltage swing across any transistor is driven into the knee region, odd harmonics will be generated, and the odd harmonic impedance terminations provided by the external circuit will cause odd harmonics of voltage to be generated in the familiar manner. The problem of designing a network to give certain prescribed odd harmonic components is however less straightforward than the loadline matching approach which serves well for the fundamental components. In the case of the third harmonic, final values of the current and voltage components are reached by a more complex interactive process. It is appropriate to state this interactive process in a mathematical form, in order to determine the possibilities of establishing a design procedure.

The current waveform shown in Figure 6.10(b) can be expressed mathematically as a dependency on two variables, the input voltage v_{in}, and the output voltage v_o:

$$i_d = I_{max} v_{in} \left(1 - e^{-\frac{v_o}{v_k}} \right) \dots v_{in} > v_t \tag{6.19}$$

$$= 0 \dots\dots\dots\dots\dots v_{in} < v_t$$

where as before, v_k represents the normalized knee voltage parameter, and v_{in} is normalized to have value between 0 and 1, representing the range from the threshold to the I_{max} current saturation level.

If the output matching network presents a resistive impedance R_3 to the device at the third harmonic, and we assume that the fundamental load is resistive, then Ohm's Law can be imposed on the third harmonic components of current and voltage:

$$\frac{1}{\pi} \int_0^{2\pi} v_o(\theta) \sin 3\theta d\theta = (-) \frac{R_3}{\pi} \int_0^{2\pi} i_d(\theta) \sin 3\theta d\theta \tag{6.20}$$

where $i_d(\theta)$ is itself a function involving $v_o(\theta)$, as given in (6.19).

In principle the relationship (6.20), along with a similar relationship for the fundamental components, does lead to unique values of the third harmonic components of i_d and v_o, for a given drive level v_{in}, and the harmonic terminations R_1 and R_3. From a design viewpoint, therefore, we need to establish an *a priori* methodology for determining the values for R_1 and R_3 which will give the desired Class F voltage waveform at a prescribed drive level. But the relationships such as that of (6.20) discourage a direct analytical approach.

The harmonic analysis can be dragged back into the analytically tractable regime by using a simpler function for the knee region characteristic. Looking again at Figure 6.10(b), it is clear that the voltage waveform is at a low level for most of the duration of the current pulse. This enables a radical simplification of the device I-V characteristic to a linear function

$$i_d = k_{scl} \cdot I_{max} \cdot v_o \cdot v_{in}, 0 < v_o < v_k, v_{in} > 0 \tag{6.21}$$

$$= 0, v_{in} < 0, v_o > v_k$$

Using a carefully selected value for the scaling parameter k_{scl}, this approximation will allow some instructive analysis to be done on the problem, but will only be a representative model for certain extreme cases[3] of practical PA devices and operating conditions; one such case would be a handset PA device operating off a very low supply voltage.

Using the classical definition of Class F as only including fundamental and third harmonics in the voltage waveform, the output voltage v_o can be written in the form

---

3.   This approximation will work well for smaller conduction angles.

$$V_o = V_{dc} - V_1 \sin\theta - V_3 \sin 3\theta \qquad (6.22)$$

and (6.19) can be recast to incorporate the simplified knee characteristic and the first few Fourier components of the current waveform, giving

$$i_d = v_{in} \cdot I_{max} \cdot \left[\frac{1}{\pi} + \frac{1}{2}\sin\theta - \frac{2}{3\pi}\cos\theta - \frac{2}{15\pi}\cos 4\theta - \frac{2}{35\pi}\cos 6\theta\right]$$
$$\cdot \left[V_{dc} - V_1 \sin\theta - V_3 \sin 3\theta\right] \qquad (6.23)$$

It is now possible to extract expressions which relate the harmonic voltage and current components. In particular, the third harmonic relationship is

$$I_3 = v_{in} \cdot I_{max} \left[-\frac{v_3}{\pi}\sin 3\theta + \frac{v_1}{3\pi}\sin 3\theta - \frac{v_1}{15\pi}\sin 3\theta - \frac{v_3}{70\pi}\sin 3\theta\right].$$

where the voltage coefficients have been normalized to V_{dc}.

So, for example, at the maximum drive level represented by $v_{in} = 1$, and normalizing I_{max} to be unity,

$$\frac{v_3}{v_1} = \frac{4R_3}{15\pi\left(1 + \frac{R_3}{\pi}\left(1 + \frac{1}{35}\right)\right)}$$

$$\approx \frac{4}{15\left(1 + \frac{\pi}{R_3}\right)} \qquad (6.24)$$

This is a remarkable result, which shows that the relative third harmonic voltage component has a very weak dependency on the value of R_3, for any values of R_3 which are significantly larger than π; this is not a high value, given that the normalized Class A loadline impedance has a value of 2 in this formulation. The third harmonic component has an asymptotic value of $\frac{4}{15}$, which is somewhat higher than the ideal value of $\frac{1}{6}$. This result would appear to be a much-sought analytical justification of the folklore notion that odd harmonics should be presented with an open circuit for best efficiency performance. But there is further quantification in this result, which gives some valuable insight on how high the impedance has to be in practice.

A similar process could be used to extract the fundamental components from (6.23), resulting in a design value for R_1, although as already commented, the much larger fundamental component can usually be satisfactorily approximated using just the transconductive component in (6.19).

The above analysis has given some useful indications on the possibility of establishing an *a priori* design method for Class F operation. The results have however been obtained using a model of marginal validity for many applications. The same procedure can be followed, but without the convenience of analytical tractability, for the more realistic exponential knee characteristic. Starting again with a voltage waveform defined as in (6.22), and a current waveform defined using the exponential knee characteristic (6.19), prescribed values for V_1, V_3, allow the current waveform to be determined for a given input magnitude, v_{in}. This waveform can be harmonically analyzed, leading to values for R_1 and R_3.

Figure 6.12 shows the resulting normalized values of R_3, plotted as a function of the fundamental voltage component v_1 for the case of $v_3 = 0.2$; the Class B case, $v_3 = 0$ is also plotted for comparison. From the previous analysis in this chapter, a zero-knee approximation gives optimum values for a Class F design as $v_1 = 1.15$, $v_3 = 0.2$ (see Table 6.1), whereby the efficiency would be 90.7%. Here we are taking account of a more realistic approximation to the knee characteristic, which not only reduces the optimum efficiency, but takes a toll on the RF power at the higher efficiency points, as also shown in the plots for $V_3 = 0.2$ in Figure 6.12. This tradeoff between power and efficiency due to knee modulation effects has been encountered in basic class AB designs in Chapters 3 and 4. In this case a good compromise would appear to correspond to a V_1 (abscissa) value of 1.0, which gives an efficiency of 77% and a relative power of –0.9 dB. But the key feature in Figure 6.12 is the plot of R_3 values, which shows a very steep characteristic. It can be seen that a wide range of R_3 values, selected on the vertical axis, intersect with this curve over a very narrow range of V_1 values; from this observation it appears that the R_3 value required to obtain the stipulated Class F performance is remarkably noncritical. Indeed, so long as the R_3 value is reasonably larger than the loadline resistance, Class F operation

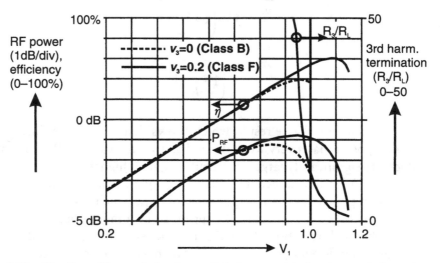

Figure 6.12 Class F amplifier analysis using realistic knee model, showing weak dependency on R_3 value (Class B performance shown for comparison).

Table 6.2 PA_Waves Class F Analysis

V_1	V_3	Rel. Power (dB)	η (%)	R_1	R_3
0.95	0.2	–0.8	74	1.1	35.3
1.0	0.2	–0.9	77	1.2	11.9
1.05	0.2	–1.3	80	1.46	5.26

can be expected at the high end of the drive range. These observations are therefore quite consistent with the previous analysis using a more idealized knee characteristic. Table 6.2 illustrates this conclusion (R_1, R_3 values normalized to ideal Class A loadline resistance), using values obtained using PA_Waves ($V_k = 0.1$; see Appendix A and Chapter 8 for further information on using PA_Waves).

Although these results may appear to be reasonably consistent with the folklore assertion that open-circuiting the odd harmonics gives Class F operation, it appears also to suggest that some useful benefits can be obtained by using a more controlled value as a design goal. Some discussion on suitable Class F matching networks will be given in Section 6.7, but it should be noted in passing that there are practical difficulties associated with the provision of a prescribed resistive impedance at the third harmonic. In practice it turns out to be much easier to design a network which has an open-circuited, or highly reactive, termination at higher harmonics. This is due primarily to the unknown nature of the termination environment that a PA product will be given in its final application. It is a common mistake to assume that a PA will behave the same way in its final impedance environment, as in the controlled lab test bench conditions, where it is usually only ever tested using broadband 50Ω attenuators. So networks which have been designed with the assumption that the 50Ω environment extends to harmonic frequencies may not perform as expected in the final application. This is also an issue with designs based on harmonic load-pull measurements, as will be discussed in Chapter 12.

6.6   The Clipping Model for the Class F Mode—Class FD

The last section considered the practical implementation of a "classical" Class F mode, whereby the fundamental voltage component can be increased by a factor of about 15% due to the auxiliary presence of a small third harmonic component. Such a design will still fall into the generic category of "quasi-linear," and can be used for amplitude modulated signals. This section considers a different category of Class F amplifier, where the output voltage is allowed to clip much more heavily, incurring higher harmonic content, heavy gain compression, and very high efficiency in practical circuits.

The conventional definitions of Class F and Class D leave something of a practical void between the simplest case of a slightly flattened sinewave using just 3rd harmonic, and the ideal case, in RF terms, of a perfect squarewave. There is also an assumption in the analysis method, which assumes that the waveforms will be "synthesized" in the frequency domain. There is an alternative formulation which avoids both of these issues, allowing a continuum of cases to be considered, and which does not rely on waveform synthesis in the frequency domain. Indeed, the formulation, shown in Figure 6.13, has already been encountered in Section 6.2, where an overdriven Class A amplifier was analyzed. Rather than adding odd harmonic components to a sinewave and performing the necessary scaling to achieve the prescribed maximum peak swing, the sinewave amplitude can be simply increased, and symmetrical clipping assumed. Once again, this analysis is essentially mathematical; we are not at this point assuming (and have yet to justify) that an actual circuit would clip symmetrically in this fashion. So using (6.3), with a clip-

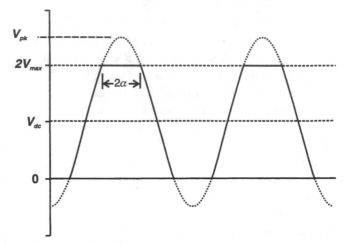

Figure 6.13   Clipped sinewave formulation.

ping region extending for a total angle of 2α, the fundamental component can be expressed as

$$V_1 = \frac{4 \cdot V_{max}}{\pi} \left(\frac{2 \cdot \sin \alpha \cos \alpha + \pi - 2\alpha}{4 \cdot \cos \alpha} \right)$$

noting that in (6.3) a factor of 2 needs to be introduced, since V_{max} is defined as an amplitude of allowable symmetrical swing, corresponding to the DC supply level, V_{dc} (I_{max} in (6.3) is a peak-to-peak value), so that

$$\kappa = \frac{V_{max}}{V_1} = \frac{\pi}{4} \cdot \left(\frac{4 \cdot \cos \alpha}{2 \cdot \sin \alpha \cos \alpha + \pi - 2\alpha} \right) \qquad (6.25)$$

Clearly, $\alpha = 0$ represents the maximum unclipped condition, and the bracketed expression has a value of $(4/\pi)$, so that $V_1 = V_{max}$. In the extreme squarewave case of $\alpha = \pi/2$, as discussed in Section 5.2, the bracketed expression has a limiting value of unity, giving the expected value for κ of $\pi/4$.

It is now possible to obtain a quantitative plot which relates the degree of squareness, represented by the value of the clipping angle, α, to the potential increase in RF power; the power increase translating directly into efficiency increase in this case, for an unchanged fundamental current. Such a plot, essentially a graphical representation of (6.25), is shown in Figure 6.14. Clearly, the waveforms for the clipped sinewaves are not identical to the harmonically synthesized cases initially considered, but they are certainly close enough for analysis purposes, as shown in Figure 6.15 for the Class F case, which can be seen to correspond very closely to a clipped sinewave with $\alpha = 45°$.

This method for representing the "squareness" of a waveform was introduced as a purely mathematical concept. It does, however, suggest a possible method of actually realizing such waveforms in practice; it may be possible to make a Class F

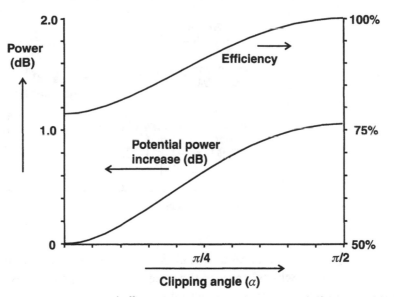

Figure 6.14   Potential power and efficiency increase for voltage-clipped Class B amplifier.

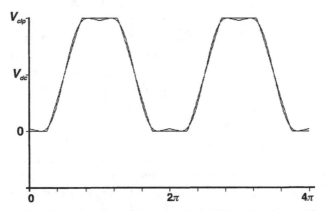

Figure 6.15   Comparison between clipped sinewave (α = 45°) and optimum Class F waveform using only third harmonic (v_3 = 1/6).

amplifier by allowing the sinusoidal voltage to clip. Figure 6.16 shows a possible circuit. The key elements are the SCSS even harmonic short, and a fundamental load resistance whose value is now made deliberately higher than the loadline values prescribed hitherto. In the first instance, it will be assumed that the fundamental load is connected to the device through a high Q series resonator, so that the device is presented with an open circuit at higher odd harmonics (this means that, for the time being, the output capacitance of the transistor will be ignored). For simplicity, it will be assumed that the device is biased and driven as for Class B operation; so that the current waveform, assuming a nonzero output voltage, will be a halfwave rectified cosinewave, with peak current I_{max}. Based on the previous discussion, the first pass at predicting the voltage waveform is to draw a sinewave, whose amplitude is equal to the fundamental component of current, $I_{max}/2$, multiplied by the load resistance R_L. This sinewave is now assumed to clip at the zero and $2.V_{dc}$ levels. As discussed

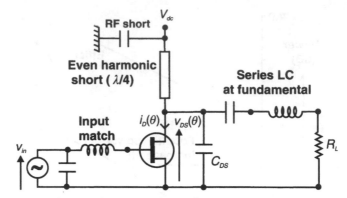

Figure 6.16 Circuit for clipped Class F amplifier analysis.

above, the even harmonic short-circuit forces the upper clipping to mirror the actual physical clipping which occurs in the knee region. Such a set of waveforms could be used as an approximation, but actually contains a serious numerical discrepancy; the fundamental component of the clipped voltage waveform does not equal the amplitude of the unclipped sinewave from which it was geometrically derived. This means that the current, based on the fundamental voltage component and the load resistor, will be lower than that based on the original current determined using the transconductance. This situation can only be resolved by adopting a more realistic model for the knee characteristic which takes account of the fact that the current itself is modified by the voltage waveform in the knee region. In fact, it becomes more appropriate to reverse the problem around, and to assume that the clipped rail-to-rail voltage is the main driving force in the circuit, rather than the transconductance of the device.

Unfortunately, although the principles of operation seem simple enough here, the interactive nature of the currents and voltages do, unfortunately, extend simple symbolic analysis beyond reasonable limits. Once again, however, some simple assumptions and approximations can be applied and the resulting analysis shows some value for *a priori* design. The following paragraph outlines such an analysis, which has been included for those who wish to pursue it in more depth for themselves; the details need not be absorbed by those who wish to examine the results, which are summarized in Figures 6.18 and 6.19.

Figure 6.17 shows a plausible strategy for analyzing a Class B stage under voltage clipped conditions. The clipping level is assumed to be at some level which is slightly lower than the amplitude represented by the DC supply level, $2V_{dc}-V_z$, where V_z has yet to be determined. The device is assumed to have a simple knee model, of the form:

$$I_{DS} = \frac{V_{ds}}{V_k} \cdot I_{sat}, 0 < V_{ds} < V_k$$

$$= I_{sat}, V_{ds} > V_k$$

where I_{sat} is the normal transconductive output current assuming no knee effects.

The analysis proceeds as follows:

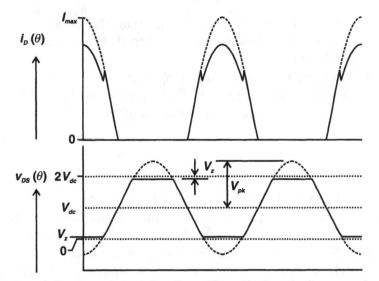

Figure 6.17   Formulation for analysis of Class B amplifier with clipped voltage waveform.

1. Determine the transconductive output current swing, ignoring the clipping effects of the DC supply levels; this is the usual simple transconductive function with selected values of quiescent bias and signal amplitude levels.

2. Using Fourier analysis, determine the fundamental component of the current waveform in (1).

3. From the fundamental amplitude determined in (2), and the selected value of load resistor, R_L, determine the ideal, unclipped, sinusoidal output voltage swing, V_1.

4. Apply symmetrical clipping, initially at clipping levels approximated by

$$V_{dc} \pm V_{clp}, \; V_{clp} = \left(V_{dc} - V_k \right)$$

5. Determine a first approximation to the clipping angle, α, using the relation [from (6.1), noting that V_{dc} is an amplitude, not a peak-to-peak value]

$$\cos(\alpha) = \frac{V_{clp}}{V_1}$$

6. Determine the fundamental component of the clipped sinewave, using [from (6.3)]

$$V_1 - V_{clp} \; \frac{4}{\pi} \left(\frac{2\sin\alpha\cos\alpha + \pi - 2\alpha}{4 \cdot \cos\alpha} \right)$$

7. Determine a value for the fundamental component of current, I_1:

$$I_1 = \frac{V_1}{R_L}$$

This value for I_1 will be smaller than the value calculated in (2) above, because in (2) no account was taken that the output voltage will be within the knee region.

8. Using the knee equation (6.19), determine a value for V_{ds}, ($=V_z$) which scales the original value of I_1 (in 2) to be equal to I_1 (in 7). This value, V_z, gives a much better approximation to the actual clipping level, V_{clp}:

$$V_{clp} = V_{dc} - V_z$$

At this point, an iteration could be performed to further refine the value for V_{clp}. Clearly, the value of V_{clp} has increased from the original estimate of $V_{dc} - V_k$ (since $V_z < V_k$) and so the voltage swing has a slightly higher value than that which the original estimate in (4,5,6) above, was based. So the procedure represented by steps (4) through (7) could be repeated. However, assuming that $V_k \ll V_{dc}$, this is possibly an unjustifiable complication, given the spirit of simplicity which this analysis pursues.

The current pulse itself will now have a modified shape. Fortunately, we have a value for the fundamental component, which enables RF power to be calculated, but the DC level poses more of a problem. Once again, it would be quite possible to recompute the current waveform, based on the original transconductance relationship with the additional modulation of the now known output voltage swing, using the knee relationship in (6.19). This could then be used as a further iteration on the exact clipping level. But again, in the spirit of simplicity, it will be assumed that the current pulse is sufficiently short that the output voltage will be at its lower, clipped, level, for most of the duration of the current flow; this means that the current scaling caused by the knee voltage function can be applied to the DC level, as well as the fundamental component. This assumption would appear to be justifiable for Class B and shorter conduction angles, but less so for moderate conduction angles encountered in Class AB operation. In practice, the approximation is quite serviceable even in these more moderate cases.

The above analysis procedure enables a quantitative analysis of a reduced conduction mode amplifier which can be driven into the voltage clipping regime (it is assumed that the current is kept within its unclipped, or nonsaturated, region). In particular, a detailed quantitative study can be made of the tradeoffs between power, efficiency, and gain compression.

Figure 6.18 shows a preliminary set of simple power transfer plots, using the analysis technique outlined above. A normalized knee voltage (expressed as a ratio to the DC supply) of 0.1 has been used for these plots; this is representative of quite a range of applications, but would have to be increased for very low voltage portable PA designs. The key parameter is the value of the load resistor, R_L. As usual in this book for this kind of analysis, the load resistor value is normalized such that a value of unity corresponds to a simple idealized linear loadline match, R_{OPT} for the device in question, so that

$$R_{OPT} = \frac{2 \cdot V_{dc}}{I_{max}} = 1$$

in the ideal linear case. The power, as usual, is normalized to the linear power,

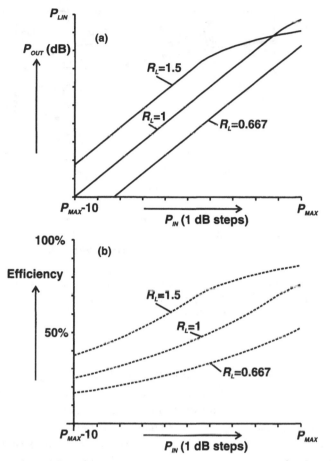

Figure 6.18 Clipped Class F analysis: (a) power transfer characteristics, and (b) efficiency, for three values of normalized load resistance.

$$P_{LIN} = \frac{V_{dc} I_{max}}{4}$$

The curve for $R_L = 1$ shows a nearly linear characteristic, with efficiency in the lower 70s and about 0.5 dB lower power than the reference linear power P_{lin}. A lower value of R_L (0.667) gives lower power and lower efficiency, but good linearity; this is the regime where the maximum drive level is limited by saturation of the device current.

Higher values of R_L are more interesting. Although gain compression quickly enters the picture, substantially higher efficiencies are possible for quite small tradeoffs in output power. Figure 6.19 shows in more dramatic form the critical effect of R_L in determining a suitable tradeoff between power, efficiency, and linearity. The magnified scales are quite justifiable; PA economics and thermal management are frequently sensitive to power variations measured in tenths of dBs and a few percentage points of efficiency.[4] This power-efficiency cartouche may look somewhat

---

4. It should be noted that in moving from an efficiency of 70% to 85% the heat dissipated in the device is halved.

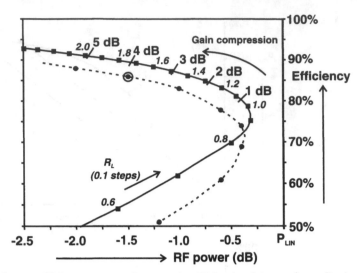

Figure 6.19 Power-efficiency contour for a varying RF load resistance, for a clipping Class B amplifier having an even harmonic output short. Solid line is derived using clipped sinewave analysis (see text); dashed line shows Spice simulation data points (waveforms for highlighted point are shown in Figure 6.21).

familiar to users of load-pull or even more basic tuning methods to explore the variations in power and efficiency of practical power amplifiers. Both experimental measurement and CAD simulation confirm the general appearance of this plot and imply that the assumptions made to derive it here are generally justified.

In order to test the general form of the characteristics of voltage clipped amplifiers designed using the approximate methods used in this section, the Spice circuit file used earlier in this chapter, can be used to plot some points, as shown in Figure 6.20. The dashed line in Figure 6.19 represents actual simulated points using this simulation, which uses a full nonlinear device model and realistic values for parasitics such as output and feedback capacitance. Other than a small offset of well under 10%, the simulated data can be seen to agree very well with the simplified theory. A set of Spice simulated voltage and current waveforms from one of the data points ($R_L = 6\Omega$, $\eta = 86\%$) is shown in Figure 6.21. It is clear that this is starting to approach a Class

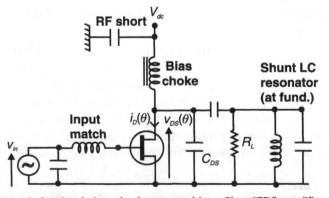

Figure 6.20 Schematic for simulation of voltage-overdriven Class "FD" amplifier.

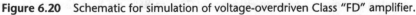

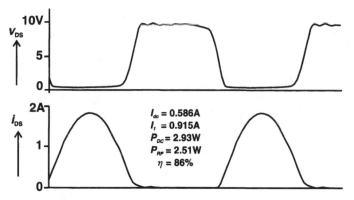

Figure 6.21   Spice simulation for $R_L = 6\Omega$ (see Figure 6.20 and Figure 5.15 for model and circuit file details).

D–like appearance, the voltage waveform looking considerably more square than a simple 3rd harmonically enhanced Class F waveform. There seems to be a case here for introducing a new nomenclature, "Class FD," which like its Class AB counterpart recognizes a continuum of useful cases between specific extremes.

The results in this section have assumed that a suitably high impedance termination can be provided at several higher odd degree harmonics. The design example illustrates the feasibility of generating voltage waveforms that do actually approach Class D squareness. This is however critically dependent on the use of a device at a frequency where its output capacitance represents a very large shunt reactance. Such a situation does have some real-world occurrence, particularly in mobile handset PAs which often use GaAs PHEMT or HBT technology even for the 830 MHz cellular band. In general, however, for higher power and higher frequency applications, it is necessary to reexamine Class F design using a device with more significant parasitic reactance. This will follow in Section 6.8.

6.7   PA_Waves

The PA_Waves program was introduced in Chapter 4 as a useful way of quantifying certain aspects of PA operation, especially including device knee effects. A general description of the program, loading and running it, is given in Appendix A. Specific files, which are set up to reproduce the figures in this section, are also included on the CD-ROM which accompanies this book.

Figure 6.22 shows a set of classical Class F waveforms, with an ideal Class B current wave and a voltage wave which shows a normalized third harmonic voltage set to be 1/6. The voltage waveform has been scaled up by a factor of 1.155, as described in Section 6.4 (see Table 6.1), in order to fill the available swing limits of zero to twice the DC supply, even harmonics being assumed to be shorted. The efficiency shows the theoretical optimum value of 90.7%, but the key issue now is the inclusion of a realistic knee characteristic. As with the Class AB analysis in Chapter 3, this has a dramatic effect on the classical results, and unfortunately it is these that always seem to form the basis of expectations. Resetting the Vk parameter in PA_Waves to a more realistic value of 0.1, a very different picture emerges. Just as

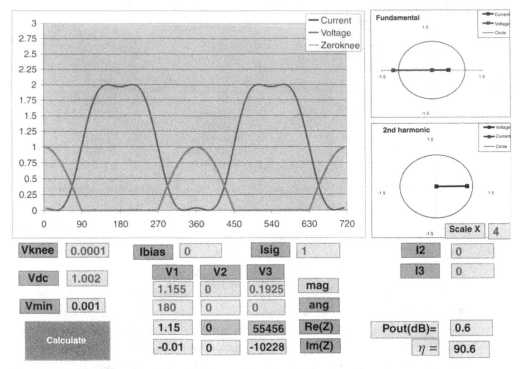

Figure 6.22 PA_Waves analysis of optimum Class F condition, zero knee assumption.

in the Class AB analysis, the voltage swing has to be reduced, or the DC supply increased, in order to prevent the current from developing deep notches which have a devastating effect on the RF power. Figure 6.23 shows that an increase in the third harmonic component from its mathematically optimum value of 1/6 can also help to recover the peak current, and hence the RF power. But the power and the efficiency (−1 dB, 82%) remain well below the values obtained using idealized analysis.

6.8 Class F Simulations

Two distinct subspecies of the Class F mode were identified in Sections 6.5 and 6.6. The heavily clipped version, described as Class FD in Section 6.6, is an ultra-high efficiency mode which requires the device to remain heavily saturated. It thus has limited applications, particularly in the communications area. It would, for example, be a candidate for use in a "LINC" system, as described in Chapter 10. The analysis in Section 6.5 however indicates the possibility that the benefits of a squared-up voltage waveform may be useful in a quasi-linear application. This section illustrates these options using a CAD simulator.[5]

Using the simulation from Chapter 5 (Figure 5.15) as a starting point, we will use some of the results from the analysis in Section 6.5 in order to obtain a clearer picture of how the odd harmonic components of voltage respond to the various

---

5. The circuit simulator used for the examples in this section is Microwave Office, from Applied Wave Research, Inc., sales@mwoffice.com. The files are included in the CD-ROM which accompanies this book.

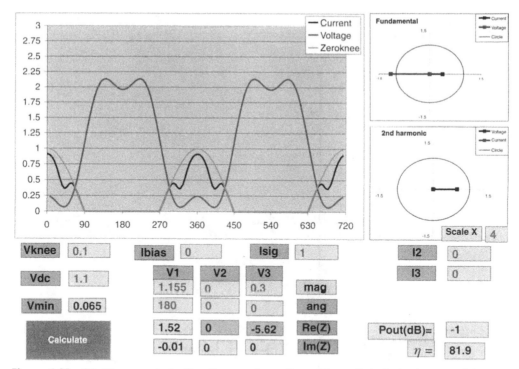

Figure 6.23   PA_Waves analysis: Class F parameters adjusted for realistic device knee model.

impedance terminations. Figure 6.24 shows the circuit which will be used for the present simulations. For comparison with other simulations in this book, the 850 MHz frequency, and the same MESFET device model, have been retained. One important difference has been incorporated for Class F simulations; the device output capacitance has been increased in value to 20 pF. This is a more representative value for higher power PA designs in the UHF and low GHz range, the X_{Cds}/R_L ratio being a factor of about 2 at 850 MHz. The simulations in this section will therefore be more representative of device technologies such as LDMOS in this frequency range. This capacitance value will be seen to have a critical impact on the practicality of Class F designs, particularly with respect to bandwidth.

At this frequency, a GaAs MESFET device gain is such that some form of lossy match is necessary on the input to maintain stability. This is the function of the input 10Ω resistor and allows for a simple low-pass input matching network. Note that the device input capacitance (C_{gs}) is brought outside the model block in the form of a linear circuit element. The voltage dependence of this capacitor can play a significant role in the design of high efficiency RF amplifiers (see Chapter 8), but for clarity this effect will be ignored in the present simulation. The use of an external C_{gs} capacitor element also allows the device drain current to be displayed using a source current meter.

The output matching consists of four sub-networks:

- The device output capacitance, now a substantially low shunt reactance at the fundamental, and lower still at harmonic frequencies;
- An ideal but realizable even harmonic short, in the form of a λ/4 SCSS;

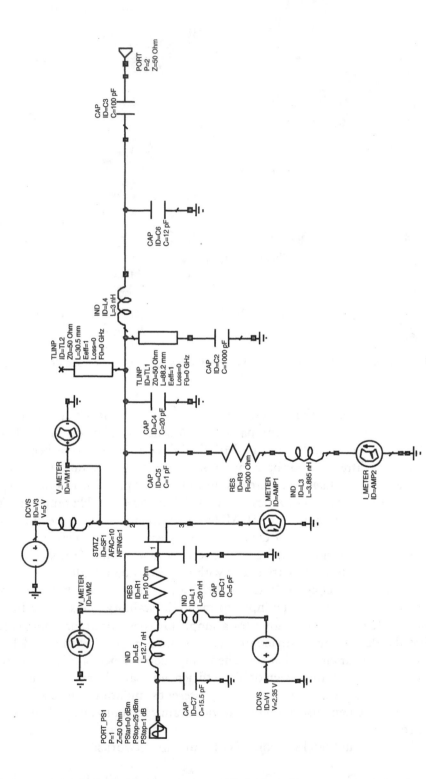

Figure 6.24   Class F amplifier; MWO simulation schematic (see file CLE_FIG_6_24 on CD-ROM).

- A low-pass fundamental matching network, which transforms the 50Ω termination down to the region of 5Ω at the fundamental, and simultaneously resonates out all of the other reactive components;
- A third harmonic network which consists of an OCSS resonator and a shunt series LCR branch.

The third harmonic termination is configured to give a variable resistive load which is essentially invisible at the fundamental, and other harmonic frequencies. The open-circuited stub essentially resonates out the device output capacitance so that the third harmonic impedance is primarily defined by the shunt LCR network.

It is already apparent that the simplistic folklore notion of "open-circuiting the odd harmonics" becomes quite a challenge to implement when the device in question has a large output capacitance, and the frequency is heading into the GHz region. Although the network described here appears to do the necessary job (see Figure 6.25) at the third harmonic, higher odd harmonics will be *short-circuited* by the output capacitance. This is consistent with *classical* Class F theory and definition, which stipulates that only third harmonic is present in the voltage waveform. The absence of higher odd harmonics does however severely limit the use of Class FD modes, as described in Section 6.6. Initially, the simulation is run using a gate bias setting in deep Class AB (–2.35v) and a drive level which runs the device into a substantially voltage-clipped condition. The resulting voltage and current waveforms are shown in Figure 6.26. An immediate observation is the modification of the current waveform due to the knee effect. In order to prevent the current from collapsing at low values of output voltage, the presence of a third harmonic component has the additional benefit of boosting the current waveform at its midpoint, much as predicted by the PA_Waves analysis in the last section (see Figure 6.23). This consideration can somewhat override the conventional calculations on the optimum value for the third harmonic voltage component in the Class F mode. Fou-

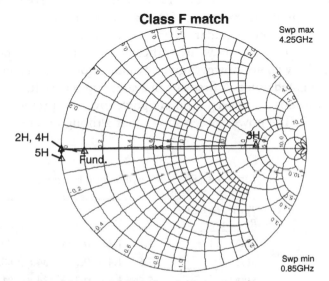

Figure 6.25   Frequency performance of Class F matching network.

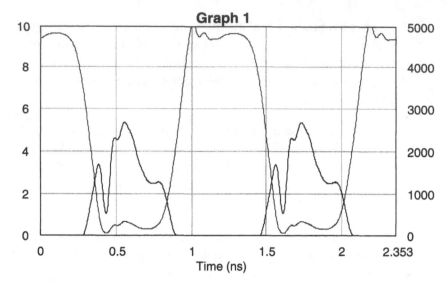

Figure 6.26 Class F simulation (using schematic and values shown in Figure 6.24). RF power output = 34.8 dBm, output efficiency = 82%.

rier analysis of the voltage waveform here shows a normalized value around 0.25 as compared to the theoretical optimum of 0.167 (1/6) obtained in Section 6.7.

Table 6.3 shows some results from varying the third harmonic termination resistor. These results show remarkable insensitivity to this value, as predicted by the analysis in Section 6.5.

The results in Table 6.3 indicate a significant departure from conventional thinking on Class F PA design. Rather than presenting the third harmonic with a nebulous "open circuit," it appears that a resistive impedance which approaches the fundamental load resistance, even to within a factor of two, does a satisfactory job. Indeed the lower limit is mainly defined by the impact on efficiency. For the 10Ω case in Table 6.3, the third harmonic power dissipated in the third harmonic resistance will be a factor of $(0.225^2)/2$ times the power developed at the fundamental, assuming a fundamental load of 5Ω. This is a nearly significant fraction (0.025), and causes an efficiency degradation of about 2% in this case.

The above results show high efficiency and acceptable power from a device of this size. They do not however give any indication of gain compression and linearity at the simulated point. A first step in evaluating the usefulness of this design for linear applications is to sweep the input power over a representative range, as shown in Figure 6.27. The power sweep shows that although there is a good range of linear

Table 6.3 Effect of Varying Third Harmonic
Termination Resistance

Third Harmonic Resistance	V_1	V_3	V_3/V_1	Pout	PAE
10Ω	5.55	1.25	0.225	34.7	79.5
50Ω	5.6	1.4	0.25	34.8	80.1
500Ω	5.6	1.45	0.26	34.8	81.2

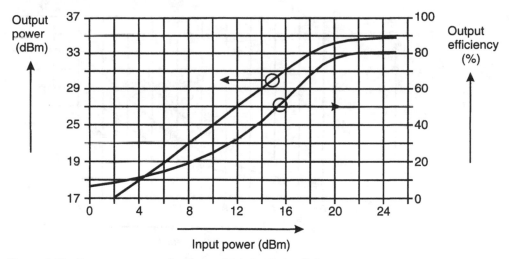

Figure 6.27   Power sweep results, Figure 6.24 circuit simulation.

performance at backed-off drive levels, the power and efficiency performance shown in Figure 6.26 (Pin = 25 dBm) represented a heavily compressed condition. Nevertheless, the efficiency at the 1 dB compression point (about 20.5 dBm drive level) still shows an efficiency of about 78%, and the linearity at lower drive levels could be expected to give acceptable spectral distortion performance in many applications.

It is useful to use the simulation to explore the effect of varying two critical design parameters, the quiescent bias setting and the fundamental load. Figure 6.28 and 6.29 show swept power and efficiency using lower, and higher, values of fundamental load. The lower load value (3Ω) causes a reduction in power gain, with higher P_{1dB}, but degraded efficiency. A higher load value (7.5Ω) causes heavy saturation at lower drive levels and lower saturated power. This latter condition represents an attempt to replicate Class FD operation, but unlike the curve shown in Figure 6.19 the efficiency does not show much sign of increasing in the clipping

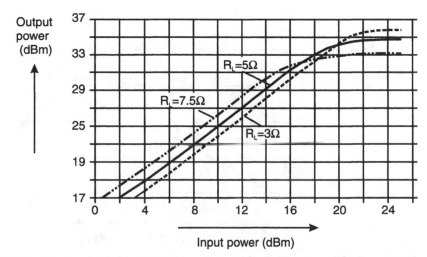

Figure 6.28   Swept output power variation with fundamental load.

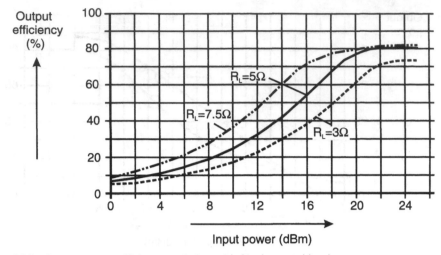

Figure 6.29 Swept output efficiency variation with fundamental load.

region. As already noted, this can be largely explained by the short-circuiting of higher odd harmonics by the device output capacitance.

Figure 6.30 shows the gain linearity as a function of quiescent bias at the median 5Ω load value. The gain characteristics are similar to those seen in Class AB amplifiers in the backed-off power regions, with gain expansion in deeper Class AB and gain compression in mid Class AB. Although there appears to be a good compromise setting, in practice there will be other considerations for obtaining best linearity with a modulated signal. AM-PM effects and the dynamic thermal characteristics of the device will significantly affect the spectral distortion. These effects will be further discussed in Chapter 9.

The simulation results appear to show that Class F has some important advantages which are realizable in practical circuits. The one remaining issue with the

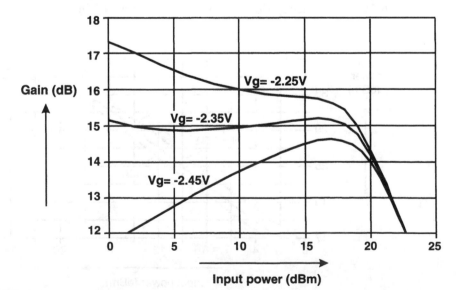

Figure 6.30 Effect of quiescent bias setting on simulated Class F power transfer characteristic.

present design is bandwidth. The device output capacitance was chosen deliberately high in order to illustrate a serious limitation of Class F operation in practice. Figure 6.31 shows the power and efficiency performance over a modest bandwidth around the design frequency, and it is clear that the bandwidth over which reasonably constant performance can be expected in this case is only a few percent. Not only that, but the actual design values for the circuit elements will show themselves to be quite critical and may require individual adjustment in production. This situation will improve rapidly as the device output capacitance is reduced. The quantification of output capacitance used in this book, the dimensionless ratio X_{Cds}/R_L, will need to have a value significantly higher than the factor of 2 used in this example. This in turn implies that even at 850 MHz, and certainly at higher GHz frequencies, a more exotic device technology such as Gallium Arsenide or Gallium Nitride is required in order to make Class F work effectively.[6]

6.9 Conclusions

The discussion and analysis of overdrive effects in this chapter have shown that considerable scope exists for trading power, or PUF, against efficiency, but this inevitably leads to higher gain compression and reduced linearity for a given power level. Results based on simple theoretical reasoning, backed up by full CAD simulations, have shown the real possibility of reaching efficiencies above 80% with actual practical circuits. This is attractive enough, in terms of heat management and battery lifetime, that it cannot be easily dismissed even when linearity is a critical issue. As will be commented in Chapter 7, where highly nonlinear switching amplifiers will

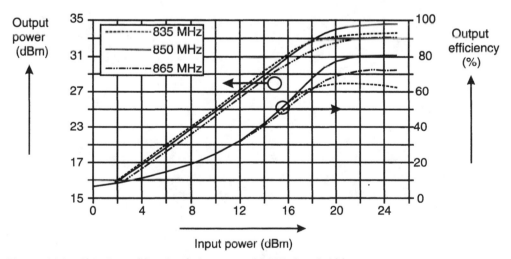

Figure 6.31 Class F amplifier simulation over a 30 MHz bandwidth.

6.   Although this design example uses a GaAs MESFET device model, the output capacitance was increased by nearly an order of magnitude in order to demonstrate the sensitivity of this parameter in Class F circuits.

be discussed, techniques such as envelope restoration and outphasing provide a possible way around the linearity issue for amplitude modulated systems.

The key to realizing higher efficiencies, or even ensuring that the best compromise has been obtained in a given design, is familiar enough at this juncture; careful selection of RF load resistance and appropriate harmonic termination. But even at lower GHz frequencies, the use of the more exotic device technologies such as Gallium Arsenide and Gallium Nitride will be necessary to achieve efficiencies in the 80% region.

References

[1] Tyler, V. J.,"A New High-Efficiency High Power Amplifier," *Marconi Rev.*, Vol. 21, 1958, pp. 96–109.

[2] Raab, F. H., "FET Power Amplifier Boosts Transmitter Efficiency," *Electronics*, June 10, 1976, pp. 122–126.

[3] Raab, F. H., "Class-F Power Amplifiers with Maximally Flat Waveforms," *IEEE Trans. Microwave Theory & Tech.*, Vol. 45, No. 11, November 1997, pp. 2007–2011.

[4] Rhodes, D., "Universality in Maximum Efficiency Linear Power Amplifiers," *Int. J. Circ. Theor. Appl.* Vol. 31, 2003, pp. 385–405.

CHAPTER 7

Switching Mode Amplifiers for RF Applications

7.1 Introduction

This chapter will consider the possibilities offered to the RFPA designer by switching mode circuits. Circuits of this kind have been used for many years in DC to DC converter applications and undoubtedly offer some possibilities for higher frequency use. This applies especially in the broader interpretation of "RF"; high power applications in the MHz and tens of MHz frequency ranges have certainly benefited from infusions of techniques and practice from the switching power converter industry.

But at GHz frequencies there remains a stubborn and irrefutable central issue, which is that RF power transistors at these frequencies cannot realistically be modeled as simple switching elements. At these frequencies, and in any currently available power transistor technology, the device will not sweep through its linear region fast enough to behave like a switch, and the behavior in the linear region must be included in the circuit simulation, just as it is when simulating conventional PA circuits. For these reasons, RF designers most frequently move on and dismiss switching modes as being unrealizable for higher frequency applications. In fact, although the problem of "slow" switching speed can never be fully resolved at higher RF frequencies, it appears that it can sometimes be judiciously out-maneuvered and that useful RF applications do exist. This can lead to some useful applications in the "gray" area, when a very high frequency technology is being used at a low relative frequency; a GaAs PHEMT or HBT below 2 GHz would be such an example.

Probably the most well-known, and certainly most widely touted, switching mode for RF applications is the Class E mode. Although in its most basic and original form, a Class E device needs to have the same unreachable switching properties as any other switching mode application, there appear to be derivatives of the classical switch-mode Class E PA which allow for slower switching but still appear to be distinct from conventional PA modes. This metamorphosis is an important topic and has been given a chapter to itself, Chapter 8; the present chapter will focus mainly on the classical switching modes which can be used quite effectively below 1000 MHz.

The chapter starts by considering a simple switching device with a broadband resistive load, and then considers the effect of tuning the load. Such an amplifier, even with an ideal switching element, has surprisingly little to offer the RF designer in comparison to the Class B or Class F modes considered in the previous chapters, but serves as a useful introduction to the subject. The Class D amplifier is considered next. In its ideal switching mode form, a "perfect" RF amplifier is created, with a halfwave rectified sinewave of current and a squarewave of voltage. This produces maximum possible fundamental RF power at 100% efficiency, but is difficult to realize at higher RF frequencies.

The bulk of the chapter considers the classical Class E switching mode in some detail. As a lead in to the following chapter, a simple design example illustrates how in the "gray area," at 800 MHz and using a GaAs MESFET device, a Class E amplifier can be realized which uses a design procedure based almost entirely on the results from the classical analysis.

Switching mode power amplifiers, without qualification, are highly nonlinear devices. This reality leads to further rejection by the wireless communications industry, on the basis that variable envelope signals cannot be passed without hopeless levels of distortion. It should however be borne in mind that the same rejection could equally be applied to a Class C amplifier, but such amplifiers were at one time commonly used in AM and SSB broadcasting. In order to overcome the distortion problem, techniques such as envelope restoration and outphasing were developed, and will be described in Chapter 10.

7.2 A Simple Switching Amplifier

Figure 7.1 shows the schematic diagram for a simple switching mode amplifier. The key element is the switch, which in this analysis will be considered as ideal. By "ideal" it means that the switch is either completely on (short circuit) or completely off (open circuit) and can switch between these two states instantaneously. Another key aspect of the idealization is that the timing of switch opening and closure is controlled by the input signal; the "conduction angle" is assumed to be discretionary in the analysis. In practice, such control of the switch conduction angle will be achieved by varying the drive level and bias point at the input of a transistor. This will usually mean that the device will be heavily overdriven in comparison to normal linear operation, and it is almost inevitable that the overall gain will be many dBs lower than a conventional linear amplifier. In any event, it will not be possible to estimate the power gain of such a circuit, without more detailed knowledge of the specific device and the drive arrangements. Useful expressions can, however, be derived for the output power and efficiency.

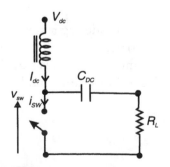

Figure 7.1   Basic RF switching amplifier.

Figure 7.1 shows a circuit which has some familiar features; a DC-blocked RF load resistor and a choked DC supply. The current and voltage waveforms are quite trivial in the symmetrical case where the switching duty cycle is 50%, and Figure 7.2 shows a more general case with a conduction angle 2α, corresponding to the switch closure period. The maximum current in the switch is now controlled entirely by the supply voltage and the RF load resistor, and the voltage toggles between zero and V_{pk}, where V_{pk} will be a function of the conduction angle α and the supply voltage V_{dc}. Before analyzing the waveforms in more detail, it is clear that this circuit converts DC energy to RF energy at 100% efficiency; at no point in the RF cycle is there a nonzero voltage and current simultaneously, so no energy is wasted as heat in the switch. This is sometimes interpreted erroneously as a perfect and final solution to RF power generation. In fact, this particular switching mode amplifier has some undesirable characteristics in that substantial energy is generated at harmonic frequencies, and the maximum fundamental efficiency is only just over 80%.

It is assumed that the DC supply voltage remains constant as the conduction angle α is varied; this will result in an asymmetrical voltage waveform whose peak value will be proportionately greater or less than twice the supply voltage as α is varied above or below $\pi/2$. This circuit can now be analyzed for power and efficiency as a function of conduction angle. With a simple broadband resistive load, the relationship between the peak voltage V_{pk} and the DC supply voltage V_{dc} is given by the mean value integral,

$$V_{dc} = \frac{1}{2\pi} \int_{-\pi}^{\pi} v_{SW}(\theta) \cdot d\theta$$

$$= \frac{1}{\pi} \int_{\alpha}^{\pi} V_{pk} \cdot d\theta \qquad (7.1)$$

$$\frac{V_{dc}}{V_{pk}} = \frac{(\pi - \alpha)}{\pi}$$

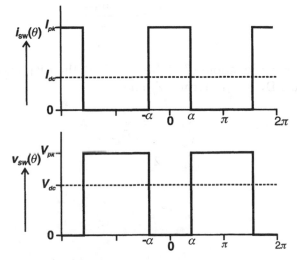

Figure 7.2   Basic RF switch waveforms.

The voltage waveform can therefore be considered to be an alternating voltage with zero mean value if it is offset by V_{dc}. So the peak-to-peak current swing will be

$$I_{pk} = V_{pk}/R_L \qquad (7.2)$$

It should be noted here that an ideal switch differs from a real device in the important respect that the peak current is entirely a function of the DC supply and the load resistor R_L; this resistor can be arbitrarily reduced in value to give any desired amount of RF power. This analysis will therefore consider primarily the power and efficiency for a given value of I_{pk}, which can then be used as a rational equivalent for comparison to a transistor having a maximum, or saturated, current, I_{max}, equal to I_{pk}.

Mean value considerations for the current waveform give, by simple inspection

$$I_{dc} = \frac{\alpha}{\pi} \cdot I_{pk} \qquad (7.3)$$

The fundamental even Fourier coefficient of current, I_1, is given by

$$I_1 = \frac{1}{\pi} \int_{-\pi}^{\pi} i_{SW}(\theta) \cdot \cos(\theta) \cdot d\theta;$$
$$\text{where } i_{SW}(\theta) = I_{pk}, -\alpha < \theta < \alpha;$$
$$= 0, -\pi < \theta < \alpha; \alpha < \theta < \pi \qquad (7.4)$$
$$= \frac{2}{\pi} \int_0^\alpha I_{pk} \cdot \cos(\theta) \cdot d\theta$$
$$\frac{I_1}{I_{pk}} = \frac{2 \cdot \sin(\alpha)}{\pi}$$

Similarly, the fundamental Fourier coefficient for the voltage waveform is

$$\frac{V_1}{V_{pk}} = -\frac{2 \cdot \sin(\alpha)}{\pi} \qquad (7.5)$$

Combining equations (7.1) through (7.5), the RF power, P_{rf}, can be expressed in terms of the DC supply terms, V_{dc} and I_{dc}:

$$\frac{V_1}{V_{dc}} = -\frac{2 \cdot \sin \alpha}{\pi - \alpha}$$
$$\frac{I_1}{I_{dc}} = \frac{2 \cdot \sin \alpha}{\alpha} \qquad (7.6)$$
$$P_{rf} = V_{dc} \cdot I_{dc} \cdot \frac{2 \cdot \sin^2 \alpha}{\alpha(\pi - \alpha)}$$

So that the output efficiency, η, is given by

$$\eta = \frac{2 \cdot \sin^2 \alpha}{\alpha(\pi - \alpha)} \qquad (7.7)$$

So in the symmetrical squarewave case, the efficiency is $(8/\pi^2)$, about 81%. Figure 7.3 shows a plot of efficiency versus conduction angle. The RF efficiency peaks at the symmetrical case of $\alpha = \pi/2$, corresponding to the generation of a maximum proportion of RF energy at the fundamental. Figure 7.3 also shows the RF output power at the fundamental and at the second and third harmonics. These power curves are calibrated relative to the RF power from a Class A linear amplifier having the same peak RF current (i.e., $I_{max} = I_{pk}$). Substituting for I_{dc} in (7.3), the expression for fundamental RF output power (7.6) becomes:

$$P_1 = V_{dc} \cdot I_{pk} \cdot \frac{2 \cdot \sin^2 \alpha}{\pi(\pi - \alpha)} \qquad (7.8)$$

and defining the linear power, P_{lin}, as

$$P_{lin} = \frac{V_{dc} \cdot I_{pk}}{4}$$

the relative power is

$$\frac{P_1}{P_{lin}} = \frac{8 \cdot \sin^2(\alpha)}{\pi(\pi - \alpha)} \qquad (7.9)$$

This function is also plotted in Figure 7.3. The striking feature of the fundamental power curve is the peak which occurs at a conduction angle of about 113° ($\alpha = 0.63\pi$). This peak represents an RF power which is 2.7 dB higher than a Class A amplifier having the same peak RF current and DC supply voltage. Although the

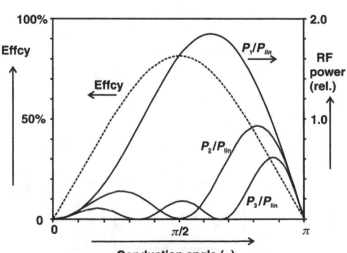

Figure 7.3   RF power and efficiency of basic RF switch.

formulation used does allow for much higher peaks of voltage than in the comparable Class A case, this condition nevertheless represents a theoretically feasible amplifier. The peak power shown in Figure 7.3 could be regarded as something of a global maximum for RF power obtainable from any kind of device using a stipulated DC supply.

The efficiency curve, shown dashed in Figure 7.3, represents something of a disappointment; despite having a device which dissipates no heat, the efficiency peaks at about 81%, corresponding to the symmetrical squarewave condition. The problem, as shown by the harmonic power plots, is that power is being wasted in harmonic generation and the next step is to eliminate this using a resonator.

7.3 A Tuned Switching Amplifier

Other than the possibility for frequency multiplication, the harmonic energy indicated in Figure 7.3 is undesirable, and can be most easily removed by placing a harmonic short across the load, as shown in Figure 7.4. This is now more familiar territory, since the voltage waveform will now assume a sinusoidal form, as shown in Figure 7.5. This makes analysis quite straightforward. In this analysis it will again be convenient to relate the current waveforms to a specific peak value, I_{pk}, in order to compare the results with that of a conventional reduced conduction mode amplifier having the same maximum peak RF current.

The relationships for current are the same as in (7.3) and (7.4),

$$\frac{I_1}{I_{pk}} = \frac{2 \cdot \sin(\alpha)}{\pi}$$

$$\frac{I_{dc}}{I_{pk}} = \frac{\alpha}{\pi}$$

The sinusoidal voltage gives the simple relationship

$$V_1 = V_{dc}$$

so the fundamental RF power is

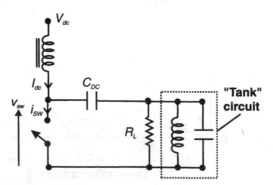

Figure 7.4 Tuned RF switching amplifier.

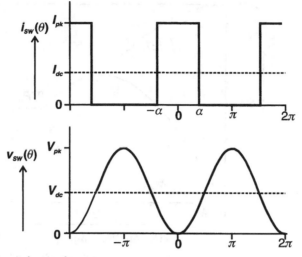

Figure 7.5   Tuned switch waveforms.

$$P_1 = I_{dc} \cdot V_{dc} \cdot \frac{\sin(\alpha)}{\alpha} \qquad (7.10)$$

giving an output efficiency

$$\eta = \frac{\sin(\alpha)}{\alpha} \qquad (7.11)$$

The relative power can be determined, as before, by expressing P_1, in (7.10), in terms of the peak current I_{pk}, using the relationship of (7.3):

$$P_1 = I_{pk} \cdot V_{dc} \cdot \frac{\sin(\alpha)}{\pi}$$

so the ratio of fundamental RF power to P_{lin} is

$$\frac{P_1}{P_{lin}} = I_{pk} \cdot V_{dc} \cdot \frac{4 \cdot \sin(\alpha)}{\pi} \left(P_{lin} = \frac{I_{pk} \cdot V_{dc}}{4} \right) \qquad (7.12)$$

The efficiency (7.11) and relative power (7.12) are plotted as a function of conduction angle in Figure 7.6. As would be expected from the analysis of a Class C amplifier, as the switched current pulse gets very short, the efficiency increases to high values, but there is a corresponding reduction in relative RF power. The peak in RF power occurs at a conduction angle of $\pi/2$, and the corresponding RF power is $(4/\pi)$ times the linear equivalent, or about 1 dB higher. But the efficiency at this point is a rather modest 63% $(2/\pi)$. Probably the most significant positive feature of the ideal switch performance in this circuit is the efficiency of about 87% at the point where the relative power crosses unity on the left side, but this still falls short of the Class F results presented in Chapter 6.

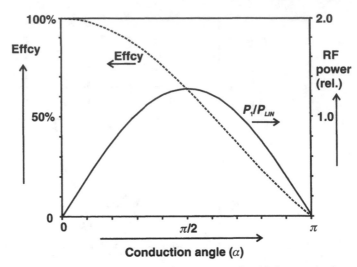

Figure 7.6   RF output power and efficiency of basic RF switch with harmonic short.

The results of the last two sections show that an ideal switch does not, by itself, give any instantaneous improvements in the efficiency of more conventional high efficiency circuits, and alternative configurations have to be sought in order to make best use of a switching device. Two such configurations will now be considered, the Class D and Class E modes.

7.4   The Class D Switching Amplifier

Figure 7.7 shows a schematic diagram for a switching mode amplifier in which a series resonant circuit is connected across a two-way switch arrangement, whereby the LCR resonator is switched between a bypassed DC voltage generator and ground for alternate half cycles. It is assumed that the repetition cycle matches the resonant frequency of the LCR circuit and that the Q factor is high.

Figure 7.8 shows the waveforms resulting from such a circuit. The key issue is that the current in the LCR branch is constrained to remain sinusoidal, with no DC offset, due to the inertia effect of the resonator and the DC blocking of the series capacitor. The result is that switch "A" conducts a positive half sinewave, and switch "B" conducts a negative half sinewave, which add together to form the neces-

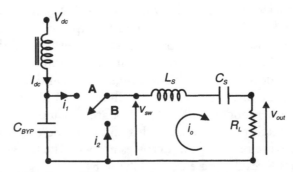

Figure 7.7   Switching Class D amplifier.

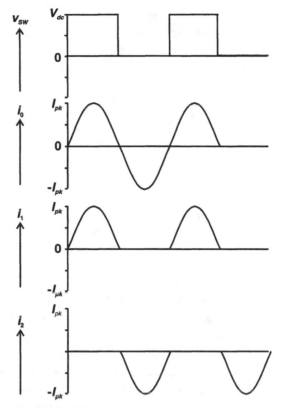

Figure 7.8   Switching Class D amplifier waveforms.

sary full sinewave of current in the LCR branch. Since the switch voltage waveform is a squarewave, it is clear that no power is being dissipated in the switch and that no harmonic energy is being generated; the half sinewaves of current flowing in each switch half-cycle contain no odd harmonics.

The analysis is therefore quite straightforward. The peak current, I_{pk}, is given by

$$I_{pk} = I_{dc} \cdot \pi$$

The fundamental RF current flowing round the LCR branch, I_1, is simply

$$I_1 = \frac{I_{pk}}{2}$$

The fundamental component of RF voltage appearing across the LCR branch is

$$V_1 = V_{dc} \cdot \frac{2}{\pi}$$

due to the the fact that the voltage appearing across the LCR branch is a squarewave of peak value V_{dc}. Once again, with ideal switches, the peak current, and consequently the power, can be made arbitrarily high by suitable reduction in the value of R_L. The RF power can be expressed in terms of I_{pk}, and V_{dc}, from these relations:

$$P_1 = \frac{V_1 \cdot I_1}{2} = \frac{V_{dc} \cdot I_{pk}}{\pi}$$

and the DC supply power is

$$P_{dc} = V_{dc} \cdot I_{dc} = \frac{V_{dc} \cdot I_{pk}}{\pi}$$

Clearly, the efficiency is 100%. The relative power, compared to a Class A device having the same I_{pk} and V_{dc}, is

$$\frac{P_{rf}}{P_{lin}} = \frac{V_{dc} \cdot I_{pk}}{\pi} \cdot \frac{4}{V_{dc} \cdot I_{pk}} = \frac{4}{\pi}, \text{ about 1 dB}$$

This switching version of a Class D amplifier has one significant difference from the RF version discussed in Chapter 5. In the switching version described here, the peak-to-peak voltage swing is equal to the DC supply voltage, and the peak-to-peak RF current is twice the peak current of each individual switching device. The switching action in some respects produces a similar overall effect to having a pair of devices in a push pull configuration. But the overall issue is realizability. The "A" switch is a high-side (ungrounded) device, which poses problems at higher frequencies both in terms of parasitic reactance and drive requirements. Such amplifiers have been reported at low RF frequencies, in the 10 MHz region [1], and offer some attractive possibilities for achieving high power and high efficiency without generating excessive voltage swing. But the possibilities for higher power microwave applications seem marginal with present technology.

7.4 Class E—Introduction

The Class E mode has been vigorously touted by its inventors [2–5] for over 30 years, to a frequently ambivalent microwave community. It has also, for most of that time, been patented, and the expiry of the patent [U.S. 3,919,656, 1975] has resulted in some renewed interest in the possibilities it offers [6, 7]. The next few sections introduce the Class E mode in its classical form, and the possibilities for higher RF derivatives are considered further in Chapter 8.

The previous sections in this chapter have already highlighted the difficulty of harnessing an ideal switch in a productive manner. Indeed, the results in Section 7.2 indicate that the sharp, rectangular pulses of current can be as much a hindrance in obtaining higher efficiency as they are beneficial. The Class E mode in some ways represents a halfway house between the analog world of conventional reduced conduction angle mode amplifiers and the digital world of ideal switches. Although most easily introduced as a switching type of amplifier, it will become apparent that the waveforms are distinctly analog in appearance, and can be approximately supported by a device with a slower switching characteristic which includes a linear region.

Furthermore, simulations and verification tests on actual amplifiers seem to support the view that Class E does truly represent an alternative to conventional reduced conduction angle operation, giving higher efficiency without the circuit complexity of more advanced Class F designs. Direct comparison is troublesome, in that Class F can, as has been demonstrated in Chapter 6, achieve very high efficiencies, but show substantially reduced power output under optimum efficiency operation. Class E has its own disadvantage in terms of peak voltage levels, and the PUF has to be traded against this limitation. But the final judgment may be that Class E appears to be easier to realize in practice using solid state transistors than a short conduction angle Class C design.

As stated in the introduction to this chapter, a Class E amplifier is unquestionably nonlinear, in the sense that variations in input power amplitude will not be reproduced at the output in any acceptable form. This also applies to a Class C amplifier, and techniques have been available for many decades to remodulate the output. For reasons that are not entirely clear, these techniques have fallen into disuse somewhere along the road of transition from vacuum tubes to solid state devices. One possible reason for this change of emphasis from remodulation to linear, or quasi-linear, RF amplification may be that there has never been a reliable method of achieving efficiencies in the 90% region using solid state devices, such as were quite normally obtained from simple tube amplifier designs, and as can still be found even in amateur radio literature. Another reason is the virtual disappearance of amplitude-only modulation systems. As will be discussed in Chapter 10, although most remodulation techniques can be adapted to preserve phase modulation, this will usually incur more system complexity. The Class E mode may just represent a possibility for achieving high enough efficiencies from solid state devices that envelope restoration and outphasing may become mainstream techniques once again.

7.5   Class E—Simplified Analysis

The analysis presented in this section is idealized, and assumes initially that the active device can be represented as a switch. This results in analytical expressions for the device waveforms. Thirty years ago it was conventional practice to pursue the analysis in symbolic form, in order to relate the external component values to the stipulated waveform parameters. This process can be followed, for example, in a classical analysis published by Raab [3]. Many authors have taken this basic analysis further, allowing for device and circuit parasitics, although the particular issue of transmogrifying a switch into an otherwise staple transconductive transistor seems to have been vigorously avoided. The continuation of the Class E theme in Chapter 8 of this book offers an alternative path to practical realization at GHz frequencies, in particular making the important transition from a switching device to a conventional transistor which operates in a linear transconductive mode above its threshold point. By way of introduction, however, we will stick to convention and analyze the Class E circuit using an ideal switch as the active element. As with the analysis of Class AB waveforms in Chapter 3, the symbolic analysis will be curtailed at strategic points, where direct computation enables a clearer path to the required goal of a quantitative design strategy.[1] Figure 7.9 shows the basic elements of a Class E circuit. An ideal switch is shunted by a capacitor C_p; the value of C_p will be seen to be an important part of the RF network and at GHz frequencies is usually neither small enough to be a device parasitic nor large enough to be a harmonic short. The RF matching network consists additionally of a series resonant circuit, which incorporates the final RF load resistor. It is assumed in this analysis that this resistor will probably be the input impedance of an additional matching section which transforms the required value up to a 50 ohm interface. As usual, the DC is supplied through a high reactance choke whose value is such that no variations of current can be supported within the timeframe of an RF cycle. Similarly, the Q-factor of the resonator is assumed to be high enough that the "flywheel" effect forces a sinusoidal current to flow in the LCR branch.

As before, it is assumed that conditions on the input can be controlled such that the open and closure timing of the switch during an RF cycle can be arbitrarily predetermined, and are effectively independent variables in the problem.

Looking now at the branch currents in Figure 7.9, it can be seen that two "immovable" currents combine to flow into the switch-capacitor combination; the DC supply current, I_{dc}, and the sinusoidal "flywheel" current in the LCR branch, $I_{rf} \cdot \cos \omega t$.

So the current flowing into the switch-capacitor combination is an offset cosine wave of current,

$$i(\theta) = I_{rf} \cdot \cos(\theta) + I_{dc} \quad (\theta = \omega \cdot t) \tag{7.13}$$

where the values of I_{rf} and I_{dc} have yet to be determined. An important normalization is now defined, writing (7.13) in the form

$$i(\theta) = I_{dc} \cdot (1 + m \cos(\theta)) \tag{7.14}$$

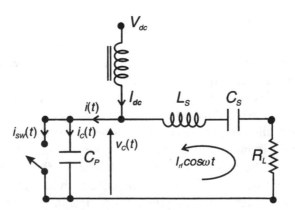

Figure 7.9 Basic Class E amplifier schematic.

1.   For better or for worse, lengthy symbolic analysis is not the flavor of the 00's. The foregoing analysis is a significant consolidation from that presented in the first edition of this book, and enables direct computation for expressions which cannot be further simplified using the symbolic approach [9].

where

$$m = \frac{I_{rf}}{I_{dc}} \qquad (7.15)$$

Figure 7.10(a) shows the offset cosine wave of current $i(\theta)$ and also shows the positions selected for switch opening and closure. The switch will be closed for the duration of the interval between the positive-moving zero crossing of $i(\theta)$, $-\alpha_1$, and an arbitrarily selected angle α_2 during the positive portion of $i(\theta)$. Clearly, when the switch is closed, $i(\theta)$ will flow entirely through the switch, and when the switch is opened, $i(\theta)$ will flow *entirely into the capacitor*. This is the definitive aspect of ideal Class E operation. The "conduction angle" ϕ can be defined as the sum of α_1 and α_2,

$$\phi = \alpha_1 + \alpha_2 \qquad (7.16)$$

The current waveforms for switch and capacitor can therefore be easily drawn, and are shown in Figure 7.10(b, c). As already noted, the key effect is the instantaneous transfer of current from switch to capacitor at the point of switch opening, but note that when the switch closes, it does so at a zero-crossing point in $i(0)$. This condition is assumed to be the optimum one for the purposes of this analysis, although a more general analysis would need to consider the effect of an impulsive current which will flow into (or out of) the capacitor at the instant of switch closure when this timing condition is not met [3, 8].

The current waveform in the switch bears some resemblance to a Class C amplifier, with the important difference that it is asymmetrical, having a slow rise and a very fast, ideally instantaneous, fall. The peak switch current, I_{pk}, will be

$$I_{pk} = I_{rf} + I_{dc} = I_{dc}(1+m) \qquad (7.17)$$

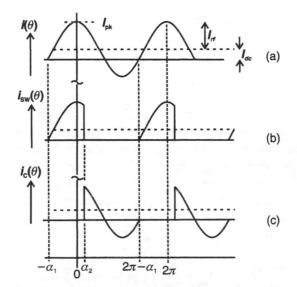

Figure 7.10 Class E current waveforms: (a) total current, (b) switch current, and (c) shunt capacitor current.

and as before, will correspond to the I_{max} of the transistor, which will ultimately have to approximate the switching function.

This is a Class E amplifier; all that remains is to integrate the capacitor current to obtain the RF voltage waveform, and to relate the dependent variables I_{dc}, m, α_1, α_2, to the independent, or input, variables ϕ, V_{dc}, ω, and I_{pk}, relationships that will lead to values for the circuit elements. The first step in this process is to trim the list of dependent variables. The first zero crossing of $i(\theta)$ gives the relation

$$I_{dc}(1 + m\cos\alpha_1) = 0$$

where

$$\cos\alpha_1 = -\frac{1}{m} \tag{7.18}$$

There is a second relationship between m and the three conduction angle parameters, which is obtained by integrating the "on" current in the switch and equating the result to I_{dc},

$$\frac{I_{dc}}{2\pi} \int_{-\alpha_1}^{\alpha_2} (1 + m\cos\theta)d\theta = I_{dc} \tag{7.19}$$

This integral can be evaluated to give a relationship between α_1, α_2, and m, and also using (7.5) simplifies to

$$\sin\alpha_1 = \frac{2\pi + \sin\phi - \phi}{m(1 - \cos\phi)}$$

and noting again that $\cos\alpha_1 = -\frac{1}{m}$ (from (7.18)), a direct dependency of m on the conduction angle ϕ can be established,

$$1 + \left(\frac{2\pi + \sin\phi - \phi}{1 - \cos\phi}\right)^2 = m^2 \tag{7.20}$$

Thus the specification of the single input design parameter ϕ leads directly to unique values for m, α_1, and α_2, which in turn lead directly to computable expressions for all of the voltages and currents, and their harmonic components, in terms of the single normalizing parameter I_{dc}. This would appear to represent a considerable simplification in the solution of this problem [9], compared to previous published attempts by various authors, this one included.

A further relationship can be obtained by integrating the switch current and equating this to I_{dc},

$$I_{dc} = \frac{1}{2\pi} \int_{-\alpha_1}^{\alpha_2} I_{dc}(1 + m\cos\theta)d\theta$$

so that

$$(\alpha_1 + \alpha_2) + m(\sin\alpha_2 + \sin\alpha_1) = 2\pi \qquad (7.21)$$

The switch/capacitor voltage waveform, $v_c(\theta)$, is given by:

$$
\begin{aligned}
v_C(\theta) &= \frac{I_{dc}}{\omega C_P} \int_{\alpha_2}^{\theta} (1 + m\cos\theta)\,d\theta \\
&= \frac{I_{dc}}{\omega C_P} [\theta + m\sin\theta - \alpha_2 - m\sin\alpha_2], \quad \alpha_2 < \theta < 2\pi - \alpha_1, \qquad (7.22) \\
&= 0, \quad -\alpha_1 < \theta < \alpha_2
\end{aligned}
$$

This voltage is shown in Figure 7.11. The voltage waveform integrates up from zero, starting at the point the switch opens ($\theta = \alpha_2$) and the entire current $i(\theta)$ instantaneously transfers from the switch to the capacitor. A peak voltage is reached at the point where the capacitor current crosses zero ($\theta = \gamma$) and in the optimum case being analyzed here, the voltage will reach zero again at the precise moment the switch closes again ($\theta = 2\pi - \alpha_1$).[2]

Looking now at the current and voltage waveforms at the switch, several important comments can be made:

1. The relationships (7.15) through (7.22) completely define the current and voltage waveforms, in terms of the selected conduction angle ϕ, and two

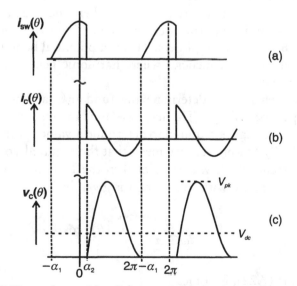

Figure 7.11   Class E RF waveforms: (a) switch current, (b) shunt capacitor current, and (c) switch/shunt capacitor voltage.

2. As already noted, this does not exclude the possibility of cases where the switch closes with a nonzero voltage across it; such cases will show an impulsive spike of current at the closure point.

normalization parameters, I_{dc} and $(1/\omega C_p)$, which will be set to unity for the ongoing analysis and then scaled to match specific designs at a later stage.

2. At no time in the cycle do current and voltage coexist at nonzero values. This implies 100% efficient conversion of DC to RF energy.

3. The only dissipative element in the circuit is the load resistor, and this is only visible to the switch at the fundamental (switching) frequency. This implies that the only energy being dissipated is at the fundamental frequency.

4. The voltage waveform is nonsinusoidal and asymmetric. This implies a mean DC level that will be lower than half of the peak value.

Item (2) above is, of course, a direct consequence of having an ideal switch element. But taken with item (3), the important conclusion is reached that this is a 100% efficient system for converting DC to fundamental, sinusoidal, RF energy, regardless of the actual value of conduction angle, ϕ, which is selected.

There is an observation to make on the reasoning presented thus far. Looking at Figure 7.11, it is clear that we have created a highly nonlinear system, with nonsinusoidal waveforms and an attendant galaxy of harmonic components. The normal reaction of the RF engineer to such a situation is to jump immediately into the frequency domain and start to analyze the necessary harmonic impedances. In fact, it will be seen that in the present analysis the only frequency domain analysis will be to determine the fundamental components of the voltage and current waveforms; the system is effectively solved in the time domain. The harmonic impedances do not play any direct part in the analysis and remain as unseen elements in the problem; the network topology itself automatically provides the harmonic terminations and we do not need to determine their values. This is not just a matter of analytical philosophy. There is a very real advantage here that we have a network which does the whole job, and not a truncated approximation to the job which we use in designing class AB or B type amplifiers. This is a potential benefit of the Class E concept, especially at lower frequencies where harmonic generation is not as inhibited by device parasitics.

It only remains to determine the fundamental and DC components of the switch/capacitor voltage. The in-phase fundamental component of the capacitor (and switch) voltage is required to determine a value for the RF power and the load resistor, R_L. By "in-phase," we mean that the required voltage is in phase with the original current which was stipulated in the LCR branch, $mI_{dc}\cos\theta$. The voltage component in-phase with this current is given by

$$V_{ci} = \frac{I_{dc}}{\omega C_p \pi} \cdot \int_{\alpha_2}^{2\pi-\alpha_1} v_C(\theta) \cdot \cos\theta \cdot d\theta$$

which, using (7.22), becomes

$$V_{ci} = \frac{I_{dc}}{\omega C_p \pi} \int_{\alpha_2}^{2\pi-\alpha_1} \{(\theta + m\sin\theta) - (\alpha_2 + m\sin\alpha_2)\}\cos\theta d\theta$$

Direct evaluation of the definite integral gives the expression

$$V_{ci} = \frac{I_{dc}}{\omega \pi C_P} \left[\cos \alpha_1 - \cos \alpha_2 - (2\pi - \alpha_1 - \alpha_2) \sin \alpha_1 + \frac{m}{4} (\cos 2\alpha_2 - \cos 2\alpha_1) \right.$$
$$\left. + m \sin \alpha_2 (\sin \alpha_1 + \sin \alpha_2) \right]$$

which using the relationship in (7.21) can be somewhat compressed to the form

$$V_{ci} = -\frac{I_{dc}}{\omega \pi C_P} \cdot \frac{1}{2m} \left\{ m \left(\sin^2 \alpha_1 - \sin^2 \alpha_2 \right) + 2 \left(\cos \alpha_1 - \cos \alpha_2 \right) \right\} \qquad (7.23)$$

There will also be a quadrature component of $v_c(\theta)$, given by

$$V_{cq} = \frac{I_{dc}}{\omega \pi C_P} \cdot \int_{\alpha_2}^{2\pi - \alpha_1} v_C(\theta) \cdot \sin \theta \cdot d\theta$$

which evaluates to

$$V_{cq} = \frac{I_{dc}}{\omega \pi C_P} \left\{ \alpha_2 \cos \alpha_2 - (2\pi - \alpha_1) \cos \alpha_1 - \sin \alpha_1 - \sin \alpha_2 \right.$$
$$\left. + \frac{m}{4} \left(4\pi - 2\alpha_1 - 2\alpha_2 + \sin 2\alpha_2 + \sin 2\alpha_1 \right) \right\}$$

and through multiple applications of (7.21) and (7.18) compresses down to

$$V_{cq} = \frac{m I_{dc}}{2\omega \pi C_P} \left\{ m(\sin \alpha_1 + \sin \alpha_2) + \tfrac{1}{2}(\sin 2\alpha_1 - \sin 2\alpha_2) + 2\sin \alpha_2 \cos \alpha_1 \right\} \qquad (7.24)$$

The mean voltage component of $v_c(\theta)$ represents the design value for the external DC supply and is given by

$$V_{dc} = \frac{I_{dc}}{2\pi \omega C_P} \int_{\alpha_2}^{2\pi - \alpha_1} (\theta + m \sin \theta - \alpha_2 - m \sin \alpha_2) d\theta$$

which, following a similar but more straightforward manipulation process than in the previous two cases, becomes

$$V_{dc} = \frac{I_{dc} \cdot m}{4\pi \omega C_P} \left\{ m \left(\sin^2 \alpha_1 - \sin^2 \alpha_2 \right) + 2 \left(\cos \alpha_2 - \cos \alpha_1 \right) \right\} \qquad (7.25)$$

which bears a striking resemblance to the in-phase fundamental voltage component V_{ci} in (7.23); these two expressions [(7.23) and (7.25)] confirm that the RF power, noting that the fundamental component of current was originally defined to be $m I_{dc} \cos \theta$,

$$P_{rf} = (-) \frac{I_{dc} \cdot V_{ci}}{2}$$

is identically equal to the DC power,

$$P_{dc} = I_{dc} \cdot V_{dc}$$

the negative sign indicating that the device is generating, rather than dissipating, RF power.

Returning to (7.24), a nonzero value of V_{cq} may appear to be in conflict with the original stipulation of a cosine current wave. In fact, this merely represents a requirement for a reactive impedance component in the series LCR branch, the LC circuit elements having thus far been assumed to be resonant at the excitation frequency ω. This can be identified in physical terms as a need to retune the LC resonator to allow for the periodic switching-in of the parallel capacitance C_p. This point can be further clarified, and some useful relationships derived prior to evaluation of the circuit element values, by considering the circuit shown in Figure 7.12.

This circuit shows a current generator, $I\cos\omega t$, supplying a circuit loop which consists of series inductive element X_L, a resistive element R_L, and a shunt element of essentially unknown impedance, but whose terminal voltage is known. This circuit can be analyzed at the fundamental frequency using linear circuit theory, but we will revert to time expressions for the various voltages and currents in order to avoid some traps which accompany the use of the ubiquitous engineering "j" artifice for representing reactive impedances. If the voltage across the shunt element is written in the form

$$v_{sh} = V_s \sin\omega t + V_c \cos\omega t$$

then the loop equation can be written

$$V_s \sin\omega t + V_c \cos\omega t + I(R_L \cos\omega t - X_L \sin\omega t) = 0$$

assuming the convention that inductive reactance, ωL, is positive, and capacitive reactance is $-1/\omega C$. Thus, by comparing coefficients of sine and cosine components,

$$R_L = -\frac{V_c}{I}, \; X_L = \frac{V_s}{I} \tag{7.26}$$

These expressions can now be applied directly to the Class E analysis, substituting V_{ci}, V_{cq}, and $m.I_{dc}$ for V_c, V_s, and I, respectively, to determine the element values in terms of the normalizing parameters I_{dc} and $1/\omega C_p$.

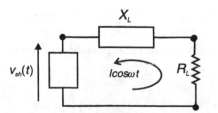

Figure 7.12   Class E fundamental impedance analysis.

The analysis of a Class E amplifier using an ideal switch is complete. For a given conduction angle the above expressions allow direct computation of all the dependent parameters in a Class E design which will initially be in normalized form, with I_{dc} and the capacitive reactance $1/\omega C_p$ being set to unity. The outstanding issue, from a design viewpoint, concerns the selection of the conduction angle, ϕ. As discussed above, the efficiency of a Class E amplifier will always be 100%, regardless of the value of ϕ. In contrast the RF power, as measured by the PUF, is a rather strong function of ϕ, as shown in Figure 7.13. This shows the power, normalized as the PUF, plotted as a function of ϕ. We will define the PUF in this case to be the ratio of the Class E power to the same device operating with optimum Class A conditions, and *with the same supply voltage*. It would seem that higher values of ϕ can give PUF factors well in excess of unity, or 0 dB, but there is an important factor which enters into the reckoning and requires a more careful compromise in the selection of ϕ. This factor is the peak voltage V_{pk}, which is determined by setting $\theta = 0$ in (7.22), giving

$$V_{pk} = \frac{I_{dc}}{\omega \pi C_p}\left\{(\alpha_1 + -\alpha_2) + m(\sin\alpha_1 - \sin\alpha_2)\right\} \tag{7.27}$$

which is also plotted in Figure 7.13, as a ratio to the DC supply V_{dc}.

Figure 7.13 shows that although smaller values of ϕ give values of V_{pk}/V_{dc} that are only slightly higher than the conventional factor of 2, as ϕ increases to give a PUF of unity, the peak voltage ratio increases toward a factor of 3, and for PUF values significantly higher than unity the peak voltage ratio climbs still higher. This is a significant downside issue for Class E operation, especially given the way in which the PUF has been defined; it could be argued that a more logical PUF definition would use the peak voltage, rather than the supply voltage, to compare the Class A and Class E powers. Clearly, in a given case it is possible to reduce the DC supply to an appropriate level to take account of the higher peak voltage. But this further reduces the available power, in comparison to a more conventional mode which has a peak voltage ratio close to 2 and would not require a lowering of the supply voltage. But there are cases when the supply voltage may be much lower, for other reasons, such as in mobile handset applications. In these cases, it is quite common to use a transistor technology having significantly higher breakdown than actually

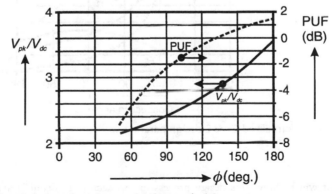

Figure 7.13   Class E design chart for conduction angle (ϕ).

required, based on the battery voltage available. Based on the plots shown in Figure 7.13, it would appear that the useful range of ϕ for normal applications is restricted to a value around 110°, plus or minus maybe about 30°, depending on whether priority is given to minimizing peak voltage or PUF.

7.6 Class E—Design Example[3]

Figures 7.13 and 7.14 plot all of the elements in a normalized Class E design, and the design process becomes somewhat analogous to the use of normalized charts for filter design. A design example will now be considered, illustrating the use and the necessary de-normalization of the final component values. Then the big step must be taken to replace the ideal switch element with a real device. In this example, the device will be the same GaAs MESFET used in previous design examples in the previous two chapters, so direct comparisons can be made. A full circuit simulation will show that most of the features predicted by the analysis can be realized, despite the nonideal nature of the device. Actual measured data on a test amplifier based on the same design will then be shown.

The first issue is to select an appropriate value for I_{max}, and a little foresight is needed in this case, since the device is to be forced into the role of a switch. It will be seen that this requires a lowering of the effective value of I_{max} of a given device, in order to serve as a suitable design value for I_{pk} in the above equations. This involves a certain amount of trial and error using a nonlinear simulator. In the case in question, the device had a working value of about 2.5A for Class A thru B design purposes, and for Class E applications, a value of $I_{pk} = 1.0$A will be taken. This may seem a big step down in current, and is indeed a foretaste of the penalties involved in making RF power transistors behave like fast switches. But it should be recalled from the results and discussions in previous chapters that pushing for rail-to-rail voltage

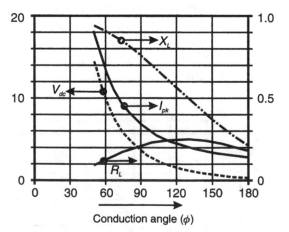

Figure 7.14   Normalized Class E design parameters.

---

3.   A spreadsheet which shows this design process in this section in detail is available on the CD-ROM: ClassE_Waves.

swings in order to obtain higher efficiencies inevitably results in much reduced peak RF currents, even for conventional reduced conduction angle modes. There are benefits which will in some cases outweigh this singular drawback.

Using Figure 7.13, a value for ϕ of 125° is selected for the conduction angle. This value aims to achieve the highest PUF (just under 0 dB), but keeps the peak voltage below a 3:1 factor on the DC supply (2.74). Figure 7.14 gives the following values for the four normalized parameters, V_{dc}, I_{pk}, R_L, X_L:

$$V_{dc} = 1.36, \ I_{pk} = 4.28, \ R_L = 0.252, \ X_L = 0.533$$

The values for V_{dc} and I_{pk} require scaling to the design values of 4.8V and 1.0A, respectively. This means that all the impedance elements have to be scaled by a factor of

$$\frac{4.8}{1.36} \div \frac{1}{4.28} = 15.1$$

giving $R_L = 3.81\Omega$, $X_L = (+)8.05\Omega$.

There is additionally the reactance of C_p which has been normalized to unity in the design charts and must now be scaled by the same factor, giving

$$\frac{1}{\omega C_p} = 15.1$$

so for a design frequency of 850 MHz, C_p has a value of 12.4 pF; the series reactance X_L will be an inductance of $8.05/(2 \times \pi \times 0.85) = 1.5$ nH.

In order to comply with the initial assumptions, a resonator at the design frequency needs to be placed in series with the series R_L/X_L combination. At this stage, these are not uniquely determined, and it is only necessary to select a reasonable value of Q, say 10, which gives an inductance of

$$\omega \cdot X_S = 40, \text{ or } L_S = 7.5 \text{ nH}$$

So the series resonant capacitor has a value of $C_S = 4.7$ pF.

At this point, it would normally be necessary to design an additional matching network to transform the 3.8Ω load resistor up to 50Ω at the fundamental frequency. The design of this network will be done as the next step, but it is instructive to simulate the circuit in its present form (Figure 7.15). The simulated waveforms for transistor output current and voltage are shown in Figure 7.16.[4] Qualitatively, the waveforms show a remarkable similarity to the idealized Class E plots in Figures 7.10 and 7.11. The current has the characteristic asymmetrical appearance, and the voltage has the predicted triangular form. The RF power is 1.05W, or 30.23 dBm; this is as much as 3 dB lower than some of the results using the same device model in

---

4.   See CD for a Microwave Office file listing, Fig_7_16.emp; note that in the simulation the input transistor parasitic elements are shown as external elements in order to avoid Cgs varactor effects.

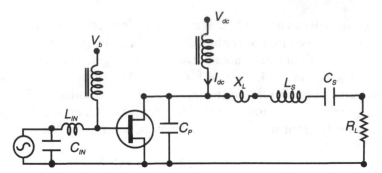

Figure 7.15 Class E amplifier design, frequency 850 MHz. $\phi = 125°$, $I_{pk} = 1$A, $V_{dc} = 4.8$v, $X_L = 1.5$ nH, (at 850 MHz), $C_p = 12.4$ pF, $R_L = 3.8\Omega$, $C_s = 4.7$ pF, $L_s = 7.5$ nH.

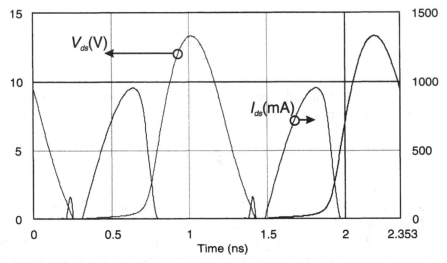

Figure 7.16 Voltage and current waveforms for simulated circuit shown in Figure 7.15; output power = 30.2 dBm, efficiency = 87% (see file CLE_FIG_7_16 on CD-ROM).

earlier chapters. But the efficiency is a borderline-spectacular 87%; this number being achieved using the ideal switch analysis element values, with no further adjustment.

In order to achieve the necessary switching action from the device, it has been biased close to its pinchoff point (Vg = −2.3, for Vt = −2.5), and the drive level has been set at a point which gives sufficient power gain (approximately 14 dB) to eliminate PAE issues. It can be seen that there is a small spike of current prior to the main conduction. In switching terms, this corresponds to a condition where the "switch" closes at a point of nonzero voltage. When using a transistor, this condition becomes less well defined, and although some minor adjustments in the circuit values can essentially eliminate the spike, it seems that better compromises between power and efficiency can be obtained with a small spike component in the device current. It is an allowable diversion at this point to adjust some of the elements to optimize the performance, given that the element values were based on an ideal switch. Figure 7.17 shows the waveforms which result from small changes in the shunt capacitor

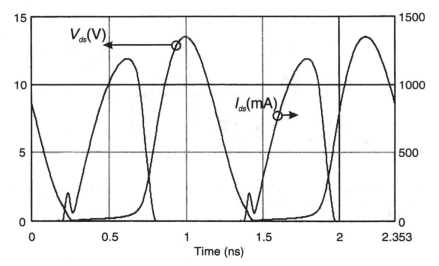

Figure 7.17   Voltage and current waveforms for simulated circuit shown in Figure 7.15; element values modified to: X_L = 1.2 nH, C_p = 13 pF, output power = 31.7 dBm, efficiency = 87%.

and the series inductor output elements. This has resulted in a significant improvement in power, at similar efficiency, 31.7 dBm at 87%.[5] The output capacitance clearly has a sensitive effect on the power and efficiency tradeoff, but also has a big impact on the peak voltage. This will be considered in more detail when a final practical circuit has been developed.

The final step in the design process can now be taken, which is to transform the load resistor up to 50Ω. This can be done using the same matching process described in Chapter 5. The series R-C part of the resonator can be transformed to a parallel R-C, where the shunt resistor is specified to be 50Ω.

Using (5.1), and referring to Figure 7.18, the transformation ratio n will be 50/3.8 = 13.15. So the value of shunt capacitance is given by

$$X_{PI} = \frac{50}{\sqrt{n-1}} = 14.3\Omega, \text{ so } C_{PI} = 13.1 \text{ pF at 850 MHz}$$

From (5.2), the series resonating inductance, X_S will now have a different value given by

$$X_S = 4 \cdot \sqrt{n-1} = 13.3\Omega, \text{ so } L_S = 2.5 \text{ nH at 850 MHz}$$

The mistuning element, X_L, will be unaffected by this transformation, so the final circuit is shown in Figure 7.18. The simulation shows very similar performance to the original plots in Figure 7.16 despite a significant lowering in the Q fac-

---

5.   It has also removed a disconcerting negative spike of device current in the original simulation, which coincides with a small negative excursion of output voltage. Depending on the device, and/or the quality of the model being used, this reverse-channel conduction region may, or may not, be physically realizable, and is best avoided in simulations.

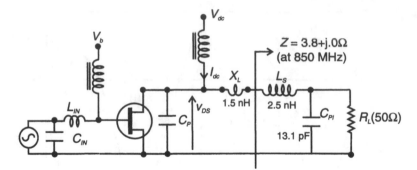

Figure 7.18 Class E design showing modification of output network to incorporate a 3.8Ω to 50Ω impedance transformation at the fundamental frequency (see file CLE_FIG_7_18 on CD-ROM).

tor of the resonance, which is incurred as a result of incorporating the impedance transformation.

A closer scrutiny of the voltage and current waveforms shown in Figure 7.19 gives some qualitative indication of how the transistor can be tricked into generating a similar set of waveforms compared to that of an ideal switch. The key is a small ramp of voltage in the knee region, which delays the full turn-on of the device, despite being in the conduction half cycle of gate voltage. This ramp gives the current pulse, which would otherwise be a half wave rectified sinewave of much greater peak amplitude (shown dashed), an asymmetrical appearance as required by the Class E equations. Thus the action of a switch can be approximated, but with a substantial penalty in peak current swing. So the main action of a Class E amplifier takes place within, and is indeed critically dependent on, the knee region of the I-V characteristic; the turn-on is largely implemented by the I-V knee characteristic, and the turn-off is more conventionally implemented by using the transconductive pinchoff. This is a subtle point which partially explains the widespread belief that Class E cannot be made to work at GHz frequencies due to the *slow and symmetri-*

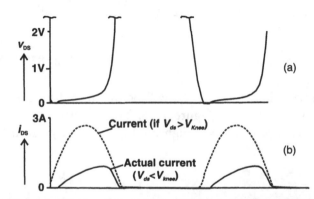

Figure 7.19 Class E quasi-switching action: (a) magnified drain-source voltage, showing ramp-up in the knee region; and (b) drain current, showing modulating effect of drain-source voltage; dashed trace is current under same bias and drive conditions but with RF voltage maintained above knee levels.

cal transconductive switching characteristics of RF power transistors. This point has been much emphasized by Sokal [2], albeit at times to unreceptive listeners and readers.

The original estimate for I_{pk} was based on some prior knowledge of the above considerations, and the simulation does show a peak current around this value. Some attempt to obtain a higher peak current, by scaling the design parameters can be attempted, but true class E operation will be compromised at higher I_{pk} values due to the slower "quasi-switching" speed. In fact, it is instructive simply to vary the value of C_p. around its design value. The results, shown in Figure 7.20, show a broad tradeoff between power and efficiency, with a promising plateau of efficiency values close to 90%. It appears that a slightly lower value of C_p, in this design, would yield almost 1 dB higher power for a minimal loss in efficiency of one or two percentage points. As the value of C_p departs more substantially away from the design value, the amplifier behavior departs from true Class E operation but shows a more favorable, and certainly comparable, power-efficiency tradeoff to the Class F results for the same transistor (shown in Chapter 6, Figure 6.19).

Figure 7.20 demonstrates the importance of the shunt capacitor value. Once again, as in Chapter 5, we find that this element has a strong effect on the PA characteristics, and yet it is an element that is rarely adjusted, and often inaccessible in practice. Its value in most of the sweep in Figure 7.20 is much higher than the transistor output capacitance, and so it is necessary to enhance the transistor output capacitance, preferably with an on-chip element, or as a minimum, a low inductance chip placed right next to the transistor die. This is an unpopular technique for RF designers, not to mention technologists who strive to minimize transistor parasitics in order to increase their high frequency performance. As with the harmonic stubs considered in Chapter 5, the placement issue can become a serious problem when using large packaged transistors, and is probably a leading cause of failure in Class E designs at higher frequencies. It should also be fairly stated that a typical load-pull tuner is unlikely to be able to simulate the combined effects of a prescribed shunt capacitor and tunable load resistor simultaneously; the shunt capacitor needs to be supplied, and then the tuner can do a much more productive job of varying the load impedance.

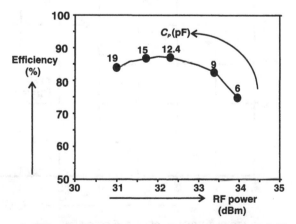

Figure 7.20   Power and efficiency simulation results for circuit in Figure 7.18, showing the effect of varying C_p from the design value of 12.4 pF.

Figure 7.21 shows some measured data from an amplifier constructed using the design parameters in Figure 7.18, with C_p about 11 pF. Power and efficiency are both plotted on linear scales to illustrate the possibility for using the supply voltage to generate, or restore, amplitude modulation. The efficiency can be seen to hold up remarkably well over a very wide range of supply voltage, although the power is significantly lower than the simulated level.

7.7 Conclusions

The previous sections have shown that Class E is a viable and potentially useful mode of RF power amplification. Simple design equations have been shown to yield useful circuit element values which can be verified in practice. Theoretical efficiencies of around 90% can be achieved, albeit with an RF output power as much as 3 dB lower than that obtainable from the same device in a conventional Class AB configuration. Class E amplifiers are highly nonlinear, but offer possibilities for envelope restoration techniques.

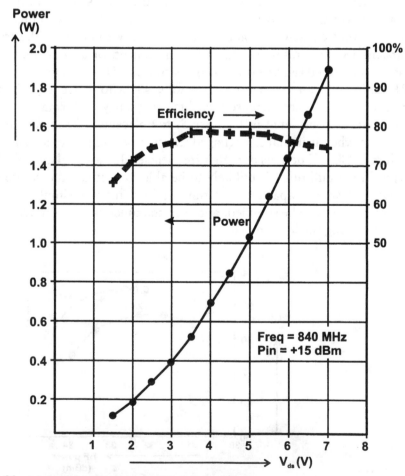

Figure 7.21   Measured performance of Class E amplifier.

On the downside, peak voltages can be as high as three times the DC supply if the power shortfall is to be kept within reasonable limits. A device with a given breakdown voltage can usually be operated in Class AB at a supply voltage half of this level. If a Class E design requires the voltage to be dropped to one third of the breakdown level, it is only fair to take account of this in the PUF calculation, as discussed (but not implemented in Figure 7.13) in Section 7.5. There is also the issue of power gain. If a typical Class E amplifier is running at 3–5 dB lower power gain, due to the overdrive requirements on the input, than a comparable linear amplifier, the upper frequency limit will be lower for a given technology. Generally speaking, 12–13 dB of linear gain would be required from a given device in order to stand a chance of useful Class E operation; however, the same can be said of conventional high efficiency modes with shorter conduction angles.

The really critical advantage of the Class E amplifier is efficiency, and with the possibility of reaching into the 90% range, the advantages of low heat generation offer some novel concepts in packaging, modulation, and power combining. As a minimum, it is a technique which deserves closer scrutiny for higher frequency applications. Throughout this chapter, little consideration has been given to the harmonic content of the voltage and current waveforms. Clearly, the central issue in implementing a Class E design at GHz frequencies is whether the higher harmonics can be sustained, and what the effects of truncating them may be. This, in essence, forms the main subject for the following chapter.

References

[1] El-Hamamsy, S., "Design of High-Efficiency RF Class D Power Amplifier," *IEEE Trans. Power Elect.*, Vol. 9, No. 3, May 1994.

[2] Sokal, N. O., and A. D. Sokal, "Class E—A New Class of High Efficiency Tuned Single-Ended Power Amplifiers," *IEEE J. Solid State Circuits*, SC-10, No. 3, June 1975, pp. 168–176.

[3] Raab, F. H., "Idealized Operation of the Class E Tuned Power Amplifier," *IEEE Trans. Circuits and Systems*, CAS-24, No. 12, December 1977.

[4] Sokal, N.O., and A. D. Sokal, "High-Effciency Tuned Switching Power Amplifier," U.S. Patent 3,919,656, 1975.

[5] Sokal, N. O., "RF Power Amplifiers, Classes A through S, et seq.," *Proc. Wireless and Microwave Technology '97*, Chantilly, VA, 1997.

[6] Imbornone, J., et al., "A Novel Technique for the Design of High Efficiency Power Amplifiers," *Proc. European Microw. Conf.*, Cannes, France, 1994.

[7] Mader, T., et al., "High Efficiency Amplifiers for Portable Handsets," *Proc. IEEE Conf. on Personal, Indoor, and Mobile Comms.*, Toronto, Canada, 1995, pp. 1242–1245.

[8] Personal communication with J. D. Rhodes.

[9] Personal communication with P. Blakey, 2005.

CHAPTER 8

Switching PA Modes at GHz Frequencies

8.1 Introduction

Chapter 7 presented the theory of switch mode PAs in much the same way that Chapter 3 presented Class AB modes; the analysis in both cases was loaded with numerous idealizing assumptions, particularly with regard to the device itself. Chapter 4 set out to reinterpret the idealized Class AB analysis in the context of GHz frequency operation. This chapter sets out to perform the same task with switching modes, and as such has the same relationship with Chapter 7 that Chapter 4 has with Chapter 3. Focusing particularly on the Class E mode, we will attempt to reinterpret classical and widely quoted theory, which is based on ideal low frequency models, in the context of GHz frequency applications.

The Class E situation is if anything more extreme than that described in Chapter 4, which referred to Class AB modes. Basically, RF power transistors are not switches at GHz frequencies, even if a finite "on" resistance is included in the switch model. The assertion that RF power transistors *can* be considered to operate as switches, "sort-of, some-of-the-time, if-you-get-things-right," has to be regarded as questionable at best, and in any case inappropriate, given that accurate nonlinear models are now available for CAD design. The dogmatic faction of the switch-model believers can even go to the extent of jettisoning the detailed nonlinear model that they, or others, use for any other kind of design using the same device, and substituting a simple switch in its place.

Even at this early stage of the chapter, the Class E community will be calling a time-out. The switch model, they are saying, is used to provide an *a priori* design approach, from which circuit topology and element values can be determined, then more detailed design optimization can be done using CAD tools and models. Fair enough; this is entirely consistent with the design philosophy promoted in this book. Unfortunately, however, lengthy papers still appear in the literature that use only switch models and do not appear to address the next step of demonstrating the same results using a real transistor model. It seems that there is still some space left in the subject, in terms of understanding what is really going on in a Class E PA circuit when "real" transistors are used. In particular, the frequency domain approach to circuit design needs to be brought into play at some point. One of the basic principles, or assumptions, of GHz frequency PA design is that only the first few harmonics matter, and the matching topologies can be designed around this simplification. The problem many GHz designers have with the classical Class E analysis is that it is performed in the time domain, and does not recognize that at higher frequencies the higher harmonic behavior will be truncated, giving slower rise and fall times which can degrade the performance down to that observed using more conventional PA modes.

Another perspective on the problem at hand can be obtained by consideration of efficiency. A classical, ideal switching device can be designed into a multitude of circuit configurations which deliver 100% efficiency. This is as much an artifact of the idealized device model, rather than the result of any particular ingenuity in the circuit design. In practice, of course, finite switching speeds and resistive losses will reduce the observed efficiency, and numbers in the literature have ranged anywhere from mid-nineties down to some more lowly numbers in the 60% region. But we have seen in Chapter 3 that a classical Class C design can give efficiencies in the mid-eighties, while incurring comparable limitations imposed by considerations of drive level, breakdown, and declining RF power output. An efficiency of 85% is not the exclusive sovereign territory of the Class E mode, although there seems to be a widespread misconception that this is so. Furthermore, some of the reasons given in Chapter 3 for the relatively low use of Class C in solid state microwave PAs, such as reduced power efficacy, high voltage swings, and high drive levels are also encountered in alleged Class E designs. Class E seems to have enjoyed something of an exempt status as far as these considerations are concerned.

This chapter does not set out to discredit the Class E mode, or those who have promoted it. At lower frequencies, which can now realistically be defined to be below 1 GHz, device technology is available that can be used to approximate classical switch-mode Class E operation quite closely. This has already been demonstrated in the last chapter, using a nonlinear device model at 850 MHz and not a switch substitute. The goal is to reinterpret and redefine a high efficiency PA mode that has most of the attributes of classical Class E, but can be designed using more familiar microwave frequency domain methods. In comparison to the vast litany of papers on switch mode RFPA analysis this is a relatively unexplored area, although was notably visited by Raab [1].

8.2   Ignoring the Obvious: Breaking the 100% Barrier

Figure 8.1 shows a very familiar waveform. It is the classical current waveform of a Class B amplifier, a half-wave rectified sinewave. This waveform has some very interesting properties. Its fundamental RF component is equal to the amplitude of a sinewave having the same peak-to-peak limits, but its DC component lies at the I_{max}/π position. It is these two figures that lead directly to the $\pi/4$ efficiency of a Class B amplifier, with the assumption that the voltage is sinusoidal, and in antiphase with

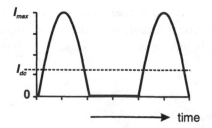

Figure 8.1   Class B current waveform.

the current. Moving from a sinusoidal current to a halfwave rectified sinewave has yielded an impressive efficiency increase of a $\pi/2$ factor.

We have seen in Chapter 6 that various tricks can be played with the voltage waveform in order to further increase the efficiency; "squaring up" the voltage can ultimately enable us to approach the 100% efficiency level. But are we not missing an important opportunity here, even perhaps ignoring the obvious? Why not engineer an amplifier with the current and voltage waveforms shown in Figure 8.2? Now we can leverage the $\pi/2$ efficiency factor twice over, giving a final efficiency value of

$$\eta = \frac{1}{2} \cdot \frac{\pi}{2} \cdot \frac{\pi}{2} = \frac{\pi^2}{8}$$

a number not just greater than unity, but quite substantially so (123%).

So of course something's wrong, and let it be clearly stated that the waveforms of Figure 8.2 *are not realizable* using a passive output termination; in particular they represent power absorption by the active device at even harmonics. It is nevertheless a very interesting and significant result, and will be used as a starting point to back into something more realizable and useful.

It is interesting to consider the actual amount of RF power such an amplifier requires, mainly at the second harmonic. The in-phase components of second harmonic current and voltage have levels of

$$I_2 = \frac{2I_{max}}{3\pi}, \, V_2 = \frac{2V_{dc}}{3}$$

giving a total second harmonic power absorption of

$$P_2 = \frac{2I_{max}V_{dc}}{9\pi}$$

whereas the fundamental power output is

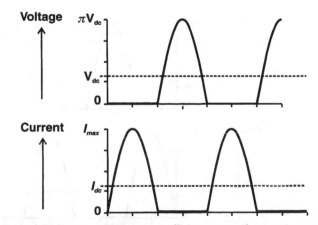

Figure 8.2 Conceptual PA having 123% output efficiency; even harmonic terminations have negative real parts (see text).

$$P_1 = \frac{I_{max} V_{dc} \pi}{8}$$

and the DC supply is

$$P_{dc} = \frac{I_{max} V_{dc}}{\pi}$$

confirming the output efficiency as $\pi^2/8$. The ratio of second harmonic input power to fundamental RF output power is

$$\frac{P_2}{P_1} = \frac{16}{9\pi^2}$$

about 18% of the fundamental power, or −7.5 dB. This second harmonic power could conceivably be supplied to the device output, and the power required to generate it could be regarded in the same light as the fundamental drive power required in any RF amplifier stage. But even a fundamental drive power requirement at this level would be regarded as marginal for a practical RFPA, due to its degrading effect on the overall efficiency. In this case the need to include an additional frequency doubling function would appear immediately to rule this out as a practical approach, although some related work at lower frequencies has been reported [2].

Returning to the waveforms of Figure 8.2, there is a more pragmatic strategy which can remove the second harmonic problem, albeit at a cost. Figure 8.3 shows the same waveforms, but with the current and voltage shifted with respect to each other by 45°, as measured at the fundamental frequency. The mutual phase shift will therefore be 90° at the second harmonic, which means that the waveforms can be realized using a reactive second harmonic termination. This has some appeal, since if the phase shift is done in the appropriate direction, the impedance termination at the second harmonic will be capacitive, and could potentially absorb the device output capacitance.

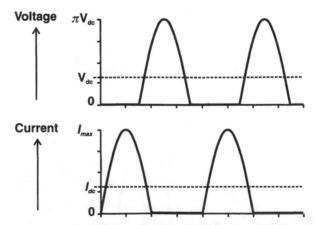

Figure 8.3   PA of Figure 8.2 with 45° differential phase offset introduced between current and voltage.

Unfortunately, this has an immediate impact on the fundamental RF power output and efficiency, both of which now have an additional factor of cos(45°):

$$P_1 = \frac{I_{max}V_{dc}\pi}{8\sqrt{2}}$$

$$\eta = \frac{\pi^2}{8\sqrt{2}}, \text{about 87\%}$$

This is still an attractive number, but there are some loose ends. The 45° fundamental phase shift will cause a 180° rotation of the fourth harmonic phase, implying a resistive termination, and the eighth harmonic will be back at square one, showing a power absorption. This problem is however about two orders of magnitude smaller than the original second harmonic issue, and can reasonably be ignored. The key conclusion in this section is that the use of substantially reactive fundamental load terminations can have some important uses in RFPA design, a fact that has not been widely recognized. The Class J PA results in Chapter 4 showed one aspect of this, and the following sections in this chapter will develop some important results for high efficiency PA modes which use the waveforms of Figure 8.2 as a starting point.

8.3   Waveform Engineering

There has, it seems, been a long tradition of designing RFPAs in a "back-to-front" manner, whereby the desired current and voltage waveforms are defined using essentially mathematical representation, and from these waveforms the circuit, and even in some cases the device, characteristics are deduced. In this section we continue that tradition, under the heading of "waveform engineering," which has been coined by other workers in the PA field [3].

The waveforms shown in Figure 8.2 demonstrate that there are hazards in this approach. We have already established that these waveforms represent power absorption at even harmonics, and as such do not represent a realizable target for a practical design. Another issue concerns the assumption, frequently used by PA theorists, that the output voltage can drop to zero without causing any modulation of the assumed current. This "knee effect" has already been discussed in some detail in connection with its impact on the classical results for Class A and Class AB amplifiers (Chapter 3). Basically, any waveforms or formulae obtained in a PA analysis using the zero-knee assumption will require the DC supply voltage to be increased by a suitable "knee factor" in order to represent a realizable result. This factor may in many cases be about 10%, but it is the variation from one device type to another, along with varying supply voltage requirements in different applications, that justify the ongoing use of the more universal zero-knee assumption.

In this chapter, the knee region will take on a new role. Rather than being a practical nuisance-factor that we would prefer to be without, it becomes an instrument by which we can modify, or engineer, the current waveforms to different shapes than those defined by the input transconductance relationship. It may not be the best tool for doing the required job, but it's a tool that comes conveniently built-in to any RF power transistor. There is another such tool, which can be used for similar purposes, and was discussed in some quantitative detail in Chapter 4: the nonlinear input capacitance. Unfortunately, this is a much more tricky tool to use. Its effects will vary more widely from one device type to another than the more easily quantified effects of the knee region. It will however emerge as an important player in bridging the technical and intellectual gap that exists between the classical descriptions of PA switching modes at lower frequencies, and their apparent ongoing utility in the GHz frequency range.

Returning to the waveforms of Figure 8.3, it is necessary to postulate a voltage waveform that has more realizable characteristics in high frequency applications; basically the sharp transitions need to be rounded off. Figure 8.4 shows some useful candidates, where an increasing second harmonic component has been added to the fundamental. The restriction of the voltage waveform to contain only fundamental and second harmonics is a justifiable assumption in higher frequency PA circuits, where very often the active device will have an output capacitance that will behave as a near-short at higher harmonics.

The device output voltage therefore has the form

$$v(\theta) = V_{dc} - v_1 \sin\theta + v_2 \cos 2\theta \qquad (8.1)$$

which can conveniently be normalized such that $V_{dc} = 1$.

Figure 8.4 shows two cases of values for v_2, $v_2 = \dfrac{1}{3}$ and $v_2 = \dfrac{1}{2}$.

The first of these cases is the familiar maximally flat case, which can be derived from the raised sine squared function,

$$v = \frac{1}{4}(1+\sin\theta)^2$$

and shows that for a v_2 value of 0.25 the DC value can be reduced to ¾ from its unity value when the voltage is a sine wave. Such a reduction in V_{dc}, while the fundamental

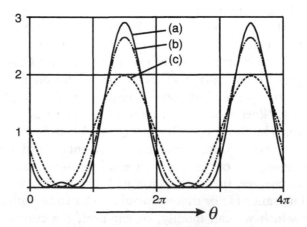

Figure 8.4   The function $1- v_1\sin\theta + v_2\cos 2\theta$: (a) $v_1 = 2^{1/2}$, $v_2 = 1/2$; (b) $v_1 = 4/3$, $v_2 = 1/3$; and (c) $v_1 = 1$, $v_2 = 0$.

amplitude is kept constant, represents a potential efficiency increase (efficiency factor); a 4/3 factor in this case. Figure 8.4(b) shows the same function scaled up in magnitude by a factor of 4/3 in order that the waveform reaches zero for the same unity V_{dc} value. Figure 8.4(a) shows that if some baseline ripple can be tolerated, the second harmonic amplitude can be increased to a higher value with a corresponding benefit in the efficiency factor. For a v_2 value

$$v_2 = \frac{1}{2}$$

the v_1 value can be increased by a factor of $\sqrt{2}$, with an equal increase in the efficiency factor, bringing the efficiency (70.7%) closer to the ideal Class B value.

This efficiency enhancement through the addition of a second harmonic component is analogous to the Class F process using a third harmonic component, described in Chapter 6. The reason the second harmonic case is less frequently considered is the phasing problem discussed in Section 8.2, and also the increase in peak voltage, clearly indicated in Figure 8.4. Once again, it has to be said that although the peak voltage may be a reliability issue in some applications, this has been another area where the "exempt status" of Class E seems to have worked in its favor. Class AB PA designers religiously divide the device breakdown voltage by a factor of two in order to obtain the maximum allowable DC supply. Only in the Class E zone, it seems, do we hear debates on how avalanche breakdown time constants maybe allow higher peak RF swings to be tolerable. We will see, however, that second harmonic voltage enhancement will play an important part in explaining the action of quasi-switching mode PAs at GHz frequencies. In truth, given that direct observation of RF waveforms is not readily possible at GHz frequencies, it is very likely that many RFPAs have nonsinusoidal voltage waveforms with peak voltages higher than $2V_{dc}$.

As in the third harmonic case, there is an issue worth solving formally at this point, which is the identification of the optimum case where the fundamental passes through its maximum enhanced value in the presence of the second harmonic component. Just as in the Class F case analyzed in Chapter 6, Rhodes [4, 5] has solved this problem by recognizing the properties of a singular condition when v_1 reaches its maximum value. Recalling (8.1),

$$v(\theta) = V_{dc} - v_1 \sin \theta + v_2 \cos 2\theta$$

and

$$v'(\theta) = -v_1 \cos \theta - 2v_2 \sin 2\theta \tag{8.2}$$

Imposing the additional condition $v(\theta) >= 0$, for all θ, we seek a value of θ, θ_z, such that $v(\theta_z) = 0$ and $v'(\theta_z) = 0$, but additionally the coefficient of v_2 in (8.1) vanishes, indicating a singular condition.

By inspection, $\theta = \frac{3\pi}{4}$ will fulfill the necessary conditions, giving

$$v_1 = \sqrt{2}V_{dc}$$

and

$$v_2 = V_{dc}/2$$

Hence the second waveform shown in Figure 8.4 represents the maximum value of v_1, and the corresponding efficiency enhancement factor of $\sqrt{2}$ is the maximum value of this property for second harmonic enhancement.[1]

Thus the waveforms of Figure 8.3 can be redrawn using the newly derived voltage function with the optimum value for v_2, Figure 8.5. Allowing for the phasing factor of 45°, the efficiency of such an amplifier would be

$$\eta = \frac{1}{2}\frac{\pi}{2}\frac{1}{\sqrt{2}}\sqrt{2} = \frac{\pi}{4}$$

a slightly disappointing result, based on the original starting point of 123% in the previous section.

The key issue with the waveforms in Figure 8.5 is the increased overlap between nonzero current and voltage, which has been caused by the rounding-off of the voltage waveform. In order to obtain higher efficiency, it is now necessary to start engineering the current waveform as well. The goal in this exercise will be to create a waveform that has a fundamental component in antiphase with the fundamental voltage component, but a second harmonic component that is in quadrature with the second harmonic voltage. Ideally, there will be no need to introduce a global dif-

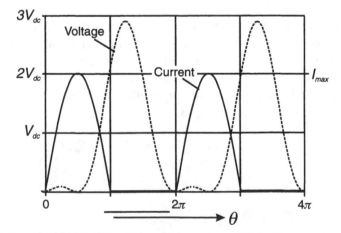

Figure 8.5 A fully realizable set of amplifier voltage and current waveforms, using second harmonic enhancement on the voltage; efficiency is back down to $\pi/4$.

---

1.   This result, which is limited here to include only second harmonic, can be extended to yield analytical expressions for the general case of multiple harmonics, as presented in [4]. The results for the specific cases of just second, or just third, harmonic enhancement, and numerical results for more generalized cases, have been presented by various authors; see [6].

ferential phase shift between voltage and current in order to achieve the second harmonic quadrature condition.

In order to understand this requirement more fully, Figure 8.6 summarizes the magnitudes and phases of the various components of current, in the simple half wave rectified sinewave case. With no differential phaseshift between the current and voltage waveforms, the fundamental components of current and voltage are in antiphase, but the second harmonics are in-phase. As the differential phasing is increased from zero (the independent variable on the abscissa in Figure 8.6), a point is reached where the second harmonics of current and voltage are in phase quadrature. At this point, however, the antiphase component of the current has reduced by a factor of 0.707 (cos 45°), and so the overall benefit of second harmonic voltage enhancement is precisely cancelled out.

It is a worthwhile effort, therefore, to seek a current waveform that requires a lower differential phaseshift to achieve the required second harmonic quadrature condition. Intuitively, this would appear to be possible by introducing some asymmetry about the vertical $\pi/2$ axis in the basic Class B waveform. Three such candidates will now be considered:

(a) Linear knee-clipped function;
(b) Smoothed Class E function;
(c) Harmonically modified input function.

Function (a) will give some quantitative insight on the expectations for power and efficiency when the knee region is used to modify the current waveform, (b) will enable a quantitative assessment of the effect of finite switching speed in a GHz frequency Class E amplifier, and (c) introduces the potential for waveshaping the current using intentional input signal distortion.

The clipping action of the output voltage while inside the knee region can be approximated to a simple linear scaling function, and thus the function

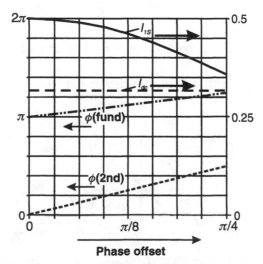

Figure 8.6   Analysis of Class B current waveform with phase offset.

$$i(\theta) = \frac{\theta}{\pi}\sin\theta............0 < \theta \le \pi$$

$$= 0......................\pi < \theta \le 2\pi$$

(8.3)

is of interest as a possible waveshaping implementation using a sloped voltage characteristic in the knee region, as shown in Figure 8.7. It will be shown in the next section how such a slope can be generated by introducing a small phase offset in the second harmonic voltage component.

As indicated in Figure 8.7, if the current waveform is engineered using the knee effect, the peak current will be substantially lower than the I_{max} of the device. This is a matter of considerable impact in the practical realization of switch mode PAs at higher frequencies. It is appropriate therefore to consider the specific Fourier components of the function (8.3). The fundamental components are

$$I_{1s} = \frac{I_{max}}{\pi^2}\int_0^\pi \theta\sin\theta\sin\theta d\theta = \frac{I_{max}}{4}$$

$$I_{1c} = \frac{I_{max}}{\pi^2}\int_0^\pi \theta c\sin\theta\cos\theta d\theta = \frac{-I_{max}}{4\pi}$$

so that the fundamental component is reduced by a factor

$$\frac{1}{2}\left(1+\frac{1}{\pi^2}\right)^{\frac{1}{2}}, (= 0.525)$$

or very nearly one-half, as compared to the original Class B waveform. This reduction is somewhat mitigated by a more favorable second harmonic phase relationship. The DC component,

$$I_{dc} = \frac{1}{2\pi^2}\int_0^\pi \theta\sin\theta d\theta = \frac{1}{2\pi}$$

is exactly half of the Class B value, a result that with the luxury of hindsight is not too surprising.

The second harmonic components, given by

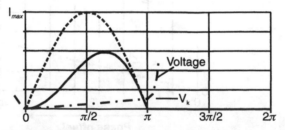

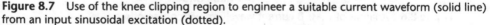

Figure 8.7 Use of the knee clipping region to engineer a suitable current waveform (solid line) from an input sinusoidal excitation (dotted).

$$I_{2s} = \frac{I_{max}}{\pi^2} \int_0^\pi \theta \sin\theta \sin 2\theta d\theta = \frac{-8I_{max}}{9\pi^2}$$

$$I_{2c} = \frac{I_{max}}{\pi^2} \int_0^\pi \theta \sin\theta \cos 2\theta d\theta = \frac{-I_{max}}{3\pi}$$

show a phase angle relative to the voltage reference which is already some way toward the required quadrature value, the actual angle being

$$\phi_2 = \tan^{-1}\left(\frac{8}{2\pi}\right) \tag{8.4}$$

or about 40°. Thus a reference fundamental differential phase shift of only 25° is required to allow the use of the √2 second harmonic voltage enhancement factor.[2] This produces much less power-factor degradation to the fundamental power than the full 45° phase shift required for the symmetrical Class B waveform. Some care is needed in the actual computation of this fundamental power in cases where the original current waveform has both sine and cosine components. The safest approach is to recompute the Fourier integral for the sinusoidal component, replacing the original current waveform $i(\theta)$ with $i(\theta-\phi)$, and putting $\phi = \phi_2/2$, so

$$I_{1s} = \frac{I_{max}}{\pi^2} \int_{\phi_2/2}^{\pi+\phi_2/2} (\theta - \phi_2/2)\sin(\theta - \phi_2/2)\sin\theta d\theta$$

this being the new antiphase current component that multiplies with the enhanced fundamental voltage component, which in turn can now have an amplitude up to the limiting normalized value of $V_{dc}\sqrt{2}$. This gives a fundamental power of

$$P_{rf} = \frac{V_{dc}\sqrt{2}}{\sqrt{2}} \cdot \frac{I_{1s}}{\sqrt{2}} \tag{8.5}$$

Putting ϕ_2 equal to the critical value obtained in (8.4) above, the fundamental power comes out to be 2.6 dB lower than the Class B baseline value, with an efficiency of 87%. It should be noted that this relative power comparison is now inappropriate unless the knee effect is taken into account also for the Class B case.

Nevertheless, the use of the knee effect to perform the current waveshaping will be quite detrimental to power, and this is usually a feature in attempts to realize Class E designs in this manner. Unfortunately, reported Class E results often omit to give the comparable power that could be obtained from the cognizant device using a conventional quasi-linear Class A or AB mode. More detailed modeling of the knee voltage, and the use of a nonzeroing voltage, can give a somewhat better tradeoff in power and efficiency. This will be covered in Sections 8.4 and 8.5.

Moving to the second candidate (b) function, a set of Class E "look-alike" waveforms can be defined by the function

---

2. The additional 50° required at the second harmonic, divided by 2 to obtain the referenced fundamental value.

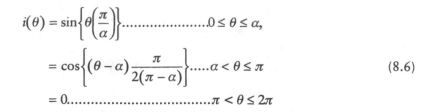

$$i(\theta) = \sin\left\{\theta\left(\frac{\pi}{\alpha}\right)\right\}...............0 \le \theta \le \alpha,$$

$$= \cos\left\{(\theta - \alpha)\frac{\pi}{2(\pi - \alpha)}\right\}.....\alpha < \theta \le \pi \qquad (8.6)$$

$$= 0.................\pi < \theta \le 2\pi$$

As shown in Figure 8.8, this function consists of a sine function up to the angle α, with the argument scaled such that from zero to α the rising half of a sine wave has been completed. From α to π, the function is the falling half of a cosine wave, with corresponding scaling and offset of the argument in order that zero is always reached at $\theta = \pi$. Thus the choice of the α parameter adjusts the time duration of the falling edge; the extreme case of $\alpha = \pi/2$ gives a Class B rectified sinewave, and $\alpha = \pi$ gives an abrupt Class E switching waveform. Intermediate values of α, such as those plotted in Figure 8.8, represent realizable waveforms whose analysis will give useful and quantitative insight on the impact of a finite switching speed in RF applications.

Figure 8.9 shows a plot of various waveform properties as a function of the parameter α. The α parameter effectively defines the sharpness of the switch-off characteristic, which becomes instantaneous when $\alpha = \pi$. As would be expected, the phase angle (ϕ_2) of the second harmonic moves back towards the in-phase value of the Class B waveform as α is reduced from π to $\pi/2$. Figure 8.9 plots the resulting power and efficiency for each α value, assuming that the appropriate phase offset is introduced and that the full $\sqrt{2}$ voltage enhancement factor is used, using the steps defined in case (a) above. It should be noted, however, that for the present purposes, the baseline zero-knee assumption is being used to compute the power and efficiency. Looking at Figures 8.8 and 8.9, it is a matter of judgment as to how, and how much, the switching speed should be reduced to represent a practical case. The median value of $\alpha = 140°$, for example, would appear to be a fairly benign attempt to replicate a fast switch, yet it can still potentially deliver an efficiency of 85%. Given that the power and efficiency are almost flat between values of 160° and 180°, these results could be interpreted as a quantitative demonstration, perhaps a vindication, that the Class E switch mode waveforms still work quite well even when the ideal switching transients are significantly smoothed out.

Actual implementation of these waveforms will have to be achieved using the two current waveshaping tools that are available: input varactor effects and IV knee clipping. This function is interesting in its obvious relationship to classical Class E waveforms, its quantification of the switching speed debate for Class E implementa-

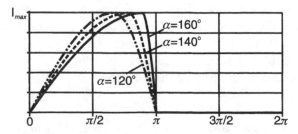

Figure 8.8 Smoothed Class E current waveforms.

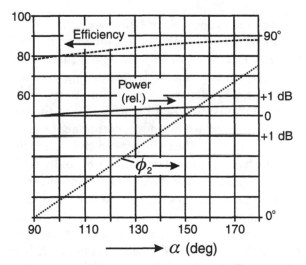

Figure 8.9   Smoothed Class E waveform properties; power and efficiency values assume appropriate ϕ_2 value and maximum second harmonic voltage enhancement factor (see text).

tion, and pointers to the kinds of waveform properties that are required for highest efficiency and practicality. Physical realization will, however, have to incorporate the knee region to some extent, and this will reduce the power and efficiency from the zero-knee values plotted in Figure 8.9. This will be addressed further in Section 8.4.

Given the potential power problems associated with I-V knee waveshaping, there is motivation to examine the alternative of input distortion. In principle, if currents such as those shown in Figure 8.8 can be engineered by the generation of a suitable input voltage, the full I_{max} of the device can still be used. Unfortunately, it is not so straightforward to model this waveshaping process as is the case with the IV knee characteristic. We have to make a general assumption that the input varactor generates harmonics, and that circuit elements will allow suitable amplitude scaling and phasing to generate specific waveforms. It is also convenient to assume an ideal transconductive device, so that the focus is on the specification and generation of a suitable input voltage waveform.

For waveshaping on the input, the voltage function

$$v_g(\theta) = \sin\theta + v_2 \sin 2\theta + v_3 \sin 3\theta \tag{8.7}$$

is of interest, since it can in principle be generated by adding suitably phased harmonic input voltage components. These components can be generated by the nonlinearity of the input varactor, with suitable harmonic tuning, or even by injecting harmonic components on the input drive signal. If this function is then truncated above the zero voltage point, due to the rectification action of the Class B bias setting, the resulting waveform (assuming the transconductance g_m to be linear and unity value)

$$i(\theta) = g_m(\sin\theta + v_2 \sin 2\theta + v_3 \sin 3\theta)\ldots\ldots 0 < \theta \leq \pi$$
$$= 0\ldots\ldots\ldots\ldots\ldots\ldots\ldots\ldots\ldots\ldots\ldots\ldots\ldots\ldots\ldots\ldots\pi < \theta \leq 2\pi \tag{8.8}$$

can have the required properties for efficiency enhancement by suitable choice of the v_2 and v_3 parameters. Three examples are shown in Figure 8.10, representing a range of magnitude for the v_3 component. The $v_3 = 0$ case (a) is of interest due to the relatively straightforward implementation, requiring only second harmonic generation in the input circuit. The intermediate v_3 value (b) represents a closer approximation to the Class E functions shown in Figure 8.8, and the third case (c) shows that the introduction of a larger third harmonic component significantly increases the scope for waveform engineering. The twin-peaked characteristic caused by higher v_3 values fits well with the basic requirements of an optimally engineered current waveform, having a high quadrature second harmonic component and a higher antiphase fundamental than in the other functions considered so far.

Figure 8.11 shows a plot of the properties of this function, over a full range of relevant values for the v_2 and v_3 parameters. Once again, these values have been computed using the procedure defined in case (a). It is important to note that the almost spectacular properties of the function for higher values of v_3 raise some issues with the underlying assumption that there will be no significant third harmonic component of voltage. There are also questions on the likelihood that such waveforms can be generated by the sole action of the input varactor.

With these reservations duly noted, Figure 8.12 shows the voltage and current waveform for such a postulated amplifier, using the (c) waveform in Figure 8.10. Only 8° differential phase shift has to be used in this case in order to obtain the quadrature second harmonic condition for current and voltage. This amplifier would deliver 0.5 dB higher power than the baseline zero-knee Class B condition, and an output efficiency of 92%. Although similar numbers have been encountered before, in connection with an ideal Class F PA, this new mode appears to represent a more realizable harmonic impedance environment. The second harmonic has a substantial capacitive termination, which may be absorbed by the device parasitic output capacitance, and the third harmonic may only require a little extra help to appear as a short circuit termination. The fundamental termination will have a substantial reactive component. This is in contrast to a Class F design, which requires a rather demanding open circuit termination at the third harmonic and a short circuit at the second. The waveforms of Figure 8.12 are a good example of the general principle that useful PAs can be postulated that do not fall into any of the established, or "named," modes. Protagonists from the school of "inverted Class F" would probably claim ownership on this example, although the waveforms shown here represent a more generalized interpretation of this classification than is usually understood. It

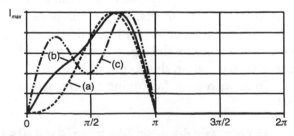

Figure 8.10 Plots of the truncated positive segments of the function $i(\theta) = \sin\theta + v_2\sin2\theta + v_3\sin3\theta$: ($\alpha$) $v_3 = 0$, $v_2 = -0.5$; (b) $v_3 = 0.15$, $v_2 = -0.3$; (c) $v_3 = 0.5$, $v_2 = -0.15$ (plots have been normalized to give equal peak values).

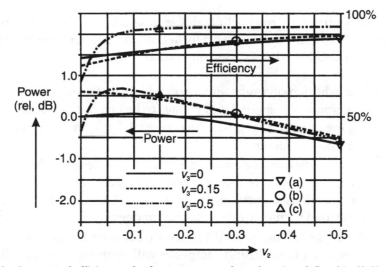

Figure 8.11 Power and efficiency plot for current waveform function defined in (8.8). Ideal second harmonic enhanced voltage assumed, with appropriate differential phase shift to give quadrature second harmonic condition (see text). Actual cases plotted in Figure 8.10 are indicated with markers (a,b,c).

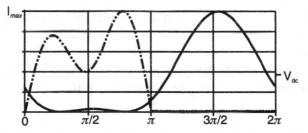

Figure 8.12 Possible PA waveforms using (c) case for current from Figure 8.10; second harmonic enhanced voltage waveform gives zero-knee efficiency of 92%.

would appear that for any stipulated restriction on harmonic content, optimum current and voltage waveforms can be identified which display similar performance, and that the optimum performance is essentially a function of the number of utilized harmonics, a concept proposed by Raab [1]. In the present examples, it should however be noted that higher harmonics are allowed in the current waveforms, since the transconductive truncation action of an RF transistor can generate such harmonics, for example, when a 40 GHz device is being used in the low GHz region.

Lower values of v_3, such as the (b) case in Figures 8.10 and 8.11, still give useful potential performance tradeoffs and represent more likely candidates for practical realization. It should be emphasized that the power and efficiency plots in both Figures 8.9 and 8.11 assume an ideal second harmonic-enhanced voltage waveform, having normalized components $V_1 = \sqrt{2}$, $V_2 = \frac{1}{2}$, $V_{dc} = 1$ with no knee effects (as shown also in Figure 8.12). As will be seen in the next section, the inclusion of a realistic model for the IV knee will degrade these ideal results.

It is appropriate to conclude this section with a brief summary, to confirm the motivations for the detailed waveform function analysis that it has presented. Taking an ideal Class B PA as a starting point:

- The addition of a substantial second harmonic voltage component allows the fundamental component of voltage to be increased by a factor of up to $\sqrt{2}$, giving the potential for power increase of 1.5 dB, and an increase in efficiency by the same $\sqrt{2}$ factor.
- This benefit *cannot be used* in a practical PA design unless the phase of the second harmonic current component is in phase quadrature with the second harmonic voltage component; a symmetrical Class AB/B/C current waveform has no second harmonic quadrature component and requires further engineering.
- A simple 45° differential phaseshift between the current and voltage waveforms of a Class AB/B/C amplifier will neutralize the 1.5 dB benefit given by the second harmonic voltage component.
- Modification of the current waveform can reduce the differential phaseshift required to enable full use of the second harmonic enhancement factor.
- Some plausible waveform candidates have been defined, which enable a useful compromise between enhanced efficiency (i.e., over 80%) and minimum fundamental power reduction.
- Optimum waveforms display considerable asymmetry, which will be a common occurrence in practice due to the effects of the IV knee and/or nonlinear effects in the device input circuit.
- Asymmetry in the RF current is a feature which is absent in most of the literature on RF power amplifier analysis, Class E being a notable exception.

8.4 PA_Waves

This section attempts to take the waveform engineering results from Section 8.3 and use them to postulate realizable PA designs. In order to assist this task, a computing tool has been created, and will be known as a "power amplifier postulator." This tool, known as "PA_waves" is available on the CD-ROM which accompanies this book. It is normal practice in the CAD design world to take an active device model and surround it with specific circuit elements. The CAD engine will then take a specified excitation, and perform the task of solving the network equations, for a range of stipulated harmonics of the excitation signal, to determine the final steady state response of the system. When the device in question is displaying nonlinear behavior, the task of the underlying software becomes increasingly challenging, and iterative processes have to be used to seek the solutions. Convergence problems are frequently encountered, and improvements on the available techniques have been the focus of much research effort over the last decade or so.

The task of the "postulator" is much simpler, but complementary and potentially useful in determining the circuit topology that should be used. The designer stipulates the device excitation and output voltage waveforms, from which the program determines the device current which is compatible with the I-V characteristics.

The I-V model includes the knee region, an important feature of the program. Then the postulator determines the circuit impedances which support the stipulated waveforms. The key aspect of using a postulator is to ensure that the circuit elements can be realized using physical components. In particular, negative resistance components have to be discarded as unrealizable. In fairness, it should be admitted that the PA_waves postulator can only handle a unilateral device. But even the interaction of a unilateral three-terminal nonlinear device with a circuit is a surprisingly difficult problem to solve. The postulation approach, as a minimum, takes the analysis of PA circuits one level higher than is possible using conventional mathematical methods, in terms of reducing the idealizing assumptions that are often made in order to make the problem analytically tractable. One such idealizing assumption that is frequently made is to ignore the effects of the knee, or device turn-on, region. The present analysis will show how important these effects can be.

Some more practical details on the installation and running of the PA_waves program are given in Appendix A. To save space, the present sequence of waveform computations will not reproduce the full PA_waves screen each time, but all of the required input parameters will be given with each figure. The CD-ROM also contains specific Excel files as set up for some of the individual figures. Figure 8.13 shows the entire screen for the starting point, which is a Class B PA using a device with a realistic model for the IV knee region. As explained in Appendix A, the current is normalized such that $I_{max} = 1$, and voltage has arbitrary units, where usually either the DC level (Vdc), or the fundamental amplitude (V1), will be set to unity, at the user's discretion. An important initial setting is the IV knee parameter Vk. As in the analysis in Chapter 3, the turn-on region is modeled using the relationship

$$I_o = g_m v_{in}\left(1 - e^{-\frac{V_o}{V_k}}\right)$$

so that when the output voltage is at V_k, the current has a scaling factor of $(1 - \frac{1}{e})$, or about 0.63. The knee scaling factor will be 90% for $(V_o/V_k) = 0.23$, so that for a device having a 90% turn-on at 1 Volt, a unity setting for Vdc would correspond to just under 4.5V. This setting is intermediate in terms of most practical PA applications; a single cell battery powered mobile PA would imply a somewhat higher value, and higher voltage applications would indicate a lower value.

In Figure 8.13, the input settings (**Ibias** = 0, **Isig** = 1) correspond to the normalized values for reduced conduction angle modes used elsewhere in this book. With the fundamental voltage component (**V1**) and the dc supply (**Vdc**) set to unity, the current waveform shows the half wave rectified sinewave form, modulated by a deep notch as the output voltage touches zero. As described in Chapter 3, a reasonable first step is to reduce the output voltage swing by a normalized factor of 0.1, in order to restore some of the current peak. This condition, shown in reduced format in Figure 8.14, will be taken as the baseline for efficiency and power using "conventional" PA techniques. It is important, however, to note that the numbers for power and efficiency (−1.4 dB, 68%) are *substantially lower* than the values frequently quoted (0 dB, 78.5%) and which come out of idealized analysis which ignores the

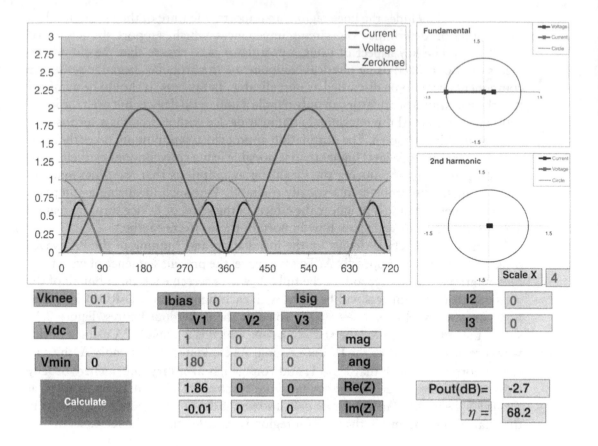

Figure 8.13   PA_waves screen with initial settings for Class B "postulation." Vk setting of 0.1 causes deep current notch where voltage grazes zero; note lower power and efficiency than "classical" values due to knee effect.

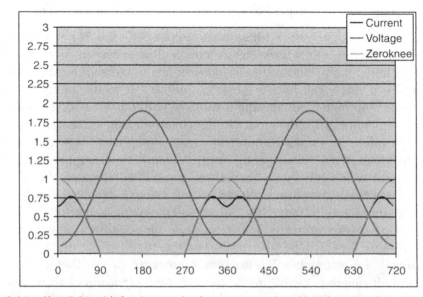

Figure 8.14   Class B PA with fundamental voltage swing reduced by Vk setting (V1 = 0.9); Power = −1.4 dB, efficiency = 68%.

clipping effect of the knee region. For convenience, the relative power, in dB, that PA_waves computes in each case refers to the power relative to the ideal Class B case, at the same Vdc setting, but with Vknee = 0. (This can be checked by setting the Vk parameter to zero.)

The next step is to observe the effect of a second harmonic component of voltage. Starting with values **V1** = 1.0, **V2** = 0.25, at a 0° phase angle, the modified waveforms are shown in Figure 8.15. The immediate benefit of the second harmonic can be seen; the voltage minimum increases and the current clipping is reduced. The real part of the second harmonic external impedance is however negative, shown in the second harmonic vector plot as in-phase voltage and current components.[3] Before taking the necessary action to remove this unrealizable component, the efficiency can be improved by increasing the voltage swing, using the second harmonic benefit. This can be implemented on PA_waves either by scaling up **V1** and **V2**, or more conveniently, reducing the **Vdc** value. The relative power computation in PA_waves uses the current **Vdc** value to establish the baseline power from which the relative power is computed, so the adjustment of **Vdc** provides a convenient way of scaling the RF waveforms. **Vknee** should, in principle, also be readjusted to main-

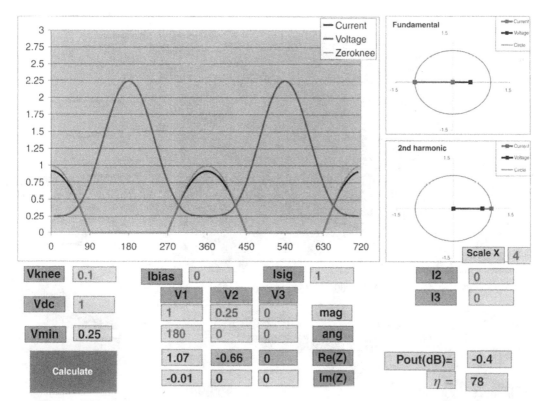

Figure 8.15 Addition of a second harmonic voltage component; note negative resistance requirement at second harmonic.

3. The PA_waves plots refer to device current and voltage; the sense of the current is reversed when considering the values of the external impedances.

tain the same ratio with **Vdc**; however, in the following set of examples, **Vknee** is kept constant. As already commented, the actual setting of **Vdc** is dependent on specific device technology being used, and as such its precise setting is a matter of judgment in specific cases.

Figure 8.16 shows the effect of reducing the minimum voltage in this manner. For the values indicated, the voltage minimum corresponds to the 0.1 value originally set in the starting-point Class B computation. The voltage waveform now dips into the knee region, and the current shows appropriate clipping. Clearly, the efficiency has increased, but the values are still of little practical significance due to the negative resistance component in the output load termination.

We now take a critical step. The voltage waveform can be phase-shifted in order to remove the negative external resistance at the fundamental; this is shown in Figure 8.17. Note that the fundamental phase has been increased by 34°, and to maintain the same waveform the corresponding second harmonic phase is increased by twice this amount, 64°. The waveforms of Figure 8.17 represent an important milestone in this sequence, since possibilities for waveshaping the current become evident. The IV knee function is causing some significant modulation on the current

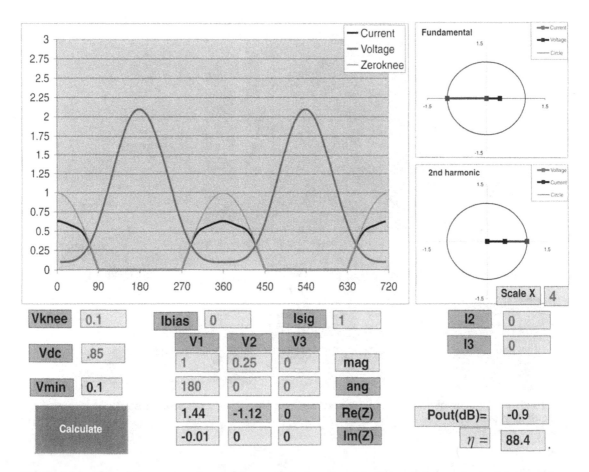

Figure 8.16 Second harmonic voltage enhancement, showing reduction in Vdc in order to reduce voltage minimum. Conditions still unrealizable using passive matching network.

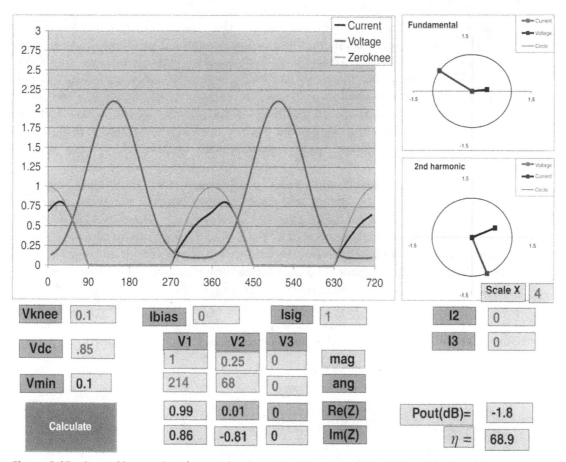

Figure 8.17   Second harmonic voltage enhancement, with reduced Vdc; differential phaseshift of 34° introduced on voltage in order to eliminate negative resistance load at fundamental.

waveform, making it become asymmetrical. Referring back to the functions analyzed in Section 8.3, it will be recalled that such asymmetry will usually cause a reduction in the phaseshift which is required to remove the negative resistance component in the fundamental load. This is shown in the PA_waves parameter values used to obtain the waveforms of Figure 8.17; the required fundamental phaseshift is reduced from the expected 45° to 34°, with a corresponding advantage in relative power and efficiency. An efficiency of 69%, however, is no real improvement over the knee-modulated Class B starting point.

There are two critical parameters which can be adjusted in order to increase the potency of the knee waveshaping action. Referring back to Figure 8.7, and the comments on the function defined in (8.3), a sloped voltage in the knee region can be used to increase the modulation of the current in the required manner. This can be implemented by introducing an offset, optimally only a few degrees, in the relative phasing of the fundamental and second harmonic voltage components. This phase offset will also interact with the chosen **V2** magnitude value, in determining the **Vdc** setting required to keep the voltage above zero.[4] The second critical parameter is the

---

4.   PA_waves analyses are invalid if the Vz parameter is negative, corresponding to a zero-crossing voltage.

Vmin value, which has an important effect on the power-efficiency tradeoff. Figure 8.18 shows the result of changing these two parameters. The voltage-current phase offset can now be reduced to 22.5°, the value used to obtain the Figure 8.18 waveforms. The efficiency has increased to 79%, a high value when a realistic model is being used for the knee region. Allowing the voltage to reach almost a zero value (0.02) takes a significant toll on the final fundamental RF power, about 1 dB less than the baseline Class B starting point. Such a tradeoff is inevitable in dealing with ultra-high efficiency modes, using realistic device models which have nonzero turn-on voltages.

All of the parameters described can be adjusted further, to obtain a surprising range of PA performance. Higher efficiencies can be obtained, but at the expense of output power. Figure 8.19 shows a condition close to the efficiency high point, for Class B bias settings. An efficiency of 87%, despite a simultaneous power reduction of over 3 dB from the baseline value, represents a milestone result and completion of the main goal of this chapter. The waveforms shown in Figure 8.19 have clear similarities to the waveforms seen in papers describing Class E PAs using a switch model for the active device. We have shown that a similar result can be obtained, according to the theory developed in this chapter, by terminating a device with a fundamental load having a substantial reactive component (specific values are given by

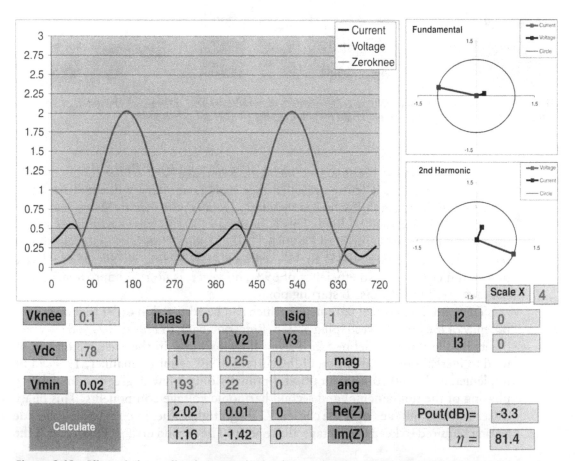

Figure 8.18 Effect of phase offset between the fundamental and second harmonic voltage components; a small slope is beneficial in engineering a more optimum current waveform (see text, and Figure 8.7).

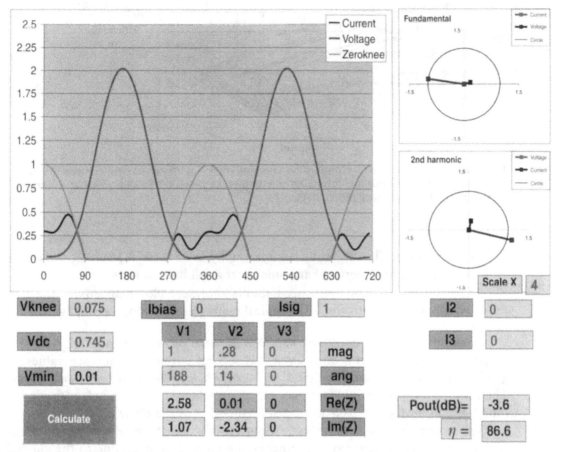

Figure 8.19   A Class E "lookalike" postulation.

PA_waves) and a second harmonic termination consisting of a specific reactance, which can conveniently be capacitive. The only idealizing assumption that may cause some issues in a practical implementation would be the absence of higher voltage harmonic components, however these will be at a low level in many practical situations, and their effects may not necessarily be detrimental.

Many more conditions can be obtained using PA_waves settings, which give the necessary quadrature conditions for the second harmonic. Independent variables which have not so far been explored are the knee voltage, **Vk**, the input bias settings (**Ibias, Isig**), the second harmonic voltage components (**V2mag, ang** at **V1ang** = 180°), and the minimum voltage (**Vmin**, itself a function of **V2mag** and **V2ang** settings). A useful summary of a wide range of realizable conditions is shown in Figure 8.20, for the Class B bias settings (**Ibias** = 0, **Isig** = 1). Here the considerable sensitivity of **Vk** and **Vmin** can be seen. A set of conventional Class B results (**V2mag** = 0), for the same range of **Vmin** and **Vk** values, is also plotted for comparison.

Clearly, efficiencies in the upper eighties are possible, but at substantially reduced power. The devil's advocate might look at these results and ask, why bother? The conventional Class B plots show up quite well alongside the new results, despite not being able to yield efficiencies in the 85% regime. The simple twofold answer is:

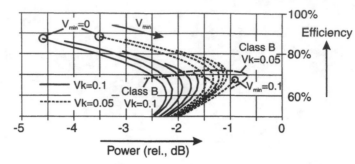

Figure 8.20 Power/efficiency contours for ranges of Vmin and Vk values. Each contour has V2 = 0.29 and represents a variation of the relative setting of V2 phase in PA_waves. Vmin contours are plotted for Vmin = 0, 0.0125, 0.025, 0.05, 0.1.

- Conventional Class B results are not usually achievable in practice, due to the requirement for a perfect harmonic short at all harmonics.
- The second harmonic enhanced results represent closer approximations to what real PA circuits do, using typical realizable matching networks.

Corresponding to a set of contours such as those shown in Figure 8.20, there is an overlay of fundamental impedance and second harmonic reactance values. A given circuit topology can only "visit" restricted areas of the power-efficiency plane. The value of the postulation approach is that impedances are specified, and suitable topology can be devised, to reach the desired region of the plane.

The higher efficiency cases clearly show that the use of the near-zeroing voltage to waveshape the current extracts a heavy premium on power, due to the clipped peak current values. In Section 8.3, it was suggested that a possible way around this problem would be to do the waveshaping on the input, either using the input varactor nonlinearity, or by physically adding suitable harmonic components to the PA drive signal. The function described in (8.5) and (8.6) is appropriate in that it can in principle be generated by simple addition of a suitably phased second harmonic component to the input voltage drive signal. Expectations for power, however, should include recognition that the fundamental component is 2 dB lower than a Class B waveform having the same peak value (Figure 8.13). Since the basic idea is to eliminate voltage clipping, it will be necessary to increase the voltage minimum level to a higher point in the knee region, or even just outside of it.

PA_waves can implement such a drive signal through the input parameters I2 and I3, which have been set to zero thus far. For example, the middle case (b) in Figure 8.10 can be implemented by setting I2 = 0.3 and I3 = 0.15 (the negative sign of the I2 component is taken care of internally by PA_waves). The **Isig** setting also has to be adjusted in order to obtain unity I_{max}, **Isig** = 0.89 in this case. Figure 8.21 shows the PA_waves spreadsheet for the particular case of **Vk** = 0.05, **Vmin** = 0.025, which indicates a relative power of –2.1 dB and an efficiency of 86.9%.

Figure 8.22 shows a summary plot, using the same parameter ranges as Figure 8.20, but using the modified current waveform. It is immediately apparent, comparing Figures 8.20 and 8.22, that significant benefits in power and efficiency have been realized, over a wide range of conditions. Most notably, the comparison in Figure 8.22 between the **Vk** = 0.05 contours and the corresponding Class B plot show dra-

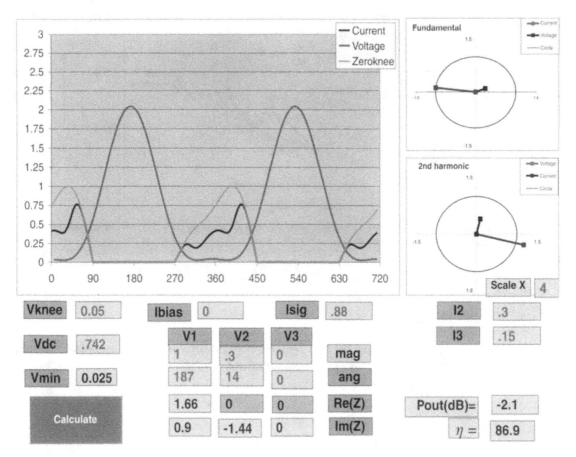

Figure 8.21   PA_waves postulation using input nonlinearity to waveshape the current (case (b) in Figure 8.10).

matic improvement in efficiency, for almost the same power levels. Clearly, input waveform engineering can play an important part in improving power and efficiency, and deserves more attention in PA design.

8.5   Implementation and Simulation

This section will show, through a specific design example, how the postulated PA examples in the last section can be transformed into actual designs. The process is an approximate one, in terms of getting exact duplication of circuit elements and waveforms. But the benefit of using a "postulator" as part of the design process will become clearer, due to its role in enabling the designer to specify the right circuit topology in the first place, before the process of simulation and optimization begins.

The example will be based on the PA_waves result shown in Figure 8.19, and applied to the same GaAs MESFET device used for Class AB and Class F simulations in previous chapters. A design frequency of 850 MHz will also be used for comparison with previous results. The PA_waves postulation gives the following impedances, normalized to the device loadline resistance:

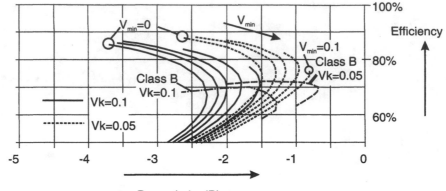

Figure 8.22   Power/efficiency contours with input waveshaping (I2 = 0.3, I3 = 0.15); other parameters same as in Figure 8.20.

- Fundamental: 2.58 + j1.1;
- Second harmonic: –j2.3.

The loadline resistance for this device has previously been obtained; for a supply voltage of 5V and an I_{max} of approximately 3.5A, the loadline value can be taken to be 3Ω. So the normalized values above are multiplied by this R_L value, giving

$$Z_1 = 7.7 + j3.3\Omega$$
$$Z_2 = -j6.9\Omega$$

We now convert the Z_2 reactance into an equivalent capacitance value at the second harmonic, which gives a value of 13.5 pF. This happens to be much larger than the C_{ds} of the device in question, but it should be noted that at higher frequencies the required capacitance value will converge toward the device C_{ds}, reaching the same value at approximately 5 GHz in this case. Figure 8.23(a) shows a suitable network for a narrowband implementation. It is a familiar network, consisting of a low-pass matching section with a capacitive input. Initially, the input capacitor would be set to a value which increases the device C_{ds} up to the design value of 13.5 pF, and the low-pass matching section would be designed to perform the necessary fundamental impedance transformation (see Chapter 5 for details on low-pass matching network design). Some adjustments can then be made to obtain a reasonably close realization of the required impedance values, as shown in Figure 8.23(b).

Using the same Spice simulation file, with suitably modified circuit elements, as used in the Class AB simulations in Chapters 4 and 5, the waveforms shown in Figure 8.24 result, with an RF output power of 34.1 dBm and an efficiency of 87%. The power should be compared with the zero-knee Class A baseline power for this device, which is $5 \times 1.75/2 = 4.375$, or 36.4 dBm. So the RF output power is 2.3 dB below the baseline value, with an efficiency of 87%, comparing remarkably well with the PA_waves prediction. The waveforms also compare quite well; the current waveform in PA_waves has some higher frequency ripples, but the fundamental and second harmonic components are close to the simulated values. The voltage waveforms are remarkably similar. Agreement this close cannot always be expected, but

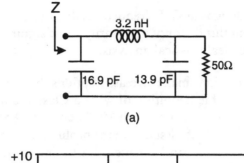

(a)

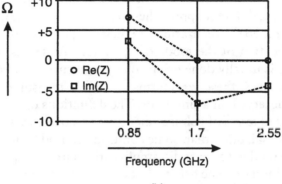

(b)

Figure 8.23 (a) Circuit topology, and (b) frequency response for design example based on Figure 8.19 postulation.

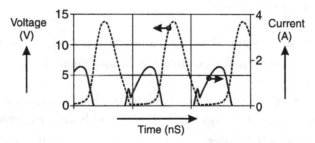

Figure 8.24 Spice simulation using GaAs MESFET and output network in Figure 8.23 (RF power = 34.1 dBm, DC current = 588 mA, efficiency = 87%).

the design process appears to be considerably streamlined through the use of the postulator.

8.6   Conclusions

It would be reasonable to summarize the results and analyses of this chapter as follows:

- Class E switching mode analysis is correct and has not been questioned.
- The Class E mode can be implemented at GHz frequencies.

- Class E designs at GHz frequencies do not have instantaneous switching transients, but this does not necessarily cause significant performance degradation from the ideal classical analysis.

These conclusions make two assumptions, however. The first is that we have a clear, unimpeachable definition of what a Class E amplifier really is; the second is that a reduction of efficiency from 100% down to 85% is "not significant."

Taking the efficiency issue first; the problem with the 85% number is that it represents a value which can be achieved, certainly in theory and quite frequently in practice, using different approaches to the design and mode of operation (e.g., Class C or Class F). So the realization of efficiency performance in this region is certainly not an indication of the mode of operation. As far as mode definitions are concerned, this is usually done using voltage and current waveforms. Unfortunately, at GHz frequencies, these waveforms are rarely measured directly, although can of course be observed on a simulator. The definitions of Class A, AB, B, C are universally agreed in terms of voltage sinewaves and truncated current sinewaves. Class D is also universal, other than some confusion caused by use of the term to represent a pulsewidth modulator in audio amplification circles. A precise Class F definition has been the subject of some debate, but only concerning the number of odd harmonics, and their respective amplitudes, that are added to the voltage waveform [6]. The Class E mode however seems to be missing such a widespread agreement in terms of an unimpeachable set of waveforms. Clearly, most papers and books on the subject (including this one, Chapter 7) reproduce the waveforms which come directly out of an ideal switch mode analysis. These waveforms could be physically drawn using the following procedure:

- Take a sinewave of current, and give it a positive DC offset;
- Pick a point in the positive region and make a vertical cut down to the zero axis;
- The portion of positive current from where it crosses the zero axis to the left of the cut, up to the cut itself, is the *Class E device current waveform*;
- Perform a time integral on the current between the right side of the cut and the previously mentioned zero-crossing point;
- Add a suitable DC voltage offset such that the waveform does not cross zero; this is the Class E device voltage.

The sticky issue is whether after doing all of this, we allow an additional step where everything gets "smoothed"; is this then still Class E? This in effect has been the main issue behind the analysis presented in this chapter. It has been shown that if the waveforms are smoothed, they can be generated in a different way, both physically and conceptually:

- Take a half-wave rectified sinewave of current, and put in some asymmetry about the central vertical axis;
- Take a voltage sinewave, add some in-phase second harmonic;

- Introduce a small phase offset between first and second harmonic components to generate a sloping baseline to the voltage waveform;
- Introduce some differential phase between voltage and current.

The first definition requires a switch to do the "cutting," and the second definition has been shown to be readily achievable using just about any kind of RF power transistor, so long as the voltage dips into the device knee region and the fundamental load has a substantial reactive component.

There has been a questionable absorption of the harmonic-limited definition into the original, more specific, switch mode Class E categorization. But as already suggested in Chapter 4, there is also an argument in favor of recognizing that second harmonic voltage enhancement is a more general technique in PA design and should, like Class F, be given a classification such as Class J.

No doubt, the debate will continue.

References

[1] Raab, F. H., "Class-E, Class-C, and Class-F Power Amplifiers Based upon a Finite Number of Harmonics," *IEEE Trans. Microwave Theory & Tech.*, Vol. 49, No. 8, August 2001, pp. 1462–1468.
[2] Telegdy, A., B. Molnar, and N. O. Sokal, "Class E/Sub M/Switching Mode Power Amplifier with Slow Switching Transistor," *IEEE Trans. Microwave Theory & Tech.*, Vol. 51, No. 6, June 2003, pp. 1662–1676.
[3] Personal communication with P. Tasker.
[4] Rhodes, J. D., "Output Universality in Maximum Efficiency Linear Power Amplifiers," *Int. J. Circ. Theor. and Appl.*, Vol. 31, 2003, pp. 385–405.
[5] Personal communication with J. D. Rhodes.
[6] Raab, F. H., "Class F Power Amplifiers with Maximally Flat Waveforms," *IEEE Trans. Microwave Theory & Tech.*, Vol. 45, No. 6, November 1997, pp. 2007–2012.

CHAPTER 9

Nonlinear Effects in RF Power Amplifiers

9.1 Introduction

Throughout the last few chapters, it has been taken for granted that any nonlinearity in the transfer characteristic of a power amplifier constitutes a problem of some kind, but so far no attempts have been made to quantify the problem. It has also been noted that the impact of nonlinearity is strongly dependent on the actual application, most specifically the modulation system in use, coupled with the relevant regulatory specifications. This chapter attempts to quantify the effects of a nonlinear power transfer characteristic, specifically the spectral distortion that is an inevitable consequence of PA nonlinearity.

In previous chapters, it was shown that a power transfer characteristic could often be derived for a particular amplifier type or mode, given reasonable assumptions about the RF load and harmonic terminations. Such transfer curves were derived using simplified models, and could also be derived using CAD simulation. Clearly, in cases where the modulation system simply varies the amplitude of a CW carrier ("Amplitude Modulation," or "AM"), it would seem to be a logical extension, or application, of these power transfer characteristics to map the input amplitude variations onto the output plane. The resulting distorted waveform can then be analyzed for its frequency components, which constitute the spectral distortion caused by the amplitude nonlinearity of the amplifier. This is the principal strategy used in this chapter, and it will be shown that much useful information about the spectral distortion characteristics can be derived in this manner.

Surprisingly, for many years this has not been the traditional method for quantifying nonlinear effects in RF amplifiers. For too long, it seems, misguided efforts have been based on the extrapolation of a power series formulation into regions of strong nonlinearity where the validity of this approach becomes highly questionable. This is not to denigrate the usefulness of power series and (especially) Volterra series analysis for these applications [1]. The issue is how many polynomial terms are required in a specific case to model the PA characteristic to the required degree of precision. The focus here is on amplifiers operating at their highest possible efficiencies, which means that the model has to be able to handle substantial amounts of gain compression, and the corresponding AM-PM nonlinearity as well. Having derived a suitable model for the PA characteristic, the methods of "envelope simulation" are used, as a means of predicting the spectral and modulation distortion under specific signal conditions. The general theory which underpins envelope simulation methods has been around for several decades [2, 3], but has only started to gain wider popularity in recent years, following the implementation of more complex digital modulation techniques. This chapter provides an introduction to the possibilities offered by envelope simulation methods, with the assumption that those who wish to pursue it in greater depth will avail themselves of a commercially available CAD package, or adopt the computational procedure described by Kenney and Leke [4]. For those who wish to do neither, this chapter, along with Appendix B, introduces a means by which envelope time domain and spectral analysis can be performed using a number of Excel files which are to be found on the accompanying CD-ROM. This approach has its uses, particularly in exploring the effects of user-specified PA characteristics.

Any amplifier, when driven into a strongly nonlinear condition, will exhibit phase as well as amplitude distortion. This is usually characterized in terms of "AM to PM" conversion, and represents a change in the phase of the transfer characteristic as the drive level is increased towards and beyond the compression point. Generally speaking, it will be seen that AM-PM in typical solid state amplifiers is a significant, but rarely a dominant, effect. The usefulness and validity of results which ignore AM to PM is therefore substantial. Indeed, the commonest manifestation of AM to PM effects is an irritating (and sometimes exasperating) asymmetrical slewing of the intermodulation or spectral regrowth display. The precise cause of this effect is key to understanding and quantifying the importance of AM-PM effects in individual cases, and it will be discussed analytically.

The techniques of nonlinear envelope simulation will initially be demonstrated using a sinusoidally modulated RF carrier, which will be compared directly with a more traditional two-tone formulation based on the "Intercept Point" concept. The envelope simulation technique will then be extended to include AM to PM effects, and more complicated amplitude modulation characteristics, which will include some simplified quantitative evaluation of different peak-to-average ratio schemes.

There follows, in Section 9.7, an introduction to digital modulation systems. This section is not intended to be a treatise on a subject which has been extensively covered in numerous books (see bibliography), but rather a need-to-know guide for the PA designer. In particular, the role of baseband filtering in converting constant-amplitude PSK and QPSK signals into variable envelope excitations which have both AM and PM will be discussed. Although a fully rigorous analysis of the distortion effects of a nonlinear amplifier on digitally modulated RF signals is beyond the scope of the present book, some general guidelines, based on the analysis techniques presented in the earlier sections, can be derived in terms of the spectral distortion levels to be expected. In fact, these general guidelines will indicate that in most cases the spectral distortion effects in most higher efficiency PAs will be unacceptable, other than in strictly constant envelope applications; detailed or more rigorous analysis is simply not required to draw this unpalatable conclusion.

The question is, then, can anything be done to leverage the promising efficiency predictions of the previous chapters in variable envelope applications. The answer is that it can, in the form of linearization and efficiency enhancement techniques, which form the subject of Chapters 10 and 14. In some respects, this chapter poses the problems caused by PA nonlinearity and attempts to quantify them, but the potential answers lie in the techniques discussed in Chapters 10 and 14.

9.2   Two-Carrier Power Series Analysis

As already commented in the introduction, two-carrier analysis, based on a power or Volterra series, represents the traditional approach to analyzing nonlinear effects in RF amplifiers. As a first step, it is necessary to review, or revisit the technique. It is assumed that most readers will have encountered it in some form already, and some of the more detailed trigonometric manipulation has been omitted. A relevant set of trigonometric formulae are given in Table 9.1 to assist in following the analysis in this and subsequent sections.

We assume a weakly nonlinear amplifier, whose output and input voltage can be related using the standard power series formulation discussed in Chapter 1:

$$v_o = a_1 v_i + a_2 v_i^2 + a_3 v_i^3 + a_4 v_i^4 + a_5 v_i^5 + \dots \tag{9.1}$$

where v_i, v_o are small, time varying quantities representing the RF input and output signals. For the time being, the a_n coefficients are taken to be scalar quantities which characterize the amplifier and are determined experimentally. Generally speaking, one set of coefficients will only apply for a single frequency and a fixed set of biasing and tuning conditions.

We now substitute for an input signal consisting of two, equal amplitude, in-band RFC signals, whose spacing is much smaller than either RF frequency.

$$v_i(t) = v \cos(\omega_1 t) + v \cos(\omega_2 t) \tag{9.2}$$

So the output voltage is

Table 9.1   Trigonometric Relationships

$$\sin(A \pm B) = \sin(A)\cos(B) \pm \cos(A)\sin(B)$$

$$\cos(A \pm B) = \cos(A)\cos(B) \mp \sin(A)\sin(B)$$

$$\sin(A) + \sin(B) = 2 \cdot \sin\left(\frac{A+B}{2}\right)\cos\left(\frac{A-B}{2}\right)$$

$$\sin(A) - \sin(B) = 2 \cdot \cos\left(\frac{A+B}{2}\right)\sin\left(\frac{A-B}{2}\right)$$

$$\cos(A) + \cos(B) = 2 \cdot \cos\left(\frac{A+B}{2}\right)\cos\left(\frac{A-B}{2}\right)$$

$$\cos(A) - \cos(B) = -2 \cdot \sin\left(\frac{A+B}{2}\right)\sin\left(\frac{A-B}{2}\right)$$

$$\cos^2 A = \tfrac{1}{2}(1 + \cos 2A)$$

$$\cos^3 A = \tfrac{1}{4}(3\cos A + \cos 3A)$$

$$\sin^2 A = \tfrac{1}{2}(1 - \cos 2A)$$

$$\sin^3 A = \tfrac{1}{4}(3\sin A + \sin 3A)$$

$$v_o(t) = a_1 \cdot v \cdot \left(\cos(\omega_1 t) + \cos(\omega_2 t)\right)$$

$$+a_2 \cdot v^2 \cdot \left(\cos(\omega_1 t) + \cos(\omega_2 t)\right)^2$$

$$+a_3 \cdot v^3 \cdot \left(\cos(\omega_1 t) + \cos(\omega_2 t)\right)^3$$

$$+a_4 \cdot v^4 \cdot \left(\cos(\omega_1 t) + \cos(\omega_2 t)\right)^4 \qquad (9.3)$$

$$+a_5 \cdot v^5 \cdot \left(\cos(\omega_1 t) + \cos(\omega_2 t)\right)^5$$

$$+ \dots \text{etc}.$$

Each line, above, of (9.3), represents a "degree" of signal distortion. Each degree of distortion generates a number of distortion products, which have either the same or lower "orders." For example, the second degree term produces second order products at frequencies $2\omega_1$, $2\omega_2$, $\omega_1 \pm \omega_2$; the fourth degree term produces fourth order products at $4\omega_1$, $4\omega_2$, $2\omega_1 \pm 2\omega_2$, $3\omega_1 \pm \omega_2$, but also produces second order products at $2\omega_1$, $2\omega_2$, $\omega_1 \pm \omega_2$. Generally speaking, the even order distortion terms will be well out of band and are of less concern than the odd order distortion products. Table 9.2 shows the results of expanding out (9.3) to include all of the distortion products up to and including the fifth degree.

Table 9.2 Two-Carrier Distortion Products, Up to Fifth Degree (Expansion of (9.3))

	$a_1.v$	$a_2.v^2$	$a_3.v^3$	$a_4.v^4$	$a_5.v^5$
1 (DC)		1		9/4	
ω_1	1		9/4		25/4
ω_2	1		9/4		25/4
$2\omega_1$		1/2		2	
$2\omega_2$		1/2		2	
$\omega_1 \pm \omega_2$		1		3	
$2\omega_1 \pm \omega_2$			3/4		25/8
$2\omega_2 \pm \omega_1$			3/4		25/8
$3\omega_1$			1/4		25/16
$3\omega_2$			1/4		25/16
$2\omega_1 \pm 2\omega_2$				3/4	
$3\omega_2 \pm \omega_1$				1/2	
$3\omega_1 \pm \omega_2$				1/2	
$4\omega_1$			1/8		
$4\omega_2$			1/8		
$3\omega_1 \pm 2\omega_2$					5/8
$3\omega_2 \pm 2\omega_1$				5/8	
$4\omega_1 \pm \omega_2$				5/16	
$4\omega_2 \pm \omega_1$				5/16	
$5\omega_1$					1/16
$5\omega_2$					1/16

The products of most interest, in terms of their possible detrimental effects, are the intermodulation (IM) products. The third order products, at frequencies $2\omega_2 - \omega_1$ and $2\omega_1 - \omega_2$, which come primarily from the third degree term and have amplitudes $a_3.v^3.(3/4)$ also have components which come from higher odd degree terms in the power series expansion. Thus the fifth degree terms, the highest included here, not only contribute close-to-carrier fifth order intermodulation products (at frequencies $3\omega_2 - 2\omega_1$ and $3\omega_1 - 2\omega_2$ and amplitudes $a_5.v^5.(5/8)$) but add a contribution to the third order intermodulation products (amplitude $a_5.v^5.(25/8)$). These higher degree contributions can be ignored when operating well below the compression level, but can become the dominant contributors in the compression and saturation regimes. A typical spectrum is shown in diagrammatic form in Figure 9.1. The intermodulation sidebands appear either side of each carrier, at a frequency spacing equal to that of the two input carriers.

This analysis assumes two separate carriers with fixed spacing. In practice, such a signal will usually be the result of the amplitude modulation of a single carrier. As will be shown in Section 9.3, the two-carrier signal would be the spectrum from a double sideband suppressed carrier AM system, with a single, fixed frequency modulating "tone."[1] In practice, even a simple AM transmitter of this kind would have a modulation band that would randomly fill a prescribed bandwidth either side of the carrier frequency. Older radio textbooks would normally show this as a typical audio, or speech, bandlimited frequency zone (Figure 9.2) but in more modern digital systems the baseband tends to exhibit a more clinical appearance, as indicated in Figure 9.3. The key point to note is that the intermodulation bands stretch out to three times the original modulation band limits, in the case of the third order distortion products, and five times these limits for fifth order, as shown in both Figures 9.2 and

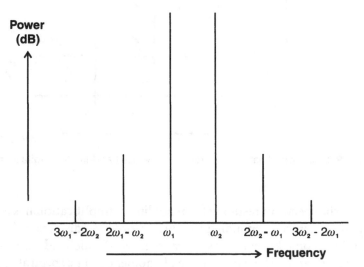

Figure 9.1 Two-carrier intermodulation spectrum.

---

1. The reason for the extensive use of a "two-carrier" test is that it is much easier to generate two cw RF signals than it is to modulate a single carrier. In practice the modulation process would generate nonlinear spectral components that would interfere with measurements on the device under test.

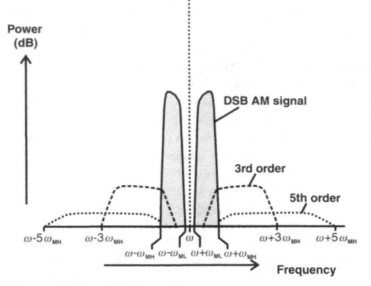

Figure 9.2 Analog amplitude modulation spectrum.

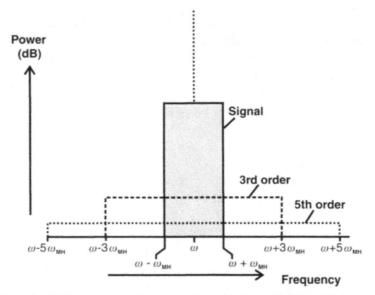

Figure 9.3 Intermodulation spectrum for a typical band-limited digitally modulated signal.

9.3. So the spectrum resulting from nonlinear amplification has a stepped appearance, with each step corresponding to a higher order of distortion. These steps have more recently become known as "spectral regrowth" sidebands, although there is nothing at all recent about the nature of the phenomenon. The spectral region of most concern in a regulated communications band is the third order step, which is the one that lies closest to the main signal. This is usually known, logically enough, as the "adjacent channel," and a specification of much importance is the integrated power which lies inside this channel, known as the adjacent channel power, or ACP. Exact definition of the ACP band will differ between modulation systems and the corresponding regulatory specifications, but the logical limit would be a frequency offset of three times the bandwidth of the undistorted signal.

Note also in Figures 9.2 and 9.3 that a portion of each distortion band lies *within* the original bandlimits of the undistorted signal. This is not a problem as far as measurement is concerned, because the distortion levels can be measured where they emerge from the main spectral zone. But the in-band "aliasing" of the distortion products and the main signal do constitute a system problem when it comes to demodulating the received signal and retrieving the information from it. Spectral regrowth, or intermodulation distortion, is often considered to be solely a problem of adjacent channel interference, due to the fact that PA linearity is usually also specified in terms of demodulation error rates.

Returning to the results of Table 9.2, intercept point (IP) is a concept used extensively by receiver designers, whose components operate well below the physical compression point. It represents the intersection between the extrapolated 1:1 slope of fundamental gain, and the 3:1 slope of the third order IM products, as indicated in Figure 9.4. If both extrapolations are done from well inside the linear region, and ideal third degree nonlinearity applies, the intercept point ("IP3") becomes a useful single-point specification from which IM levels can be easily calculated at different signal levels. It is nevertheless a highly conceptual point which can never be reached in practice, and even well below the intercept point the actual measured IM3 data points can wander substantially from the extrapolated 3:1 line at higher power levels, as indicated in Figure 9.4. These deviations from the assumed straight-line plot are caused by higher degrees of nonlinearity, and are most evident in the higher power regions, which are of most interest for a PA user. Thus intercept point is of limited use in PA design and specification, despite still being held in high esteem by receiver designers.

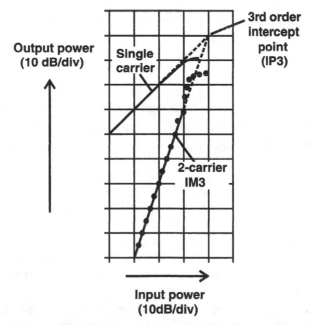

Figure 9.4 Single-carrier power transfer and two-carrier IM3 power amplifier characteristic (solid circles represent typical measured data).

One well-known derivative of the intercept point concept is a relationship between the third order intercept point, IP3, and the 1 dB compression point, P1 dB. This "proof" assumes that the power series can be approximated to include terms only up to the third degree. In this case, the intercept point plot will indeed be a straight line having a 3:1 slope, and the gain compression will be solely caused by third order effects.

If we first consider an amplifier driven with a sinusoidal signal which causes a gain compression of 1 dB, then the fundamental output voltage is given, from (9.1), as

$$v_{oc} = a_1 \cdot v_{ic} \cdot \cos \omega t + a_2 \cdot v_{ic}^2 \cdot \cos^2 \omega t + a_3 \cdot v_{ic}^3 \cdot \cos^3 \omega t$$

From Table 9.1, we recall that

$$\cos^3 \theta = \frac{1}{4}(3\cos\theta + \cos 3\theta)$$

so that the output voltage amplitude at the fundamental frequency is

$$v_{oc} = a_1 \cdot v_{ic} + \frac{3}{4} \cdot a_3 \cdot v_{ic}^3$$

With no compression or nonlinear terms, the output voltage amplitude would be simply $a_1 . v_{ic}$; if this is reduced in power by a factor of 1 dB, the output voltage will become

$$v_{oc} = a_1 \cdot v_{ic} \cdot 10^{-0.05}$$

so combining the two expressions for v_{ic} gives

$$a_1 \cdot v_{ic} \cdot 10^{-0.05} = a_1 \cdot v_{ic} + \frac{3}{4} \cdot a_3 \cdot v_{ic}^3 \qquad (9.4)$$

It is clear from (9.4) that the sign of a_3 must be negative for gain compression to occur. So at this point for clarity, a negative sign will be introduced for a_3, and (9.4) rearranged to give

$$v_{ic}^2 = \frac{4 \cdot a_1 \cdot \left(1 - 10^{-0.05}\right)}{3 \cdot a_3} \qquad (9.5)$$

Equation (9.5) gives a value for the input amplitude of a sinusoidal signal which causes 1 dB gain compression, in terms of the two power series coefficients, a_1 and a_3. It should be emphasized again that higher order terms in the power series have been ignored.

We now consider the same amplifier, having the same bias and tuning conditions, but with a two-carrier signal applied to its input, both signals having equal amplitude. The amplitude of each carrier, v_{ip3}, at the third order intercept point will, by definition, be given by

$$a_1 \cdot v_{ip3} = \frac{3}{4} \cdot a_3 \cdot v_{ip3}^3$$

using Table 9.2 to determine the amplitude of each third order intermodulation product. Note that the definition of third order intercept point, by convention, relates the amplitude of a single intermodulation carrier to the amplitude of a single input carrier. Rearranging this expression gives a relationship between the input third order intercept signal amplitude and the two power series coefficients:

$$v_{ip3}^2 = \frac{4 \cdot a_1}{3 \cdot a_3} \tag{9.6}$$

and combining (9.5) and (9.6),

$$\left(\frac{v_{ip3}}{v_{ic}} \right)^2 = \frac{1}{1 - 10^{-0.05}} \tag{9.7}$$

which evaluates to a ratio of 9.2, or about 9.6 dB.

So the input third order intercept point is about 10 dB higher than the cw, or single tone, 1 dB compression point. This will be equivalent to 13 dB difference if the total output power is used to represent the two-carrier case. It should also be noted that if the reference parameter is taken as input power, then 1 dB allowance has to be made for the gain compression. So, for example, if an amplifier has a measured cw output 1 dB compression point of +30 dBm, the two-carrier intercept point would be expected to be at +40 dBm output for each carrier, or 43 dBm total power output. Of course, a 1 watt amplifier will never actually deliver 20 watts; herein lies the weakness of the concept. But if it is assumed that as the power is backed off from this point the third order products fall at a rate of 3 dB for every dB of output power reduction, or 2 dBc per dB of backoff, it is easy to calculate the IM3 level at any specified power level.

As rules of thumb go, this relationship has its uses. Indeed, so long as the intercept point is used to extrapolate well down into the linear region, it can frequently give answers, in terms of IM3 levels, that are within a few dB of measured values, for a unit that may have only been specified originally in terms of its 1 dB compression point. But the perils, not to mention lost revenues, in assuming that IM specs can be met on the basis of a P_{1dB} spec alone, are only too well known in the merchant microwave amplifier community. The need to include fifth, and higher odd degree, terms as the compression point is approached has already been stressed. Other deviations from the simple rule occur in multistage amplifiers where several intermediate stages may generate additional contributions to the intermodulation products. These additional components have a strong tendency to be in phase with the IM products generated in the output stage, and can alter the final IM result quite significantly.

Some other relationships can be proved using a similar approach. One of particular interest in this chapter, and for variable envelope PA analysis in general, is the 1 dB compression point of a two-carrier signal. This is simple enough to derive. If the 1 dB compression amplitude of *each carrier* is v_{ic2}, then using Table 9.2, the amplitude of either fundamental carrier will be

$$v_{oc2} = a_1 \cdot v_{ic2} - \frac{9}{4} a_3 \cdot v_{ic2}^3$$

so that the two-tone 1 dB compression point is given by

$$v_{ic2}^2 = \frac{4 \cdot a_1 \cdot \left(1 - 10^{-0.05}\right)}{9 \cdot a_3} \tag{9.8}$$

and by comparison with (9.6), the ratio between the single-carrier and two-carrier signal amplitudes for 1 dB compression is

$$\left(\frac{v_{ic1}}{v_{ic2}}\right)^2 = 3 \tag{9.9}$$

This corresponds to a power ratio of about 5 dB, but it should be noted that the ratio in (9.8) compares the *individual* carrier amplitudes. So if the comparison is made in terms of total average signal power, the two-carrier compression point is 2 dB lower than the single carrier case.

This result is subject to all of the same caveats as discussed above for the 1 dB compression—IP3 relationship. It does, nevertheless, provide a sobering introduction to the subject of distortion effects with variable envelope signals. Here we have a simple, but quite relevant case of a signal which consists of two carriers, closely spaced and equal in amplitude, but which require a mean power backoff of 2 dB in order to reach the same compression level of a cw signal having the same undistorted mean power level. This should not be too much of a surprise, since clearly the two-carrier signal will display amplitude reinforcement beats having an instantaneous peak power 3 dB higher than the average level. But given the unsatisfactory nature of the approximations in reaching this result, it would seem logical at this point to look at distortion from a time domain, or "envelope" viewpoint.

9.3 Two-Carrier Envelope Analysis

The analysis presented in Section 9.2 clearly has limitations, mainly due to the exclusion of higher degree terms in the power series, which become important as the compression point is reached. The analysis so far has essentially been done entirely in the frequency domain. It therefore pays to look at the same problem in the time domain to get a more direct intuitive feel for what happens to an amplifier when driven with a variable envelope signal. Starting once again with the two-carrier signal,

$$v_s(t) = v \cdot \cos(\omega_1 t) + v \cdot \cos(\omega_2 t) \tag{9.10}$$

this can be rearranged, using Table 9.1 again as trigonometric guidance, in the form

$$v_s(t) = 2 \cdot v \cdot \left\{ \cos\left(\frac{\omega_1 - \omega_2}{2}\right) \right\} \left\{ \cos\left(\frac{\omega_1 + \omega_2}{2}\right) \right\} \tag{9.11}$$

If we now define $\omega = \dfrac{\omega_1 + \omega_2}{2}$; $\omega_m = \dfrac{\omega_1 - \omega_2}{2}$, (9.11) can be written as

$$v_s(t) = 2 \cdot v \cdot \cos \omega_m t \cdot \cos \omega t \tag{9.12}$$

This is now recognizable as a double sideband, suppressed carrier, amplitude modulated signal, having a carrier frequency of ω and a sinusoidal baseband modulating signal of frequency ω_m. In order to generate such a radio signal it is necessary for the modulating signal $\cos \omega_m t$ to reverse the phase of the RF carrier by 180° at each zero crossing, that is to say the modulation is "case-sensitive" to the sign of the modulating function, $\cos \omega_m t$. This phase reversal is not really visible when the resulting envelope of the signal is drawn as in Figure 9.5. Drawing envelopes of this kind presents a challenge to the printer; inasmuch as it is not possible to resolve the individual RF cycles, the phase variations cannot be indicated either. It is therefore conventional to construct lines which connect all the RF peaks, much as the signal

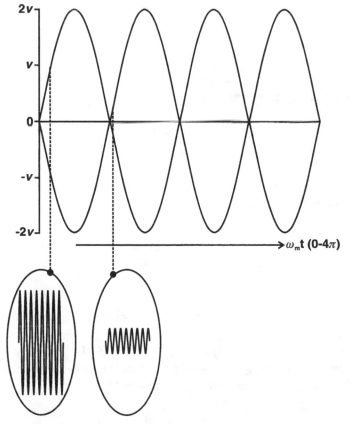

Figure 9.5 Two-carrier signal envelope.

would appear on an oscilloscope using a peak detector. This is somewhat unsatisfactory, since these lines do not appear on the real signal, and the detector is unable to detect the phase reversals each half cycle of the modulation envelope.

It is therefore important to remember that a magnified section of such a plot would reveal the much faster, fine-grain, RF waveform. This is indicated in Figure 9.5 at two places on the modulation cycle. It should also be noted that, conversely, such magnified time samples show an RF sinewave with no *perceptible* cycle-to-cycle amplitude or phase variation. It is a fundamental assumption in envelope analysis that the modulation frequency is sufficiently slow compared to the RF variations that the conditions at any instant can be considered to be quasi-static, and can be determined from a static, cw measurement at the same input power. In most present day wireless communications systems, this is usually still a reasonable assumption. Essentially, the amplifier has to have sufficient bandwidth that the performance characteristics do not change over the bandwidth of the modulated signal. This is almost certainly true for single channel applications, although multichannel, spread spectrum signals may start to show the first signs of violation on this basic assumption.

Some important power relationships need to be established for the waveform in Figure 9.5. It is clear that the RF signal sweeps between an amplitude of zero and a peak voltage amplitude of $2v$, twice the amplitude of each original carrier in the two-tone representation. So the peak envelope power (PEP) is given by

$$P_{pk} = \left[\frac{2v}{\sqrt{2}}\right]^2 = 2v^2 \tag{9.13}$$

assuming, for convenience, a 1Ω resistive load.

In this particular case, the mean power can be most easily determined using the two-carrier representation of (9.10). Each carrier has a power given by

$$P_n = \left[\frac{v}{\sqrt{2}}\right]^2$$

and the total power is the sum of individual carrier powers, so in this case the overall mean power P_m is given by

$$P_m = 2 \cdot \left[\frac{v}{\sqrt{2}}\right]^2$$

$$P_m = v^2 \tag{9.14}$$

A very important parameter in envelope analysis, and PA analysis in particular, is the ratio of P_{pk} to P_m, the "peak-to-average" power ratio, or "PAR," which in this case is clearly a factor of 2, corresponding to 3 dB. Note also that a constant amplitude cw signal having the same power as the two-carrier signal under discussion, would have a voltage amplitude v_{cw}, given by

$$v_{cw} = v \cdot \sqrt{2}$$

In some cases, where the modulated signal does not lend itself to mathematical spectral analysis quite so readily, it will be necessary to find the mean power by direct integration of the square of the RMS RF amplitude over a complete modulation cycle,

$$P_m = \frac{\omega_m}{2\pi} \int_0^{\frac{2\pi}{\omega_m}} \tfrac{1}{2}\{f(w_m t)\cdot v\}^2 \, dt \tag{9.15}$$

where the signal is defined as

$$v_s(t) = v \cdot f(\omega_m t) \cdot \cos \omega t \tag{9.16}$$

and the modulating function $f(\omega_m t)$ is a scalar function with a periodicity of ω_m.

We are now in a position to examine the effect of a nonlinear amplifier on the variable envelope signal shown in Figure 9.4. Clearly, if the mean power level of the variable envelope signal is set to be equal to the conventional, single-carrier 1 dB compression power, the peaks of the envelope will incur much greater than 1 dB of compression, and will saturate the amplifier for a significant portion of the modulation cycle. On the other hand, for an equal portion of the modulation cycle the signal will be well down in the linear region and incur no distortion at all. One might therefore speculate that the damage might equal itself out. Unfortunately, the power series analysis in Section 9.2 has already given us warning not to be so optimistic.

In order to analyze this problem quantitatively, a typical P_{in}-P_{out} amplifier characteristic is needed. For the purposes of this chapter, it will be convenient to model the characteristic using an abstract describing function, rather than to base the PA characteristic on a nonlinear simulation at component level. Such a characteristic can then be fitted, either to the RF simulated data, or to actual measured points. Figure 9.6 shows a suitable function. In this case, the "softness" of the compression characteristic can be varied by choosing two tangible parameters, P_{COMP} and P_{SAT}. P_{COMP} represents the difference (in dB) between the 1 dB compression point and the

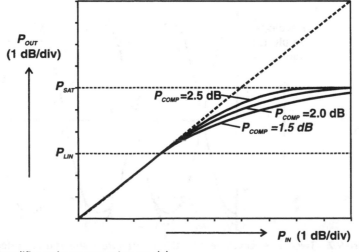

Figure 9.6   Amplifier gain compression model.

maximum linear power point, and P_{SAT} represents the difference between the saturated power point and the maximum linear power point. The amplifier is assumed to be perfectly linear below the linear power point, P_{LIN}. The characteristic up to P_{LIN} is a straight line of the form

$$y = x$$

where y and x represent linear points on equal dB scales.

Above the P_{LIN} point the characteristic takes the form

$$y = x - a \cdot x^n$$

With some simple algebraic manipulation, it is possible to determine the peak value of this function, which is set to represent the specified value of P_{SAT}. Above this value, the characteristic is defined to be ideally flat, or saturated. This gives one relation between the parameters a and n, and the other relationship can be determined using the 1 dB compression point. The resulting equations relating a and n cannot be solved analytically, but a simple numerical computer routine quickly yields the required pair of parameters. Figure 9.6 shows a set of typical compression characteristics for three devices having the same values for P_{SAT}. For convenience, the analysis will be normalized such that the linear power gain of the amplifier is unity, and the average cw power corresponding to P_{LIN} is 0 dBm.

The signal waveform of Figure 9.5 and the amplifier compression characteristics of Figure 9.6 can now be combined to show the effect of the amplifier amplitude nonlinearity on the shape of the signal envelope. This is shown in Figure 9.7. Clearly, since the envelope has a peak-to-average ratio ("PAR") of 3 dB, if the average input power level is set to a value 3 dB lower than the linear power point (P_{LIN}), on the amplifier compression characteristic, no distortion will occur. As the level is increased, the peaks of the envelope encroach into the compression region, and show a symmetrical flattening. When the mean envelope power equals the cw 1 dB compression power, very extended limiting is evident, this is the case shown in

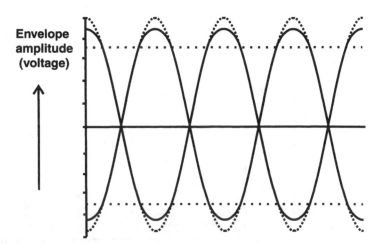

Figure 9.7 Two-carrier envelope distortion (dotted trace shows undistorted signal). Dotted line indicates maximum linear power level.

Figure 9.7. It might be thought that if the same time magnifier could be used, such as in Figure 9.4, the actual RF peaks in Figure 9.7 would also be showing similar clipping behavior. But it is assumed here that the amplifier is tuned and filtered such that only a narrow band of frequencies in the vicinity of the RF carrier are transmitted to the output load. The key assumption in envelope distortion analysis, which requires justification [2–4], is that if a bandpass amplifier characteristic is assumed, then the output waveform can be assumed to be an RF sinewave with amplitude and phase modulation as defined by the static AM-AM and AM-PM characteristics.

The distorted envelope shown in Figure 9.6 can be assumed to have the form

$$v_o(t) = A \cdot \{ \mu_1 \cos \omega_m t + \mu_2 \cos 2\omega_m t + \mu_3 \cos 3\omega_m t + \dots \} \cdot \cos \omega t \qquad (9.17)$$

and the third order IM products are the result of the multiplication of the term $\mu_3 \cos 3\omega_m t$ and the RF carrier $\cos \omega t$,

$$\mu_3 \cos 3\omega_m t \cdot \cos \omega t = \frac{A \cdot \mu_3}{2} \{ \cos(\omega + 3\omega_m)t + \cos(\omega - 3\omega_m)t \} \qquad (9.18)$$

So the IM3 levels can be determined by performing a Fourier analysis on the distorted output *envelope*, which will give a value for μ_3. The result of performing such an analysis, using direct computation in this case, is shown in Figure 9.8. This plot shows the IM3 levels which correspond to the distorted time domain envelope characteristics in Figure 9.7. For both of these plots, the amplifier characteristic has been the intermediate compression gain model shown in Figure 9.6.

Figure 9.8 shows that the IM3 vanishes when the peak envelope power drops below the onset of gain compression. In this case an overly optimistic amplitude compression model has been assumed, whereby the amplifier reverts to perfect linearity below the P_{LIN} level (0 dB). In practice, weakly nonlinear effects will take over

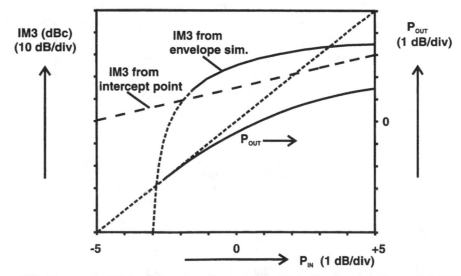

Figure 9.8   Two-carrier IM simulation, based on envelope distortion shown in Figure 9.7. IM3 distortion vanishes at the 3 dB backoff power, in accordance with the gain compression model shown in Figure 9.6 ($P_{COMP} = 2$ dB case).

at lower power levels, and will cause IM3 levels to be asymptotic to a line with 3:1 slope. Such a line, shown dashed in Figure 9.8, is drawn such that it passes through an intercept point which is 13 dB higher than the cw 1 dB compression power (+15 dB in this case, based on P1 dB = +2 dBm). Experimental points would be expected to fall closer to the dotted line for power levels well below the compression point, and closer to the solid line at higher power levels, where strongly nonlinear effects will dominate.

In interpreting the IM3 levels shown in Figure 9.8, some generalization is necessary in terms of what level of IM3 is acceptable. Clearly, this is a specification which will vary widely between applications, but for the present analysis it will be assumed that 20 dBc will usually be unacceptable for most applications. A representative range of systems may tolerate a level of 30 dBc, so a 25 dBc level could be used as a meaningful standard of comparison between different amplifer characteristics, understanding that some systems will have much tighter requirements. For example, Figure 9.9 shows three IM3 plots, each corresponding to the three different compression characteristics shown in Figure 9.6. Despite having the same saturated power level, there is a substantial difference between the three IM responses, which is summarized in Table 9.3.

Clearly, the harder-saturating device shows lower IM3 levels than the softer devices at any given power level. Significantly, however, the power output level for the established comparison level of −25 dBc shows a fairly close correlation to the value of P_{COMP}; in other words, the cw 1 dB compression point does give a useful relative indication for device IM3 comparison. Unfortunately, the −25 dBc criterion represents an IM3 level that is much higher than many system applications will tolerate.

9.4   Envelope Analysis with Variable PAR

So far, the envelope amplitude distortion has only been analyzed using a two-carrier system as the excitation signal; this limits the peak-to-average RF power ratio to a single case of 3 dB. In this section, the analysis will be extended to explore the variation of distortion characteristics with more complicated envelopes, which have different peak-to-average power ratios. This is an important intermediate step to take, before considering some actual practical modulation systems such as OQPSK and DQPSK.

Table 9.3 Summary of Two-Carrier Envelope Simulation for the Three Device Compression Characteristics Shown in Figure 9.6 (All Powers in dB Relative to P_{LIN})

P_{SAT}	P_{COMP}	P_{1dB} (Two-Tone)	P_{OUT} for −25 dBc IM3 level
3	2.5	0.7	0.0
3	2.0	0.2	−0.5
3	1.5	−0.3	−1.0

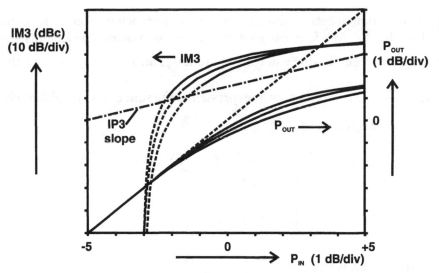

Figure 9.9 Two-carrier envelope IM simulation, showing effect of compression model (see Figure 9.6 for model details).

The easiest way in which the peak-to-average power ratio can be changed is to include one or more higher harmonics in the modulating signal,

$$v_s(t) = v \cdot \left\{ \cos(\omega_m t) + m_2 \cos(2\omega_m t) + m_3 \cos(3\omega_m t) + \ldots \right\} \cdot \cos(\omega t) \qquad (9.19)$$

so that by suitable choice of the number of harmonics of ω_m, and the value of the m_n coefficients, various kinds of envelopes can be investigated. [Note that now the m_n coefficients represent the envelope of the original signal, as distinct from the μ_n distortion coefficients in (9.17) and (9.18).]

One issue, which is often the subject of idle speculation, is the case of envelopes which have very high PARs due only to the random and very occasional occurrence of the high peaks. This is a common situation in dense multicarrier communications systems. The speculation concerns the appealing concept that if the high peaks are fast enough and rather infrequent, the heavy clipping which they will incur will not be "noticed." This issue will be dispatched in due course, but initially a few cases will be considered where just the second harmonic of ω_m is added. This extends the possible range of PAR up to a factor of 4 (6 dB). So the signal under consideration can be represented as

$$v_s(t) = v \cdot \left\{ \cos(\omega_m t) + m_2 \cos(2\omega_m t) \right\} \cdot \cos(\omega t) \qquad (9.20)$$

which has a peak voltage amplitude of $v \cdot (1 + m_2)$, so the corresponding peak power is

$$P_{pk} = \left(\frac{v^2}{2} \right) \left\{ 1 + m_2 \right\}^2 \qquad (9.21)$$

As before, the mean power can be determined either by direct integration of the squared RMS voltage amplitude or by expanding the time domain function into its frequency components. In this case, the latter approach is taken. Inspecting (9.20), it is clear that this is a four-carrier system, with symmetrical sidebands appearing at frequencies $\omega \pm \omega_m$ having voltage amplitude $\dfrac{v}{2}$, and at $\omega \pm 2\omega_m$ having voltage amplitude $\dfrac{v \cdot m_2}{2}$. So the mean power, expressed as the sum of the individual powers at all four frequencies is

$$P_m = \frac{1}{2}\left\{\left(\frac{v}{2}\right)^2 + \left(\frac{v}{2}\right)^2 + \left(\frac{v \cdot m_2}{2}\right)^2 + \left(\frac{v \cdot m_2}{2}\right)^2\right\}$$

$$= \left(\frac{v^2}{4}\right)\left(1 + m_2^2\right)$$

(9.22)

So the PAR is given by

$$\frac{P_{pk}}{P_m} = 2 \cdot \frac{\left(1 + m_2\right)^2}{1 + m_2^2}$$

(9.23)

A representative set of values for the PAR, in dB, is given in Table 9.4.

Some of these envelopes are plotted out, as voltage magnitudes, in Figure 9.10. All four envelopes shown in Figure 9.10 represent the same mean RF power; clearly, those with higher peak values have amplitude functions which spend proportionately more time below the mean value.

The amplifier transfer characteristic can now be applied to these envelopes, using a chosen mean power level. Figure 9.11 shows the case which corresponds to the intermediate amplifier compression characteristic in Figure 9.6, and a mean power level which corresponds to the cw 1 dB compression point. Envelope (a) in Figure 9.10 has a peak power level which is 6 dB above the mean level, and will thus swing 6 dB into the compression region. The resulting envelope, shown in Figure 9.11(a) has most of its peak clipped away, but the remainder of the envelope, comprising about two thirds of the envelope cycle, is essentially undistorted. Lower peak values incur less peak distortion, but the distortion occurs for a longer proportion of the cycle.

Table 9.4 Representative Set of Values for Peak-to-Average Ratio (Signal Defined by (9.20))

m_2	$\dfrac{P_{pk}}{P_m}(= \text{PAR, dB})$
0	3 dB
0.13	4 dB
0.25	4.7 dB
0.32	5 dB
1.0	6 dB

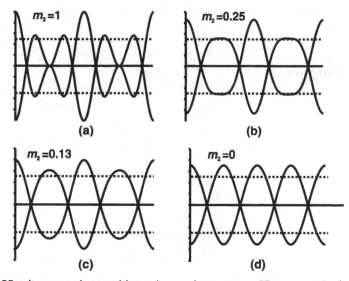

Figure 9.10   RF voltage envelopes with varying peak-to-average RF power ratios but equal mean power levels: (a) 6 dB; (b) 4.7 dB; (c) 4 dB; and (d) 3 dB. Dotted lines indicate voltage amplitude corresponding to mean power level.

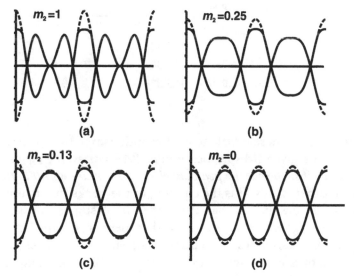

Figure 9.11   RF voltage envelopes from Figure 9.10, showing gain compression on envelope peaks.

Figure 9.12 shows the corresponding third order intermodulation levels, where it can be seen that despite the shorter distortion time, the higher peak envelopes show higher IM3 at the same measured mean power levels. Unfortunately, this contradicts the speculation that short, high peak envelope excursions will somehow go unnoticed. The key issue is that waveforms with high PARs have more power locked up in the high peaks, so conversely if the peaks are clipped then it has to be expected that the effect will be substantial. Table 9.5 summarizes the amplitude distortion results in Figure 9.12.

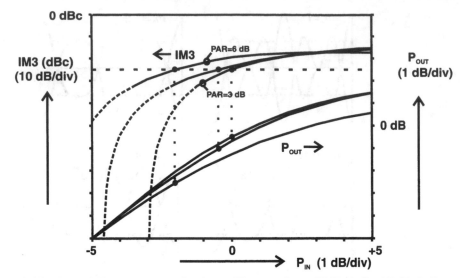

Figure 9.12 Two-carrier IM response for three different values of PAR (6 dB, 4.7 dB, 3 dB cases from Figure 9.11).

Table 9.5 Summary of Envelope Simulations for Variable P_{pk}/P_m

m_2	Peak/Av	IM3	P_{IN}	P_{OUT}	Compression
0	3 dB	−25 dBc	0.0	−0.5	0.5
0.25	4.7 dB	−25 dBc	−0.4	−1.0	0.6
1.0	6.0 dB	−25 dBc	−2.0	−2.6	0.6

Once again, there is little scope for optimism in these results. A basic rule seems to apply for a given IM specification: *as PAR varies, the mean power output has to be backed off such that the peak envelope power swings up to the same compression point.* In other words, it is the clipping of the peaks that mainly determines the IM levels. These results are based on the simplest possible simulation of variable peak-to-average envelopes and show that there is perhaps 1 dB of leeway in making power backoff estimates; the 6 dB peak signal has to be backed off by a further 2 dB in order to obtain the same IM3 level as the 3 dB peak signal having the same mean power.

9.5 AM to PM Effects

AM-PM effects seem to have something of a poor-relation status in the RF power amplifier world. Amplitude distortion is well understood, and has an established litany of rules of thumb and design guidelines. But phase distortion effects seem more inscrutable, despite their obvious potentially disruptive effects in phase modulated systems. One factor is that until recently, AM-PM was a tricky measurement to perform on an RF power amplifier. Older vector network analyzers had to be calibrated at very low fixed power levels, and a change in the level could produce a phase error greater than the amplitude dependent phase change of the device under test. Fortunately, modern instruments are much less susceptible to these effects, and the phase of the transfer characteristic can be confidently swept over the same power range as the P_{in}-P_{out} characteristic. But there is still an almost complete absence of calibration points, such as intercept or 1 dB compression points, which provide a rough roadmap for amplitude distortion effects.

There is also a good deal of mystery about exactly what causes AM-PM distortion in the first place. Clipping on supply rails can, as we have shown, do quite a reasonable job of explaining gain compression, but it is not clear where phase distortion comes from, even when looking at well-clipped voltage and current waveforms. A good starting point, therefore, is to look at some measured AM-PM data. Figure 9.13 shows data taken for the actual amplifier designed and built in Chapter 2. Although originally designed as a Class A amplifier, AM-PM data has been taken for two Class AB conditions.

As might be expected, the best behavior is exhibited by the Class A condition. The phase characteristic remains almost constant, or within a degree or so (which may be the measurement limit), and only starts to climb when the drive level is at about the 1 dB compression point. Although the maximum AM-PM excursion is shown to be about 25°, this corresponds to a fully saturated condition. As the device is biased down into deeper Class AB operation, the AM-PM phase change starts at a lower power level, well back into what would appear to be the linear region based on the amplitude characteristic. Indeed, the deep AB condition (AB$_2$) shows an almost linear phase change versus dB drive level up to the compression point, where it starts a more rapid reversal. So here is a final twist in the reduced conduction angle PA story; it seems that deep AB operation may cause substantial AM-PM problems in the precompression zone. Such curves as are shown in Figure 9.11 are quite typical for a device such as a MESFET which has a varactor-like input junc-

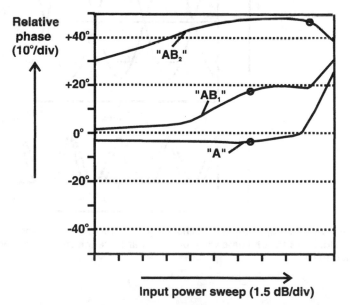

Figure 9.13 Measured AM-PM for 1 watt 1.9 GHz PA. (Circles indicate input drive level for 1 dB compression.)

tion. Insulated-gate type devices such as LDMOS will typically be better behaved, although can show significant AM-PM which is derived from a larger nonlinearity in the output capacitance. It pays to perform such measurements using a source-pull test setup, especially during device evaluation and characterization, since unlike the output compression characteristics, AM-PM behavior usually shows a strong dependency on the input impedance environment (see Chapter 12).

A detailed analysis of the causes of AM-PM behavior, such as shown in Figure 9.12 is outside the present scope.[2] In general, AM-PM effects can be traced to the signal-level dependency of several key transistor model elements. For FETs the input capacitance and both the depletion and junction resistance of the gate-source diode can be primary culprits. It should be noted that nonlinear resistance, in the presence of linear reactance, can cause AM-PM effects, just as much as nonlinear reactances. For BJTs, the nonlinear base-collector capacitance adds an important additional source of drive-dependent phase shift. All of these effects are detailed, interactive, and highly complex in themselves, and pose great challenges for physical modeling, such as that which is more successfully employed for compression and clipping effects. So the only way that AM-PM can be treated in a concise manner is to resort to empirical describing functions fitted to physical measurements.

Figure 9.14 shows, in a conceptual manner, the important features of the AM-PM behavior for a device under variable amplitude envelope excitation. Even with no attempt to model the precise shape of the phase characteristic a key feature emerges. The phaseshifts which occur at envelope amplitude maxima have twice the fundamental frequency of the double sideband modulation, as indicated by the

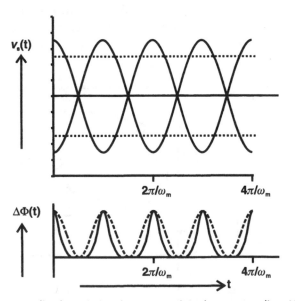

Figure 9.14 Envelope amplitude variation (upper trace) and corresponding AM-PM phase variation (solid, lower trace). Phase variations occur at twice the amplitude modulation frequency (dashed trace).

2.   See Chapter 4 for further discussion on the effects of nonlinear input varactor effects.

dashed sinewave. For simplicity, therefore, the initial analysis will assume an RF output signal which has the form

$$v_s(t) = \cos(\omega_m t) \cdot \cos\left\{\omega t + \frac{\phi}{2} \cdot (1 + \cos 2\omega_m \cdot t)\right\} \tag{9.24}$$

where ϕ is the peak magnitude of the AM-PM phaseshift under the conditions being considered, and it is assumed that $\phi \ll 1$ radian. So as the envelope amplitude reaches either a positive or negative amplitude peak, the phase of the RF carrier shifts by ϕ radians. Equation (9.24) can therefore be expanded to give

$$v_s(t) = \cos(\omega_m t) \cdot \{\cos \omega t \cdot \cos \Psi - \sin \omega t \sin \Psi\}$$

$$\text{where } \Psi = \frac{\phi}{2} \cdot (1 + \cos 2\omega_m \cdot t)$$

so using the approximation that $\cos \Psi \approx 1$, $\sin \Psi \approx \Psi$

$$\begin{aligned}
v_s(t) &= \cos(\omega_m t)\left\{\cos \omega t - \frac{\phi}{2}(1 + \cos 2\omega_m t)\sin \omega t\right\} \\
&= \cos(\omega_m t)\left\{\cos\left(\omega t - \frac{\phi}{2}\right) + \frac{\phi}{2}\cos 2\omega_m t \cdot \sin \omega t\right\} \\
&= \cos(\omega_m t)\left\{\cos\left(\omega t - \frac{\phi}{2}\right) + \frac{\phi}{4}\sin(\omega + 2\omega_m)t + \frac{\phi}{4}\sin(\omega - 2\omega_m)t\right\} \\
&= \cos(\omega_m t)\left\{\cos\left(\omega t - \frac{\phi}{2}\right) + \frac{\phi}{4}\sin(\omega + 2\omega_m)t + \frac{\phi}{4}\sin(\omega - 2\omega_m)t\right\} \\
&= \frac{1}{2}\cos\left\{(\omega + \omega_m)t - \frac{\phi}{2}\right\} + \frac{1}{2}\cos\left\{(\omega - \omega_m)t - \frac{\phi}{2}\right\} \\
&\quad + \frac{\phi}{8}\left\{\sin(\omega + \omega_m)t + \sin(\omega + 3\omega_m)t + \sin(\omega - \omega_m)t + \sin(\omega - 3\omega_m)t\right\}
\end{aligned} \tag{9.25}$$

So the most significant effect of the AM to PM distortion is to generate additional terms at the same frequencies as the normal third order intermodulation products, $\omega \pm 3\omega_m$. These additional terms, having a magnitude of $\frac{\phi}{8}$, will be added vectorially to the third order intermodulation products, generated by amplitude distortion. The amplitude distortion terms can be determined, following the formulation of (9.24), by adding odd order amplitude distortion terms, giving

$$\begin{aligned}
v_s(t) &= \{\cos \omega_m t + \mu_3 \cos 3\omega_m t + \mu_5 \cos 5\omega_m t + \dots \} \\
&\quad \cdot \cos\left\{\omega t + \frac{\phi}{2} \cdot (1 + \cos 2\omega_m \cdot t)\right\}
\end{aligned} \tag{9.26}$$

Expanding just the term for the third order amplitude distortion, $\mu_3 \cos 3\omega_m t$, gives

$$
\begin{aligned}
v_s(t) = {} & \frac{1}{2}\cos\left\{(\omega+\omega_m)t - \frac{\phi}{2}\right\} + \frac{1}{2}\cos\left\{(\omega-\omega_m)t - \frac{\phi}{2}\right\} \\
& + \frac{\phi}{8}\left\{\sin(\omega+\omega_m)t + \sin(\omega+3\omega_m)t + \sin(\omega-\omega_m)t + \sin(\omega-3\omega_m)t\right\} \\
& + \frac{\mu_3}{2}\cos\left\{(\omega+3\omega_m)t - \frac{\phi}{2}\right\} + \frac{\mu_3}{2}\cos\left\{(\omega-3\omega_m)t - \frac{\phi}{2}\right\} \\
& + \frac{\phi\mu_3}{8}\left\{\sin(\omega-3\omega_m)t + \sin(\omega+5\omega_m)t + \sin(\omega-3\omega_m)t + \sin(\omega-5\omega_m)t\right\}
\end{aligned}
\tag{9.27}
$$

Assuming that the cross term can be ignored in comparison to the others, the third order intermodulation products each have a magnitude which is represented by

$$
v_{im3} = \frac{\mu_3}{2}\cos\left\{(\omega \pm 3\omega_m)t - \frac{\phi}{2}\right\} + \frac{\phi}{8}\sin(\omega \pm 3\omega_m)t
\tag{9.28}
$$

This shows the principal components of the third order IM products, with representations from both amplitude and phase distortion mechanisms. Note that under the ongoing assumption of small ϕ, the phase distortion contributions will remain approximately in quadrature with the amplitude ones. So there is no possibility of significant cancellation occurring between the components generated by the two distortion mechanisms. The AM-AM and AM-PM components will always combine to produce IM3 sidebands which have a higher amplitude than either of these individual parts. Experimental measurement of the relative contributions of amplitude and AM-PM contributions requires the measurement of the relative phase of the IM sidebands; this is not a straightforward task.

Equation (9.28) also enables a quantitative estimate to be made on the relative impact, or importance, of the AM to PM distortion. With AM distortion acting alone, a C/I (carrier-to-intermod ratio) of 25 dB would correspond to a value of $\frac{\mu_3}{2}$ = 0.056, so that an equal AM to PM contribution would indicate a value of $\theta = 0.45$ radians, or about 25°. As can be seen in Figure 9.13, this is very much in line with the kind of AM to PM phase variations which can be measured in practice, even in the precompression region. If AM effects are ignored completely, it is worth noting that even one degree of AM to PM distortion can cause IM3 at a level of −53 dBc, and this will rise according to a law of 6 dB per doubling of the phase distortion. In practice, this can give rise to a situation in the precompression region where AM to PM effects can have an important impact on IM3 levels, and can easily dominate higher order IM levels. We will return to these computations in Chapter 14, where the point is made that in order to reduce IM3 levels by any more than about 10 dB, a linearization device must correct AM-PM as well as the more tangible AM-AM effects, and this correction needs to be effective at well backed-off drive levels and for phase errors measured in terms of just a few degrees.

Equation (9.28) also implies a departure from conventional wisdom concerning the frequent observation of asymmetrical IM sidebands. This phenomenon is correctly attributed to the presence of both AM to PM and AM to AM, but the explanation frequently goes astray in asserting that the upper and lower AM to PM sidebands have opposite polarity. Careful examination of (9.27) and (9.28) does not reveal any difference, or asymmetry, between the upper and lower IM3 sideband coefficients, even though the analysis takes account of both amplitude and phase components. The actual explanation of this effect requires an additional factor that has not been included. Referring back to Figure 9.14, it has been assumed all along in this analysis that the AM distortion and PM distortion mechanisms occur with perfect time synchronism. Both are described in (9.24) as cosine functions at the modulation frequency ω_m, with no phase offset between them. Due to the low frequency represented by ω_m, it seems a reasonable assumption that even though the AM and PM distortion effects arise through different physical mechanisms in the device, any time delays will be measured on a timescale that is negligibly small in comparison to a modulation cycle. In some cases, and for some types of device, this may actually be incorrect. Surface states and dynamic thermal effects can sometimes introduce a significant time hysteresis measured in microseconds or even longer. This can cause the idealized amplitude or phase characteristic in Figure 9.14 to show some asymmetry, which in turn causes both in-phase and quadrature terms to be generated in the subsequent analysis. So a more general version of (9.26) would have to be used to include these effects,

$$v_o(t) = \left[a_1 \cos(\Omega t + \Delta) + a_3 \cos 3(\Omega t + \Delta) + \ldots \right]\left[\cos\{\omega t + \Phi \cos(2\Omega t)\} \right] \quad (9.29)$$

where Δ represents the effective phase difference between the mechanisms causing AM and PM. The full expansion of (9.29) gets a little involved, and we quote here the result (see [5], Chapter 3 for the full analysis), which gives the magnitude of the upper ($IM3_{USB}$ at frequency ω_{3U}) and lower ($IM3_{LSB}$ at frequency ω_{3U}) IM3 sidebands:

$$
\begin{aligned}
IM3_{USB} &= \left\{ \frac{a_3}{2} \cos 3\Delta + \frac{a_1 \Phi}{4} \sin \Delta \right\} \cos \omega_{3U} + \left\{ \frac{a_1 \Phi}{4} \cos \Delta - \frac{a_3}{2} \sin 3\Delta \right\} \sin \omega_{3U}, \\
IM3_{LSB} &= \left\{ \frac{a_3}{2} \cos 3\Delta - \frac{a_1 \Phi}{4} \sin \Delta \right\} \cos \omega_{3L} + \left\{ \frac{a_1 \Phi}{4} \cos \Delta + \frac{a_3}{2} \sin 3\Delta \right\} \sin \omega_{3L}
\end{aligned}
\quad (9.30)
$$

Evaluation of (9.30) shows that quite small values of Δ will cause significant asymmetry. Even if Δ is negligible at the device level, there are additional mechanisms in the external circuitry of most amplifiers which can contrive to make Δ have a nonzero value. The most important of these is remodulation of the amplifier output caused by supply rail voltage variations. These are induced by the varying current drawn from the DC supply by a Class AB type of amplifier, and have time constants which are significant at envelope modulation timescales. These effects have some important practical ramifications for bias network and decoupling design, and will be discussed in detail in Chapter 11. It is also worth noting that the above analysis is restricted to a special case, where AM-AM and AM-PM distortion are caused by unique and orthogonal physical mechanisms. It can be shown more generally that different distortion mechanisms, operating with different time constants, can cause asymmetrical IM sidebands whether the distortion is uniquely AM-AM or AM-PM in nature.

9.6   PA Memory Effects

Any RFPA will show some dynamic deviations from its static characteristics. Such deviations have become known, for better or for worse, as "memory effects." These effects are very troublesome for the process of predistortion, as will be discussed in Chapter 14. Memory effects are an additional source of nonlinear behavior that is typically not accounted for in PA models, and represent a source of error in attempts to simulate the distortion characteristics of any PA. Memory effects can be traced to three main causes:

- Dynamic thermal effects;
- Unintentional modulation on supply rails;
- Semiconductor trapping effects.

The second of these is probably the most common cause of asymmetrical IM sidebands, and can be potentially cured by more attentive design of the biasing networks to the PA stage. Inasmuch as a whole chapter is devoted to this subject (Chapter 11), this effect will not be discussed here. Suffice it to say that a circuit simulator should be able to predict such "memory effects," provided that the designer can be bothered to include sufficient details of the bias circuitry, as well as the RF circuit itself, in the simulation file.[3]

"Trapping effects" is another unsatisfactory generic term, which at worst can be used to categorize just about any anomalous effect which is observed in a semiconductor. The external manifestation is usually most apparent when observing the current in a device whose input voltage is switched from below cutoff to a value which places it into its linear conduction region. The corresponding current can show a response which is characterized by several time constants: the expected "fast" response from an RF device which may be measured in picoseconds, but some additional much longer time constants which can stretch the timescale required to reach a final steady state current into microseconds, milliseconds, or even seconds in extreme cases. A useful signature when making such a measurement is a strong sensitivity of the response to light.

The more exotic the semiconductor, the more likely the effects will be troublesome. GaAs device manufacturers have struggled with this problem for three decades and although the reputable suppliers seem to have eliminated most of the visible manifestations, it is still a very clear and present danger. Trapping effects defy analytical, or even behavioral, modeling, and can thus cause major headaches in the development of predistortion algorithms. The best strategy, it seems, is to use a device technology, and a supplier, who appear to be free of this affliction.

Dynamic temperature effects undoubtedly play a part in PA distortion mechanisms [6, 7], yet the topic seems to generate perennial controversy. The central theme in the ongoing debates seems to be whether heat moves around fast enough to cause changes on a timescale that will give observable nonlinear effects. The short answer seems to be yes, it can, in particular when the timescale is being measured in

---

3.   This is actually where time domain simulators such as Spice, show some benefits over the slicker harmonic balance engines; this in turn may explain why designers seem frequently not to include the bias network in their simulations.

a typical signal modulation domain of anywhere from tens of nanoseconds up to milliseconds. This debate is best cut short by doing a simple calculation.

Figure 9.15 shows a one-dimensional model of an RF power transistor chip. The heat equation can be solved for the case when a source of heat H, which is assumed to be uniformly distributed across the upper surface, is turned on at time $t = 0$. We assume that the underside of the chip is perfectly attached to an infinite heatsink which is maintained at ambient temperature, for convenience this temperature will be normalized to zero. Taking a good engineering undergraduate mathematics textbook [8] as a fast-track path to the solution, we obtain the following expression for the temperature $\theta(z, t)$ at time t and distance z from the heatsink:

$$\theta(z,t) = H\frac{z}{k} - \frac{8HL}{k\pi^2} \sum_{nodd} \frac{(-1)^{(n-1)/2}}{n^2} \sin\left(\frac{n\pi z}{2L}\right) \exp\left(\frac{-k\pi^2 n^2 t}{4L^2 s\rho}\right) \quad (9.31)$$

where k is the thermal conductivity, s is the specific heat, and ρ the density of the semiconductor material. Clearly, there is a time constant τ, where

$$\tau = \frac{4L^2 s\rho}{k\pi^2}$$

which characterizes the dynamics of the heat diffusion. The temperature finally settles out to the steady-state solution of a linear temperature ramp, after a few units of τ, as shown in Figure 9.15. These parameters are listed in Table 9.6, for GaAs, Si, and SiC[4]:

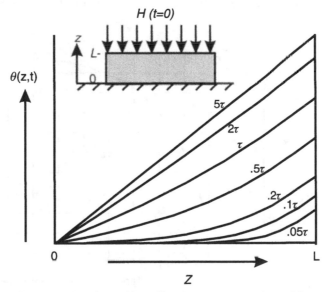

Figure 9.15   One-dimensional analysis of heat flow in a power transistor chip.

---

4.   Gallium Nitride (GaN) devices are usually fabricated using either Si or SiC substrates, so the SiC column would be applicable.

Table 9.6   Semiconductor Material Properties

	GaAs	Si	SiC
k (thermal cond. cal/cm-sec-K)	0.1	0.35	0.8
s (specific heat, cal/g-K)	0.1	0.17	0.7
ρ (density, gm/cc)	5.3	2.3	3.2

Taking a die thickness of 50 microns for GaAs, τ comes out to be about 50 *micro*seconds. Silicon works out to be about 5 times faster, at 10 μsec, and SiC somewhere between, around 25 μsec. The square-law dependency on die thickness can significantly reduce these values; even 20 microns is common for GaAs power transistor manufacturers when through vias are used. An alternative way of plotting (9.31) is shown in Figure 9.16. Here the temperature at the heat source level is plotted against time, in units of τ. This curve has a significantly sharper turn-on than an exponential characteristic in the t /τ <1 region. For example at a time t = τ/ 10, the temperature has already risen to 20% of its final steady-state value. The key issue in RF applications is the statutory assumption that thermal effects are very *slow* in comparison to the variations in the heat source H. For once, faster means more trouble. Figure 9.16 shows that based on the above estimates of τ, there is a possibility of maybe 10° of temperature variation on a timescale of about 1 microsecond, which in turn may cause about 0.1 dB of gain variation for most varieties of RF semiconductor [9].[5] Once into the microsecond region, we are well into the modulation time domain of typical communications signals.

It is therefore relevant to consider the other half of the problem, which is the expected time variation of the heat source H. It is a feature of Class AB power operation that the power dissipated in any RF power transistor will vary considerably

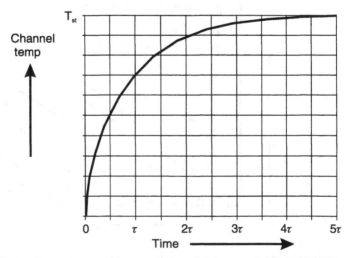

Figure 9.16   Dynamic temperature characteristic at heat source [$z = L$ in (9.31)].

---

5.   This estimate is based on the assumption that the power dissipation operating limits for most transistors allow for an increase of at least 50°C above ambient for the active region.

with the input signal envelope amplitude, and on the same timescale. It is a straight-forward modification to the classical Class AB analysis presented in Chapter 3 to plot the variation of power dissipation versus input drive level, for various Class AB settings. This is shown in Figure 9.17, which can be considered to be a direct deriva-tive from Figure 3.11. As would be expected, the extremes of Class A and Class B show the power dissipation heading in different directions as the drive level is increased. There is however an intriguing condition, corresponding to a quiescent setting of about 10%, where the dissipation remains virtually constant for any drive level.

It should be noted that the power dissipation scale in Figure 9.17 has been nor-malized to the ideal RF output power in Class A or Class B operation. This is a con-venient normalizing parameter, but in practice this amount of dissipation would take most higher power RF transistors well beyond their specified maximum rat-ings.[6] So the lower half, or even third, of the total plotted dissipation range will still represent a dissipation level which may cause some tens of degrees of temperature increase under steady state conditions. There appears to be a sensitive range of qui-escent settings in the deep Class AB region where the dissipation function can devi-ate rapidly from a flat characteristic.[7] It is still regarded as a matter of conjecture as to whether the linearity "sweet spot" which is frequently observed on the test bench is caused by this effect. It is likely to be an important component in the linearity behavior of most RFPAs, and will certainly come into play in more stringent appli-cations which require ACP and IM levels to be below the −50 dBc levels.

One popular technique for assessing the presence of memory effects is to per-form a dynamic gain measurement, where the gain of the amplifier is measured on the fly with the relevant modulated RF signal as the input excitation. This turns out to be quite a lot more difficult than it may at first appear, and such results need to be

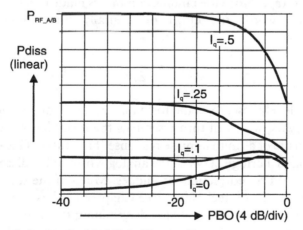

Figure 9.17   Device dissipation for ideal Class AB operation.

6.   In other words, high power RF transistors cannot be assumed to run safely at full Class A bias; check this out at your peril.
7.   The characteristics in Figure 9.17 represent ideal Class AB behavior, and the actual value of the critical set-ting will vary from one device type to another, and also with frequency band.

treated with some caution. Commercial vector signal analyzers (VSAs) are typically not capable of performing a simultaneous two-port measurement and have limited bandwidth. In order to see nonlinear effects that cause IM levels below the −40 dBc level, the resolution of the gain measurement has to be at the 0.01 dB level, which requires a more innovative approach than the simplest application of a commercial VSA.[8] There is an element of "Catch-22" in this area. Such high resolution typically requires the averaging of a large number of measurements, and this is difficult when the excitation itself has a random nature. Spectrum analyzers can resolve down to this degree of resolution because they are in effect performing a massively averaged measurement through the use of narrowband IF and video filters.

Some further comments on dynamic gain measurement and some data will be presented in Section 9.8. The impact of memory effects on the "predistortability" of an RFPA has been the main reason for the recent emergence of these effects as a research topic, and will be further addressed in Chapter 14. As already stated, unintentional bias modulation is probably the most common cause of the most serious memory effects observed in practice, and will be given more detailed attention in Chapter 12. This section has so far addressed the various possible physical causes of memory effects in PAs. From a modeling viewpoint, bias modulation and thermal effects would clearly lend themselves to a physically based modeling approach, but anomalous semiconductor effects are a very different matter. Possibly for this reason the "top-down," or behavioral, approach seems to have been much favored for modeling and correcting PA memory effects. It is a fairly straightforward matter to extend the power series representation, discussed in some detail in the early sections of this chapter for memoryless PA modeling, to include the possibility of memory effects. Indeed, it is the inclusion of memory effects into the power series formulation that elevates it into a fully fledged Volterra series. Essentially, the device output is expressed as a sum of polynomial series, each using a progressively delayed value of input voltage. Notwithstanding comments made in the preface about double Σs, it is actually worth the effort of mentally unraveling this more generalized formulation,

$$v_o(t) = \sum_k \sum_n a_{k,n} v_{in}^n (t - k\Delta)$$

This closely resembles the formulation for a convolution integral over discrete time intervals, which in turn resembles the formulation used in DSP to define the characteristic of a finite impulse response (FIR) filter. This expression thus serves as something of a bridge between the analog world of PA distortion effects and the digital world of DSP correction techniques. Much of the recent development work on digital predistortion techniques (see Chapter 14) has been focused on streamlining the number of a_k terms, and also determining the appropriate choice of the "tapping" interval, Δ. An obvious starting point is to use a value for Δ that corresponds to the sampling rate used to generate the baseband modulation signals. This can do a reasonable job in modeling bias modulation, but poses problems in representing the

---

8. Clearly, VSAs are just fine if the effect being explored causes gain variations of the order of a few tenths of a dB or higher; this can occur very easily in cases of supply rail modulation.

longer time constants which characterize thermal and semiconductor effects. The recognition that memory effects start to dominate PA nonlinear behavior at the extreme levels of linearity that modern standards demand has been a major roadblock in the otherwise rapid progress of digital linearization methods. Minimization of PA memory effects, through semiconductor process and circuit development, has thus become an important area for ongoing research.

9.7 Digital Modulation Systems

9.7.1 Introduction to Digital Modulation

As already stated in the introduction to this chapter, this section is not a treatise on digital modulation. This subject has been well covered elsewhere and is the subject of many dedicated books (see bibliography). The goal here is to draw attention to the key issues which affect the design and operation of typical PA designs using a few standard modulation schemes encountered in wireless communications systems. The most important of these are the GSM (GMSK) system used throughout Europe, the 2.5G EDGE derivative which is finding more widespread use than the GSM replacement for which it was originated, and the various generations of CDMA implementation, including the 3GPP WCDMA system. Orthogonal Frequency Division Multiplex, or OFDM, is a more recent addition and is now standardized in various 802.11 and 802.16 variants aimed at Wireless Local Area Networks (WLAN) and high speed data access (WiFi, WiMax). Unfortunately, the standards associated with all of these systems are frequently updated, and whole new systems are being created at a bewildering pace.

When making the step from analog to digital modulation, it is usually necessary to make a conceptual step from mathematical functionality and analytical certainty to random sequences and statistical probabilities. In principle, it is still possible to employ the same envelope approach to estimate the distortion levels in RF power amplifiers as was used earlier in this chapter to analyze simpler modulation formats. The key word, however, is *estimate*. This can be a little frustrating to the RF designer. Given a power transfer characteristic, and an envelope function (such as in Figure 9.9), it is possible to state, with full mathematical certainty, what the IM distortion levels, and even phases, will be; the envelope repeats every modulation cycle. But a QPSK signal based on pseudo-random bit sequences will never repeat, no matter how long the timescale.[9] One enters a world where normal RF engineering concepts of certainty and causality become unsatisfactorily blurred. The result of this is that most discussions involving nonlinear amplification of these signals seem always to retain a qualitative element. Simple questions such as how much power backoff is needed to meet a given spectral mask specification are never given hard, definitive answers.

Needless to say, CAD tools which purport to *give* such definitive answers are widely available. In performing the most critical function of predicting the ACP

---

9. This is not quite true of course when using a commercial signal generator, which will have a finite length to its internal pseudo random bit generators, and for this reason some care has to be exercised in assuming "real world" randomness will give the same results.

response of a PA using a stipulated modulation format as the excitation, the key problem is how to model the PA characteristics. Most commercial packages still use the envelope simulation approach, discussed and demonstrated in Section 9.3 for two-carrier, and multicarrier, signals. These packages also, for the most part, make the quasi-static assumption, that the PA can be entirely characterized by its static AM-AM and AM-PM response, usually cast into a power series format. The power series will have maybe around nine complex coefficients for a ninth degree model. This has been shown to have sufficient complexity to model most PA characteristics [5].

Given these PA characteristics, and the quasi-static assumption, it is actually quite a straightforward process of computation to determine the spectral distortion. The problem, and a wonderful source of income for software and instrument vendors, is the creation of the signal itself. RF engineers are reluctant to spend many hours digging into the vast mountains of regulatory documents that actually define the modulation standards. One of the features of this chapter is to demonstrate the use of some Excel spreadsheets, which have a representative data listing of a few important modulation systems, from which useful quantitative tasks and calculations can be performed. These spreadsheets are not intended as a substitute, or competition with, the various professional software packages and CAD simulators which are now available, which can perform some of the same tasks with greater precision. It should however always be recognized that the chain is only as strong as its weakest link, and the weak link in this zone is the PA model and the quasi-static assumption.

9.7.2 QPSK Modulation Systems

Quadrature Phase Shift Keying, or QPSK, is a modulation system which uses the phase of the RF carrier to define a number of states, each of which is assigned a binary symbol. A Binary Phase Shift Keying (BPSK) system has just two phase states, denoted by a 180° phase shift, and in this simplest case each state would be a binary 1 or binary 0, respectively. QPSK systems, by definition, use four phase states, so each state can denote a 2-bit symbol. 8PSK and 16PSK systems are also widely used in the various mobile communications systems.

Digital modulation systems are normally defined using an IQ diagram, such as that shown in Figure 9.18. Each point on the diagram represents a "constellation point," having a designated magnitude and phase, and is designated as a digital "symbol." This defines an important class of modulation systems which use constant amplitude symbol states. The identity of each symbol is then entirely determined by its phase angle. Since the phase modulation is normally implemented using a quadrature combining scheme, it is conventional to represent the Cartesian components of any point on the constellation diagram as a pair of "I" and "Q" values, representing the in-phase and quadrature components.

The trajectory which is followed during a symbol transition is critical in determining both the spectrum of the final signal and also the envelope amplitude variation. A diagram such as Figure 9.18 does not give a realistic indication of these trajectories, or the time taken to make a transition in relation to the time spent at a constellation point. Clearly, if the transitions were made very quickly, and the signal

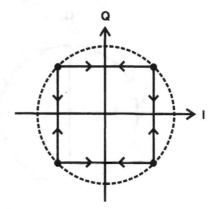

Figure 9.18 QPSK modulation.

spent most of the time sitting at points on a constant amplitude circle, the overall envelope variation would be minimal. This would be a most desirable situation from the PA viewpoint, but unfortunately such a system would have a very broad spectrum and would be most inefficient in terms of channel capacity.

Figure 9.18 represents, with the limitations already discussed, an "Offset Quadrature Phase Shift Keyed" system, or OQPSK. The significance of the "offset" term is that the phase transitions are constrained to move around the square of constellation points in either direction, but are not allowed to cross to a diametric point; this is achieved in practice by offsetting the I and Q channels by one half of a clock cycle. It should also be noted that successive symbols can be duplicated; movement does not necessarily take place every clock cycle. The "nonzero-crossing" restriction is of some benefit to the PA designer in that it restricts the dynamic range of the final RF envelope. This might mean, for example, that a deep Class AB condition could be considered for such a signal, since the small signal dropout of the transconductance (see Chapter 3) may be below the low point of the envelope. This benefit of nonzero crossing was seen as a major concession to the nonlinear devices in a system when digital modulation systems were being originally developed. Unfortunately, the advantage is quickly wiped out if there are multiple carriers present.

Figure 9.19 shows a different system, in which the phase transitions are constrained to move along a constant amplitude circle. So regardless of the speed of the transitions, constant envelope amplitude will be maintained. This system is clearly very favorable to PA design, since it is insensitive to amplitude distortion. With some minor additional qualifications, such a system can be regarded as a frequency shift keying system (FSK), and (with some further qualifications which are not detailed here) is representative of the modulation scheme used in the European GSM (Global System for Mobile Communications) mobile phone network. In principle, even the most highly nonlinear, saturated PA designs can be used in this system. There are actually some caveats in practice, such as imperfections in the low-cost implementation of the handset modulators which introduce some AM onto the signal, and the need to control the rise and fall times of the time-duplexed transmission bursts. Both of these can cause AM-PM in highly nonlinear amplifiers which in turn causes spectral distortion. The benefit of constant envelope also comes at a price in that the GSM system is less efficient than, say, the NADC (North

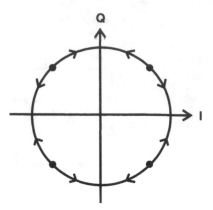

Figure 9.19   Constant amplitude QPSK ("GMSK").

American Digital Cellular) system in terms of channel capacity. As a result, the luxury of constant envelope operation has been short-lived and the next generation along this path, the EDGE (Enhanced Data Rates for GSM Evolution) system, is highly sensitive to PA linearity, as we will see later in this section.

A system which is something of a compromise between the conflicting requirements of high channel capacity and low envelope amplitude variation is the $\pi/4$ DQPSK system, shown in Figure 9.20. This is, in effect, two interlaced OQPSK signals having a $\pi/4$ offset between their phase constellations. Each successive symbol transition jumps from one constellation to the other, so there will always be a phase change at each beat of the symbol clock. Figure 9.20 shows the four possible transitions from each constellation point. The net result is that, as in the OQPSK system, there are no zero-crossing transitions; there will be either $\pm\pi/4$ or $\pm3\pi/4$ transitions. This system was used extensively in the first generation of cellular telephone systems, notably the NADC network, and the Japanese Personal Handi-Phone (PHP).

It has already been commented that if the phase transitions in all of these QPSK systems were abrupt, the signals would have very wide bandwidth. In practice it is necessary to restrict the bandwidth of the signal, using a suitable filtering process. The trick is to limit the bandwidth without compromising the information content. Such a trick makes its inventor justifiably famous, since it is rare indeed that Nature

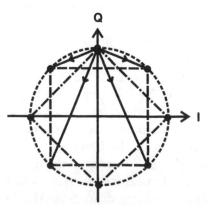

Figure 9.20 $\pi/4$ DQPSK (NADC) constellation diagram.

allows such an apparent contradiction in the so-called "Conservation of Grief" principle. The underlying theory behind the filtering of such signals was originally derived by Nyquist, and can be found in almost any communications theory text-book. The bandwidth can be restricted without any loss of information content, so long as certain prescribed filter types are used. An example of how this works can be seen in Figure 9.21. If the data stream is represented as a series of impulses, which are either "1" or "0," and this signal is fed into a filter whose impulse response has an undulating characteristic that crosses zero at each symbol point, it is clear that the filtering process will not generate any "inter-symbol interference," or ISI. In practice, this filtering is applied separately to the I and Q components of the final phase-modulated signal. The digital elements are more likely to be continuous, rather than impulsive, but the basic theory can be shown to remain intact. Although the filtering process will modify the waveform of the original signal quite consider-ably, the information remains untouched inasmuch as it can still be retrieved in its prefiltered form.

Such filters are called, reasonably enough, Nyquist filters. The filtering is nor-mally performed on the baseband data streams prior to upconversion to the RF car-rier frequency, so the filters will be low-pass types. The rolloff frequency characteristic of this filter is clearly critical in determining the bandwidth of the final signal, and hence the capacity of the system. It is also equally critical in determining the amount of envelope amplitude variation in the final signal. Figure 9.22 shows a typical result of Nyquist filtering. The sharp trajectories of the original QPSK tran-sitions have been smoothed out, but most significantly there is now *amplitude mod-ulation* on the signal. Even worse, the peak amplitudes, which result from certain critical combinations of symbol sequences, are significantly higher than the unity circle amplitude of the original signal.

The occurrence of high peaks is a statistical issue, and will extend the discussion started in Section 9.4 where some guidelines about the impact of clipping signal peaks were proposed. Here we have a more difficult problem, due to the noncyclic nature of the higher peaks. It is a feature of different modulation systems that some may display very high PAR peaks, but the peaks occur infrequently enough that it may be acceptable to allow the PA to clip them. Some versions of the WCDMA sig-nal formats come into this category, as does the 802.11/16 OFDM format. Other cases show lower PAR, but the occurrence of peaks is much higher. The EDGE sys-tem is such a case, and is worthy of further consideration.

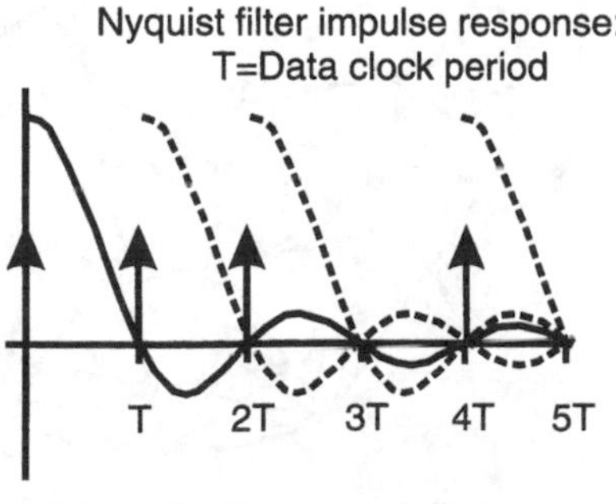

Nyquist filter impulse response:
T=Data clock period

Figure 9.21   Impulse response of Nyquist filter.

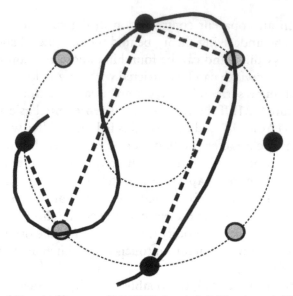

Figure 9.22   Effect of Nyquist filter on $\pi/4$ DQPSK signal. Solid line is post-filter trajectory, compared to ideal prefilter (phase-only) transitions (dashed line).

Figure 9.23 shows the prefiltered EDGE constellation. It is somewhat analogous to the $\pi/4$ DQPSK system, in that it is a pair of 8QPSK systems. Zero crossings are eliminated by inserting a $\pi/8$ phase rotation at each symbol transition. An important distinguishing feature is the use of a Gaussian filter whose characteristics are carefully prescribed so that an EDGE spectrum cannot be distinguished from that of a GSM signal. This has the effect of spreading the original constellation points all over the constellation space, as shown in Figure 9.24, which compares the effects of the Nyquist and Gaussian filter responses on the NADC and EDGE constellations. An

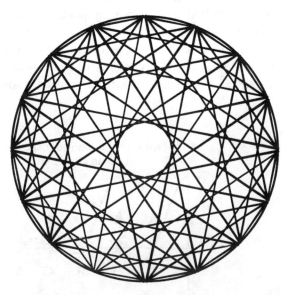

Figure 9.23   EDGE constellation.

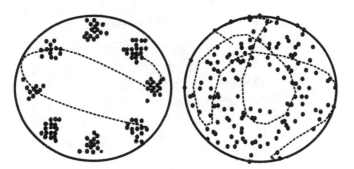

Figure 9.24   Effect of filtering on constellation points: root raised cosine filter in $\pi/4$ DQPSK system (left); and Gaussian filter in EDGE system (right).

actual EDGE IQ trajectory plot, shown in Figure 9.25 cuts a rather striking pattern, which has various void areas that are never visited. Such a plot indicates the cumulative journey that the signal vector makes over the IQ plane, but gives little indication of which areas are visited more than others. A "CCDF" (Cumulative Complementary Distribution Function) plot is the traditional format in which such signal statistics are quantified. Figure 9.26 shows a CCDF for an EDGE signal, where the probability of the signal having an instantaneous power higher than the selected x-axis value is plotted.[10] There is however a case for using the more humble bar chart for plotting the same information, as shown in Figure 9.27. Either way, it becomes clear that the EDGE signal is unusual in that the peak amplitudes show the highest statistical occurrence.[11] This makes linear EDGE PAs more challenging to

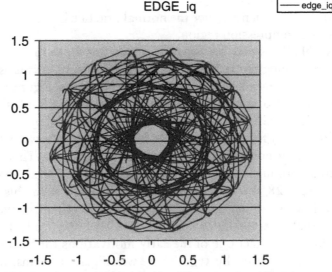

Figure 9.25   EDGE IQ trajectory (400 symbol transitions; see Excel file Edge_stats).

10.   CCDFs sometimes plot the vertical scale logarithmically; here a linear scale is used.
11.   The plots shown in this section were obtained using the Excel spreadsheet Edge_stats, available on the accompanying CD-ROM.

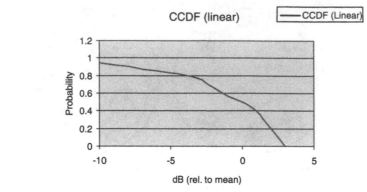

Figure 9.26   EDGE "CCDF" plot.

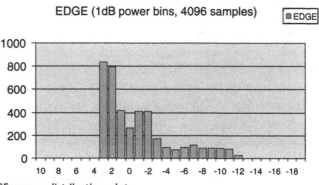

Figure 9.27   EDGE power distribution plot.

design for high efficiency, since the normal practice is to place the peak power levels well into the compression region.

Most QPSK systems use filters which do not move the constellation points. For example, returning to the NADC (π/4 QPSK) system, this uses a raised root cosine filter, giving a constellation trajectory plot shown in Figure 9.28. Here the amplitude statistics, also shown in bar chart form in Figure 9.28, show a marked difference from the EDGE plot, with a much lower statistical probability of a value in the PEP region. Partly for this reason, PAs for NADC handsets can be designed such that 1–2 dB range of power lies in the compression region. This gives good efficiency (commercial products were often quoted above 40%) while maintaining acceptable ACP. Figure 9.28 shows another key observation with this modulation system, which is that the actual peaks always occur directly in between the constellation points, so that a demodulator does not "see" the amplitude distortion on the signal peaks. This was in fact one of the early motivations for the development of QPSK modulation schemes of this type. There was an era, it seems, when the problems of nonlinear PAs were treated with higher priority.

9.7.3   CDMA and WCDMA

The introduction of the IS95 standard in the United States, its subsequent rapid displacement of the burgeoning NADC system, and its evolution into a worldwide

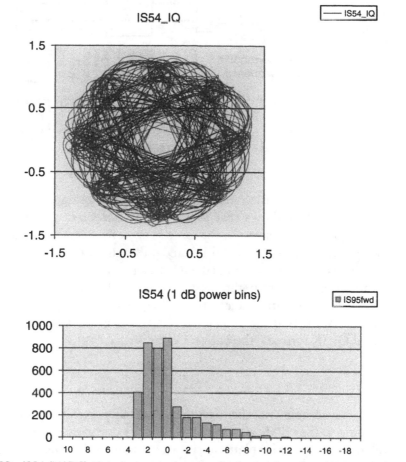

Figure 9.28 IS54 (NADC) IQ trajectory and power distribution bar chart (400 symbols, see Excel file IS54_stats).

mobile 3G ("Third Generation") communications system is one of the more remarkable stories in the recent history of radio communications. How a modestly clever technical idea from academia was used to plan and execute such a wide-ranging *coup d'etat* has been told elsewhere, and is not repeated here. The impact of this revolution has been felt in the RFPA field just as much as any other, but from the PA designer's viewpoint it is not essential to have a detailed understanding of all the digital magic that goes into the formulation of a 3G signal, and a detailed treatment of that subject is well beyond our present scope. It is however within the stipulated boundary conditions of this section to quantify some of the key technical aspects of CDMA ("Code Division Multiple Access") techniques, inasmuch as they have a direct bearing on PA design.

The basic underlying principle of a so-called "direct sequence" CDMA multiplexing system is illustrated in Figure 9.29. Each data stream is multiplied by a pseudo-random "chip" sequence. The chip rate will typically be many times higher than the original data bit rate, and can thus be considered to "spread" the data stream over a much larger bandwidth than its original form. At any point in the process, the original data stream can be retrieved by a suitable inverse process, provided the pseudo-random "key" sequence is known. The chipping process is, in effect, a

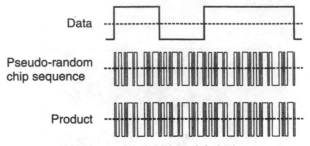

Figure 9.29   Direct sequence CDMA concept (BPSK modulation).

form of digital modulation and is usually encoded using one flavor or another of QPSK. Figure 9.29 shows that to an outside observer, a single channel chipped CDMA signal looks essentially the same as any other digital signal, albeit running at a much higher bit rate than the original data stream. But the key point in the CDMA process is that the various chipped data streams can now be literally added together, with full confidence that the digital keys will be able to unlock the individual components at any later stage in the process.

If we assume that the chipped data streams going into the summer each have standard QPSK formats, the output can be written as

$$S_{mult} = \sum_n \beta_n$$

where β_n is a complex number having one of four possible values: $+1, -1, +j, -j$, corresponding to the QPSK symbol states. The result can be represented as a matrix of possible states on the IQ plane, as shown in Figure 9.30. The signal jumps randomly around the points of the matrix, but the visits to the extremities of the matrix are rare, due to simple statistical considerations. Taking a typical value of $n=16$, the chances of a visitation to one of the extremities ($+16, -16, 16j$, or $-16j$) will be 2^{-30}, which can only be expected in a data stream at least 10^9 symbols in length. This statistical bias on the frequency of visits to different parts of the matrix strongly differentiates this system from the amplitude-phase symbol states of a QAM modulation scheme. Thus as the number of channels is increased, the probability of registering the highest possible peak values are very low. As with any digital modulation sys-

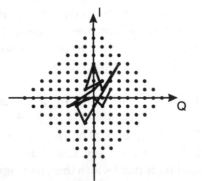

Figure 9.30   Constellation for eight summed CDMA chip streams.

tem, the digital signal is passed through a suitable bandlimiting filter which smoothes out the transitions, but retains the intersections at constellation points.[12]

So despite all the digital magic going on, the signal which gets finally upconverted to the RF band has an envelope amplitude which shows an *increasing PAR* as the number of digitally multiplexed data streams is increased. Indeed, it could be speculated that the final PAR using a single carrier, modulated with a multiplexed CDMA signal, may be comparable to that observed using a more traditional multicarrier approach. Such a comparison is fairly academic, and will not be pursued at this point. The benefits of CDMA, it seems, have found great approval in the communications world, but they do not necessarily include lower PAR for a given information rate.

Figure 9.31 shows the I-Q plane trajectory of an IS95 uplink (mobile, or handset) signal, along with the amplitude distribution bar chart. In the IS95 CDMA system, mobiles only transmit a single data channel, and thus the signal has the outward appearance of an OQPSK signal. It can be seen to have negligible zero crossings, and the trajectories are smoothed out by the Nyquist filter response. The

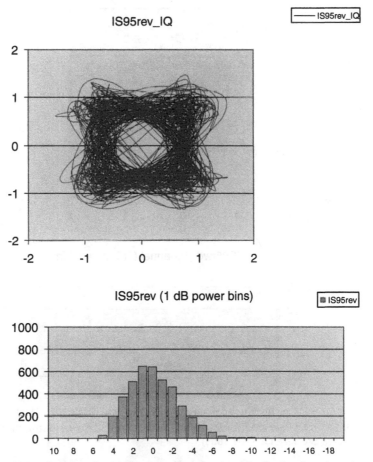

Figure 9.31   IS95 uplink (handset) IQ trajectory and power distribution bar chart (see Excel file IS95rev_stats).

12. In practice, this analysis will be further complicated by amplitude control on the individual data streams.

bar chart shows that although the signal has a 5 dB PAR, the incidence of the peaks is low. This has resulted in some surprisingly efficient handset PA products, showing over 40% efficiency. It should however be noted that in most cellular systems, the ACP requirements for mobiles are considerably more forgiving than the basestation specifications, due to the nonuse of the mobile transmit channel in adjacent cells.[13]

On the downlink, or basestation end, things quickly get more complicated. Figure 9.32 shows the I-Q trajectory and bar chart for a downlink transmitter which is only sending a single data channel. It can again be recognized as a straight QPSK format, with zero crossings allowed and appropriate filtering. Figure 9.33 shows the same plots for a case of eight data channels. The I-Q trajectories have been plotted using about 150 symbols, which is not enough to ensure that all permitted points on the IQ plane are represented. It is worth noting that within the timeslot structure of a 3GPP signal, these data burst lengths may in some cases be representative.[14]

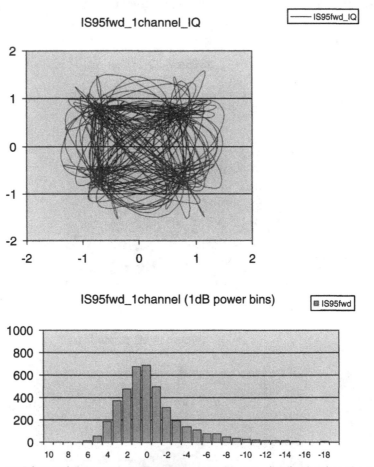

Figure 9.32 IS95 forward (basestation) IQ trajectory and power distribution bar chart; single channel (see Excel file IS95fwd_stats).

---

13. Hence the origin of the "ACP" term.
14. In comparing the "real" data shown in Figures 9.32 and 9.33 to the conceptual plot in Figure 9.31, it should be noted that for convenience the QPSK constellation points were rotated by 90°.

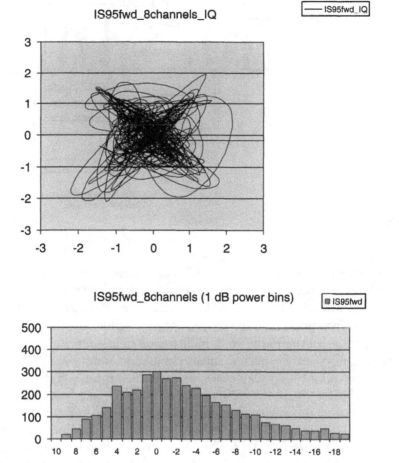

Figure 9.33 IS95 forward (basestation) IQ trajectory and power distribution bar chart; eight channel (see Excel file IS95fwd_stats).

In its basic form, "wideband" CDMA (WCDMA) follows much the same principles and implementation as the IS95 system, the main change being the use of a higher chip rate, now mainly standardized at 3.84 Ms/sec. This was the starting point for the evolution of the international "3GPP" (Third Generation Partnership Project) standard. The 3GPP standard has already evolved into a sumptuous cocktail of digital radio transmission techniques, an evolution which still continues. Figure 9.34 attempts only to give a general impression of the complexity of a 3G signal, rather than a detailed definition which can be found elsewhere [10]. The main additional feature, in comparison to a basic CDMA system, is the introduction of a time multiplexing dimension, so that a basestation transmits a 10 millisecond "frame" which consists of 15 "timeslots." Within a single timeslot, the signal comprises a burst of CDMA-encoded data, which is further partitioned into user and system management sub-timeslots. The system management data provides information on synchronization and amplitude control for the mobile user, and is typically sent in a "midamble," which is about a third of each timeslot. Individual channels can have numerous different formats, which allow for different spreading codes and data

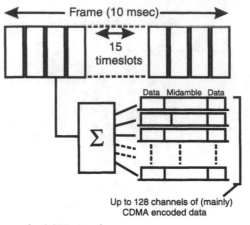

Figure 9.34   Basic structure of a 3GPP signal.

rates. There is a so-called "HSDPA" (High Speed Download Packet Access) format, which reduces the spreading factor to enable data download rates up to 14 Mb/sec.

The difficulty a PA designer faces here is that the actual signal the PA has to handle, in a given physical emplacement, has a bewildering range of possibilities, which can change substantially on a millisecond timescale. It becomes less relevant to examine IQ plots for specific timeslot configurations when they change all the time. The timescale of likely changes is very significant, since it connects back to the thermal analysis in Section 9.6. Now that we have the possibility of a signal changing on a millisecond timescale, dynamic thermal effects will come into play, especially given that the linearity of the PA is being maintained down to an equivalent gain discrepancy of maybe 0.001 dB. This has important implications for linearization using digital predistortion.[15]

Figure 9.35 shows a portion of the IQ trajectory for a timeslot which has just two DPCHs (Dedicated Physical Channels). This is not a typical signal, but it shows the basic underlying constellation pattern. Figure 9.36 shows how this develops as more channels are added, eight in this case. The underlying diamond structure of constellation points, as shown in Figure 9.30, can still be seen in both plots. These signals have high PARs, and it is clear that some *a priori* peak clipping is desirable. In many cases, it can be demonstrated that removal of the highest peaks will have a negligible impact on bit error rates; the trick is to remove the peak in such a way that there is also negligible impact on the spectral distortion. Crest Factor Reduction, or CFR, has thus become a research topic in its own right [11]. In general, it is better to remove the peaks on an *a priori* basis using DSP, rather than letting the RFPA clip the final signal, but as always there's never a completely free lunch.

The difficulty of specifying and testing PAs for 3GPP applications has been recognized in the form of test standards, which define configurations that are representative of typical expected operation. Unfortunately, early experiences with pilot 3G installations has caused the concept of "typical" operation to change, to the ongoing frustration of PA manufacturers.

15. It should be clarified that this issue is addressed in the 3GPP standards, and considerable care has been taken to maintain constant average power from burst to burst, but there are exceptions such as the HSDPA mode.

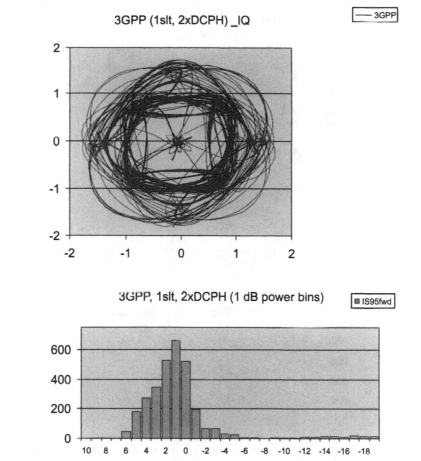

Figure 9.35   IQ trajectory and statistics of a two-channel DPCH timeslot burst.

9.7.4   OFDM Modulation, 802.11/16 Standards

Orthogonal Frequency Division Multiplexing, or OFDM, was originally devised as a very secure method of encoding and transmitting data, and not surprisingly was developed for military communications applications. Its ability to reduce sensitivity to multipath effects and interference has launched it into mainstream use for the higher data rates required in WLAN, and other related applications. Whether in fact these newer wireless communications systems actually require all the attributes of OFDM is something of a moot point, given the additional factor of powerful commercial interests and the patented monopoly on the use of CDMA techniques. However, the 802.11b WLAN standard has come into widespread use, and the same capabilities over greater distances ("WiMax") are promised from implementation of the 802.16 standard.

Figure 9.37 shows a starting point for a qualitative explanation of OFDM. In a conventional FDM scheme, several separate signals, using nonharmonically related carrier frequencies, and possibly different modulation systems, are spaced in the frequency domain such that the individual spectra do not overlap. This is a necessary requirement when it comes to the separation of the signals in the receiver, which will use IF filtering to perform this task. Figure 9.38 shows the same set of signals

3GPP

3GPP (1slt, 8xDCPH) _IQ

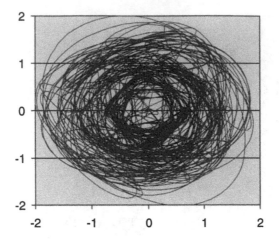

3GPP, 1slt, 8xDCPH (1 dB power bins)   IS95fwd

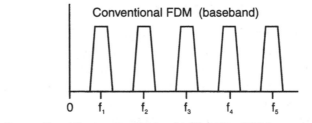

Figure 9.36   IQ trajectory and statistics of an eight-channel DPCH burst.

Conventional FDM (baseband)

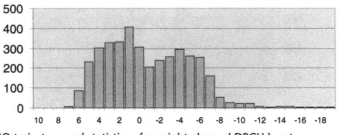

Figure 9.37   Conventional Frequency Division Multiplexing (FDM).

$f_n = n \times f_0$

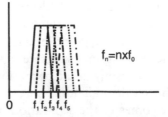

Figure 9.38   Orthogonal Frequency Division Multiplexing (OFDM).

with much closer spacing. In an OFDM scheme this is acceptable because the individual carriers are generated using a common synthesizer, and are harmonically related. This enables each individual carrier, along with its own modulation, to be creamed out using a suitable FFT algorithm.

In the first instance, this is clearly a more spectrally efficient system than that shown in Figure 9.37. The 802.11 and 802.16 standards exploit a further facility, which is to distribute each individual data stream across the many carriers (52 in the case of 802.11a). This frequency diversity reduces sensitivity to fading effects, which are of much greater concern in high data rate systems. It also reduces the individual carrier symbol transmission rate, by a proportion equal to the number of subcarriers.[16] From the PA designer's viewpoint, the multicarrier basis is troublesome, since it inevitably causes high PAR, significantly higher than other single carrier QPSK systems so far encountered. This is illustrated in the standard format in Figure 9.39. The PAR is in the region of 10 dB, but the statistics indicate that the

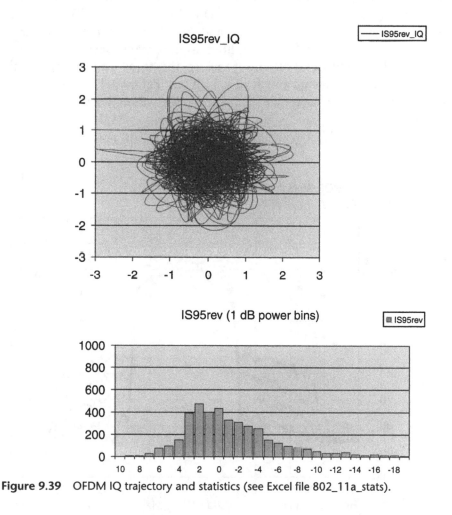

Figure 9.39 OFDM IQ trajectory and statistics (see Excel file 802_11a_stats).

16. The 802.16 standard will allow for a number of multiple users to be allocated an individual smaller group of subcarriers, in order to lower the FFT processing time.

upper 2 dB or so of the range can probably be clipped without significant increase in ACP or demodulation errors.

9.8   30 Watt LDMOS Test Amplifier Measurements

It seems that a chapter devoted to the nonlinear effects in RFPAs would not be complete without showing some actual measurements. The difficulty is what to measure, given the multitude of modulations standards and regulatory specifications, both of which are in any case in a continuous evolution. When dealing with a real PA rather than the more in-vogue virtual variety, there will always be effects which are visible, do not comply with theoretical expectations, and demand further explanations. One can thus quickly lose sight of the original objectives. The goal in this section is to show a few typical measurements which illustrate some of the more theoretical concepts discussed in this chapter, but it is not possible to do this comprehensively. The sample amplifier is a two-stage 2 GHz LDMOS PA. At the 1 dB compression point the output power is 45 dBm (30 watts), and the efficiency of the output stage is 55%. These are typical numbers for an amplifier using LDMOS devices at this frequency.

Figure 9.40 shows a set of output spectra, measured using a single channel IS95 forward link signal, having a PAR of 6.5 dB (see Figure 9.32). The maximum drive condition has been set such that the highest peaks cause the amplifier to reach its 1 dB compression point, causing an ACP level of about −35 dBc. The mean efficiency is about 17% at this level, which represents about a 7 dB power backoff from the cw

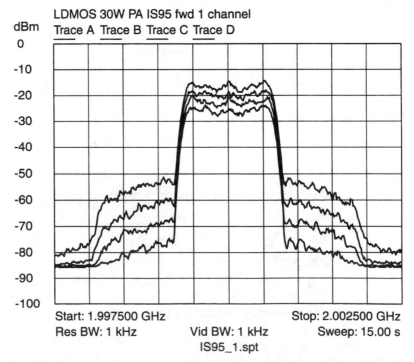

Figure 9.40   Test PA spectrum: IS95 forward, single channel; 3 dB backoff steps from 1 dB compression at PEP.

1 dB compression point. The subsequent plots show the effect of successive 3 dB backoff power settings. Figure 9.41 shows the same conditions for an eight-channel IS95 signal. The starting condition shows a higher ACP level, about −30 dBc, due to the higher PAR.

Figure 9.42 shows a summary of power, ACP, and efficiency for the two cases, and also includes a third case of 64 channels. For a basestation application, it is most likely that the ACP level will be specified at −50 dBc or even lower. Clearly, in order to reach this level by power backoff alone even the unrepresentative single channel case would yield an efficiency barely above 10%. The multichannel cases do not even give a convincing impression that the ACP level would ever drop below −50 dBc, whatever the level of backoff. These results underline the need to engage efficiency enhancement and linearization techniques in basestation applications.

Figure 9.43 shows the equivalent spectral plots for a single carrier EDGE signal input to the same test amplifier. Due mainly to different characteristics of the Gaussian filter, the in-band and out-of-band ACP levels are not as clearly differentiated as in the IS95 plots. Of the three drive levels shown, only one clearly breaks the spectral mask. Figure 9.44 shows the dynamic gain plots for the same three cases, showing that the case which breaks the spectral mask has a heavily compressed dynamic characteristic, with the peak power almost at the saturation level. This may seem a more favorable result than expected, given our previous comments on the EDGE signal statistics. The point here is that the criterion we are using, in the form of one particular release of a regulatory specification, is not especially demanding. It turns out that only the lowest drive case in Figure 9.43 meets the

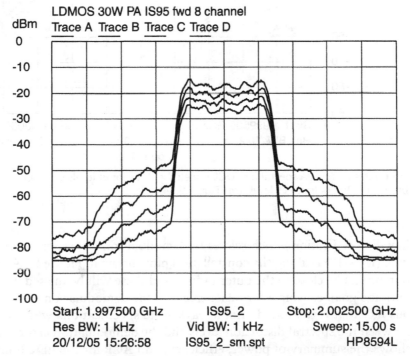

Figure 9.41   Test PA spectrum: IS95 forward, eight channel; same mean power levels as in Figure 9.40.

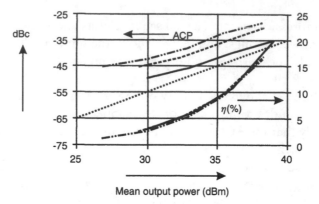

Figure 9.42 ACP and efficiency of test PA: IS95 single channel (solid), IS95 8 channel (dashed), IS95 64 channel (chain). Dotted diagonal shows ideal IM slope.

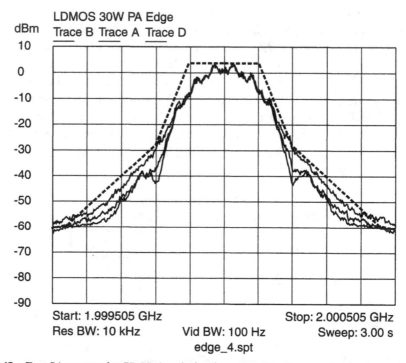

Figure 9.43 Test PA spectra for EDGE signal: the three input levels cause increasing gain compression at the output PEP levels (see Figure 9.44).

EDGE specification which relates to demodulation errors, or Error Vector Magnitude (EVM). Recalling the constellation plot shown in Figure 9.24, many constellation points lie close to the outer radius, and these will be missed by the compressed output signal, causing a serious increase in the demodulation error rate. It is a curious fact that regulators do not always write specifications which have consistency between the spectral distortion levels and the demodulation error rates. Figure 9.45 shows the summary of power, efficiency, and ACP for the EDGE measurements.

The dynamic gain plots in Figure 9.44 were obtained using a pair of high speed logarithmic detectors (AD8318) and a digital sampling oscilloscope (DSO). Each

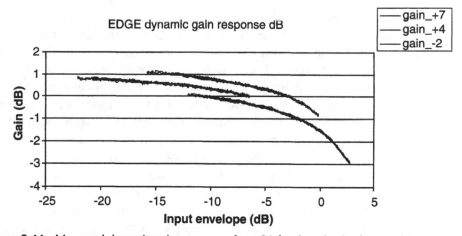

Figure 9.44 Measured dynamic gain response of test PA for three levels of mean drive power (see Figure 9.43 for ACP spectra). (Gain levels for the three drive ranges have changed due to case temperature differences.)

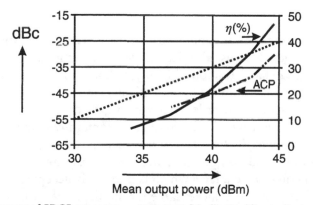

Figure 9.45 Summary of EDGE measurements on test PA. (Dotted line indicates 3:1 IM3 slope.)

gain trace consists of 10,000 points taken at a rate of 100 samples per symbol. The gain offsets can be assumed to be caused by different case temperatures for the three measurements. It would appear that for the cases shown there is no visible looping or hysteresis on the individual gain traces. The thickness of each trace is about 0.1 dB and represents the limit of the measurement resolution. There is however some sign that the curves do not trace exactly the same path when the offsets are removed.

It will be noted that in this section we have slipped quietly out of the modeling and simulation party and show results from physical measurements.[17] This chapter has addressed the two basic elements required for the simulation of the results presented in this section: a behavioral model for the PA and an envelope simulator. There are many such commercial software packages available, either as stand-alone items or adjuncts to more general mathematical analysis packages. It would seem a logical conclusion to put these elements together and show that the test PA measure-

17. It has been suggested that the physical measurement approach, in keeping with modern inclinations, should be known as the "analog simulation" option.

ments can be predicted. Unfortunately, obtaining a suitable model for a given physical PA is the weak link in the process of predicting the kind of detailed measured results shown in this section. Experience with digital predistortion has underlined this problem. Although it seems quite a straightforward process to model the nonlinearities of any PA for a given excitation, any significant change in the excitation will require some modifications to be made to the modeling coefficients. Such is the process used in adaptive predistortion linearization (see Chapter 14), and this applies in particular to changes in the mean power level of signal. Simple models are however quite useful for comparative purposes. Given the signal itself as a starting point, the computational process for determining the output spectrum is fairly straightforward. This can be illustrated using the Excel IQ files used in this chapter, along with the resident Excel FFT routine, to perform some basic spectral distortion computations (see Appendix B). In general, it seems that the success of techniques for PA modeling and simulation at the system level have not advanced as satisfactorily as component level simulators for RF design.

9.9 Conclusions

As stated in the introduction to this chapter, treating the subject of PA nonlinearity is a challenge which usually delivers less than it might promise. Typical scenarios can be analyzed to death, but the bottom line is much the same as that which can be deduced using the simplest of heuristic arguments: any signal whose peak envelope power exceeds the power level at which gain compression or AM-PM occurs will cause distortion, which in the frequency domain will appear as intermodulation or spectral regrowth. If the mean power level of the signal is backed off to the point where about 1 dB compression occurs at the peak envelope power, the spectral regrowth will drop to a level which is frequently acceptable for single channel applications, but probably unacceptable for multichannel applications. In the latter case, external linearization will enable a compromise in the amount of additional power backoff required in order to satisfy the regulatory spectral requirements.

This basic backoff guideline can be used for setting mean power levels with different peak-to-average ratio signals. In a case such as a CDMA uplink having 5 dB PAR, this would mean operating the PA at a mean power level 5 dB below the 1 dB compression point; a 3 watt transistor would be needed to deliver 1 watt of CDMA power, at an efficiency down by a factor of 3 on the efficiency measured in a cw test at the 1 dB compression point; this may typically be well below 30%. Such simple rules of thumb can be questioned and refined, using more detailed analysis. There is, simply, no holy grail to be found in analyzing distortion effects, no clues as to how the mathematics of distortion can be defeated.

Chapter 10 will describe a range of techniques, many dating from the earliest era in RF broadcasting, whereby higher efficiencies can be obtained from amplifiers which operate in backed-off conditions with variable envelope signals. Chapter 14 will describe linearization techniques, which can enable amplifiers to meet regulatory specs without the maximum literal application of the backoff rule. These techniques have become the new key to the development of high efficiency amplifiers for modern wireless communications systems.

References

[1] Maas, S., *Nonlinear Microwave Circuits*, Norwood, MA: Artech House, 1988.

[2] Kaye, A. R., D. A. George, and M. J. Eric, "Analysis and Compensation of Bandpass Nonlinearities for Communications," *IEEE Trans. Commun.*, October 1972, pp. 965–972.

[3] Blachman, N. M., "Band-Pass Nonlinearities," *IEEE Trans. Inform. Theory*, Vol. IT-10, April 1964, pp. 162–164.

[4] Kenney, J. S., and A. Leke, "Simulation of Spectral Regrowth, Adjacent Power, and Error Vector Magnitude in Digital Cellular and PCS Amplifiers," *Microw. Jour.*, October 1995.

[5] Cripps, S. C., *Advanced Techniques in RF Power Amplifier Design*, Norwood, MA: Artech House, 2002.

[6] David, S., et al., "Thermal Transients in Microwave Active Devices and Their Influence on IM Distortion," *IEEE MTT Symposium Digest*, Phoenix, AZ, Vol. 1, 2001, pp. 431–434.

[7] Parker, A., and J. Rathnell, "Contribution of Self Heating to Intermodulation in FETs," *IEEE 2003 Intl. Microw, Symposium*, Fort Worth, TX, WE4C-5.

[8] Riley, K. F., M. P., Hobson, and S. J. Bence, *Mathematical Methods for Physics and Engineering*, 2nd ed., Cambridge, U.K.: Cambridge University Press, 1997.

[9] Cripps, S. C., "A New Technique for Screening and Measuring Channel Temperature in RF and Microwave Hybrid Circuits," *IEEE Semitherm Symp.*, 1990, pp. 40–42.

[10] 3GPP Web site, http://www.3gpp.org.

[11] Sperlich, L., et al., "Power Amplification with Digital Predistortion and Crest Factor Reduction," *IEEE 2003 Intl. Microw, Symposium*, Fort Worth, TX, WE5B-1.

CHAPTER 10

Efficiency Enhancement Techniques

Introduction

The power amplification of amplitude modulated RF signals has two inherent problems. The first is that the envelope, and hence the modulating signal, will be distorted to some degree, if the power amplifying device is used at its full rated RF power level. The second is that conventional power amplifier designs only give maximum efficiency at a single power level, which is dependent on the circuit design but usually near the maximum rated power for the device. As the drive power is backed off from this point, the efficiency drops sharply and the heat dissipation can increase, even though the RF output power is also decreasing. The overall effect is therefore to measure a mean efficiency which is much lower than the efficiency at the maximum power, or PEP, level.

Solutions for the second of these problems have been available for many years and can be quite effective; they will be termed "efficiency enhancement" techniques. The first problem, which essentially requires the transfer characteristic of the amplifier to be made more linear over a given power range, represents a bigger challenge, and the solutions always have some limitations. In particular, higher linearity will always come at a price, usually in the form of lower efficiency but also in baseband frequency limitations. Such techniques are nevertheless widely used, and are quite indispensable in many transmitter applications. Logically, the term "linearization" should only be applied to this latter category. Inevitably, some implementations use both efficiency enhancement and linearization techniques together, and conversely some specific techniques can be implemented as either efficiency enhancement or linearization as the primary goal. This chapter focuses on the techniques which primarily improve or enhance the efficiency under modulated signal envelope conditions, without specifically improving the linearity. Linearization techniques are covered in Chapter 14.

Several efficiency enhancement techniques were invented in the early era of radio broadcasting. The motivation at that time was thermal management and running costs. A shortwave broadcast transmitter delivering tens of kilowatts could consume a large amount of electrical power, which constituted an important part of the running costs. Three classical techniques fall into this category; the *Doherty Amplifier*, the *Outphasing Amplifier* initially proposed by Chireix, and the *Envelope Elimination and Restoration (EER)* technique demonstrated by Kahn in the early days of single sideband (SSB) transmission.

This chapter will consider all three classical techniques, along with some more recent derivatives. In particular *Envelope Tracking* (*ET*) has emerged as another important candidate under this heading. The basic underlying theory behind these techniques will be analyzed in some detail in this chapter. It has been a matter of experience in this industry over the last 10 years or so, that although promising conceptually at the outset, specific implementation of any of these techniques typically poses many problems, especially in high volume applications. This has led to a situation where there are many research papers but not many commercial products. Digital signal processing looks poised to come to the rescue in some of the troublesome areas, although this introduces a new dimension of complexity and cost. It should be noted that this has been a very active area for patents. Readers intending to pursue commercial implementations for any of the techniques discussed in this chapter are strongly advised to conduct a thorough patent search before releasing a product.

10.1 Efficiency Enhancement

A logical starting point for a discussion on efficiency enhancement is to return to the idealized analysis of a classical Class B amplifier given in Chapter 3. Figure 10.1 shows the case normally considered, where the device is biased to its threshold point and the RF drive has the precise magnitude required to swing the current up to the maximum linear value for the device, I_{max}. The output load consists of a load resistance at the fundamental frequency, shunted by a resonator which provides a short circuit at all harmonics. The load resistor is carefully chosen so that the resulting sinusoidal RF voltage has a prescribed maximum amplitude. In the ideal case, where zero knee voltage is assumed, the value of the load resistor is given by

$$R_{opt} = 2 \cdot \frac{V_{dc}}{I_{max}}$$

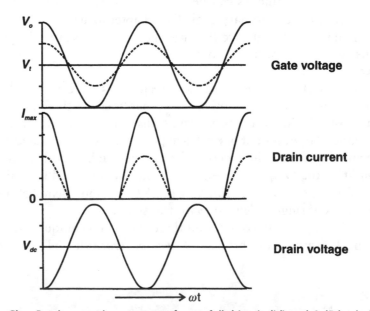

Figure 10.1 Class B voltage and current waveforms, full drive (solid) and 6 dB backoff condition using a load resistor having twice the normal loadline value (dotted).

noting that the fundamental component of a halfwave rectified sinewave having a peak value of I_{max} is $I_{max}/2$. The RF output power is given by

$$P_{RF} = \frac{V_{dc} \cdot I_{max}}{4}$$

and the DC supply power is given by

$$P_{DC} = \frac{V_{dc} \cdot I_{max}}{\pi}$$

so the output efficiency is $\pi/4$, or about 78.5%.

Let us now consider what happens when this amplifier has a lower incident RF input level. Suppose that the voltage amplitude of the input RF signal is reduced by a factor p from the ideal maximum level used in the design equations given above. The assumed linearity of the device transconductance means that the output RF current will still be a halfwave rectified sinewave, but with a peak value reduced by the factor p. So the fundamental component of RF current will be

$$I_1 = \frac{I_{max}}{2p}$$

and, with the original design value of load resistance, the output voltage swing will have an amplitude given by

$$V_1 = \frac{I_{max}}{2p} \cdot R_{OPT} = \frac{V_{dc}}{p}$$

so the RF output power is

$$P_{RF} = \frac{V_{dc} \cdot I_{max}}{4p^2} \tag{10.1}$$

and the DC supply power is

$$P_{DC} = \frac{V_{dc} \cdot I_{max}}{p\pi} \tag{10.2}$$

and the corresponding efficiency is

$$\eta = \frac{\pi}{4p} \tag{10.3}$$

So, for example, a 3 dB reduction in drive power, corresponding to $p = \sqrt{2}$, results in a 3 dB drop in RF output (from 10.1), but a corresponding reduction in efficiency from 78.5% to 55.5%. At 6 dB power backoff, the efficiency drops to 39%. As noted in Chapter 3, the ideal Class B amplifier does have the virtue of remaining linear as the power is backed off, although in practice the fading of the transconductance as the cutoff point is approached will cause a roll-off of gain at low drive levels. This effect will be ignored in this analysis, which is mainly concerned with maintaining efficiency at higher drive levels.

The reason for the drop in efficiency at reduced drive levels can be clearly seen in Figure 10.1; the dotted waveforms represent a 6 dB backoff condition and the problem is that the selected value for the load resistor is now too small to allow a full rail-to-rail voltage swing. If the load resistor value could be increased to $2.R_{OPT}$, just for the 6 dB backed-off condition, the efficiency would return to the optimum value of 78.5%. In general, if the load resistor has a value R_p at a drive level backed off by the voltage factor p, then the efficiency will return to 78.5%, if

$$R_p = \frac{V_{dc}}{I_{max}} \cdot 2p \tag{10.4}$$

The RF ouput power with the load resistor reset to a value of R_p will be

$$P_{RF} = \frac{V_{dc} \cdot I_{max}}{4p} \tag{10.5}$$

The DC supply power will be invariant to the value of RF load resistance,

$$P_{DC} = \frac{V_{dc} \cdot I_{max}}{p\pi} \tag{10.6}$$

so that, combining (10.5) and (10.6), the efficiency can be seen to remain constant at $\pi/4$.

So the immediate speculation is that if the load resistance could somehow be made to change its value dynamically, according to (10.4), maximum efficiency could be maintained. In fact, the problem is yet more challenging, because (10.5) shows that even if such a physical challenge could be met, such an amplifier would be nonlinear; the RF output power would drop in proportion to p, not p^2. Such a response is shown in Figure 10.2. One could speculate about how useful such an amplifier may be, but the central issue concerns the manner in which the load resistor, or the matching network elements used to transform the load resistance from a 50Ω termination, could be dynamically altered on the same timescale as the amplitude modulation carried by the input signal. It so happens that both the drive-dependent resistor and the linearity issue can be solved simultaneously, using a configuration of two separate amplifiers, originally proposed by Doherty in a classic paper dating from 1936 [1]. Doherty's invention, in its original form, only changes the load resistor value optimally for a few dB of backoff from the maximum power level, but it solves the linearity issue elegantly at the same time.

Before considering the Doherty amplifier in more detail, there are some other observations which should be made concerning efficiency enhancement techniques. The key issue, when dealing with a signal which has a time-varying envelope amplitude, is that the full power capability of the PA is only needed at the peaks of the signal envelope. When the envelope dips down to much lower amplitudes, the capability of the PA is wasted, to a greater or lesser extent. Reduced conduction

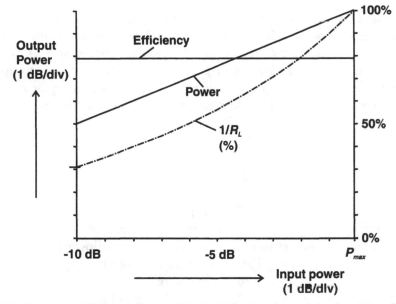

Figure 10.2 Power and Efficiency characteristic for PA having dynamic load variation (load varied inversely with input voltage, see text).

angle modes, even in their original basic form, do alleviate this problem in that the current peaks adapt to the amplitude of the incoming signal envelope. Due to the rectified nature of a Class B, or deep Class AB current waveform, the DC supply will also automatically reduce under low envelope excitation. This will not happen in a Class A amplifier, and even a lighter Class AB mode will be particularly susceptible to low efficiency during envelope minima, as shown in Figure 10.3.

Another general possibility that often comes to mind is somehow to control, or adapt, the DC supply as a means of reducing the power capability during lower envelope points. Such a strategy is sometimes introduced as a weaker implementation of Kahn's envelope restoration technique, but is not in fact at all the same thing;

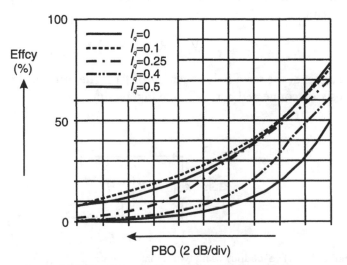

Figure 10.3 Power backoff (PBO) efficiency characteristic for Class AB PAs.

it falls under a sub-category which will be known as *envelope tracking* and will be considered in Section 10.6. Conversely, it may even be possible to modify dynamically the effective periphery of the RF power transistor, so that only a smaller portion of the whole device is active during low points of the envelope. This would clearly not be possible for a single packaged device, where the internal cells are all hard-wired together, but may be an option to consider in a circuit-combining scheme.

10.2   The Doherty Amplifier

The Doherty amplifier was first proposed in 1936 [1], and is primarily an efficiency enhancement, or power conservation, technique. The original paper was concerned with very high power tube amplifiers generating kilowatts of RF power in a short wave broadcast station. This may seem a rather remote application, both in terms of chronology and relevance to modern wireless communications systems, and indeed the modern wireless communications industry did seem a little slow on the uptake in the early 1990s. The uplink, or handset, mobile transmitter in systems which have an amplitude modulated signal and a peak-to-average power ratio anywhere between 3 and 6 dB, is an obvious candidate for this technique. Despite a few orders of magnitude reduction in power level from the original application, the key problem of maintaining efficiency during lower points of the envelope is solved, at least to a very useful extent.

One of the interesting aspects of the Doherty configuration is that it uses what would today be termed as an active load-pull technique. The concept that the resistance or reactance of an RF load can be modified by applying current from a second, phase coherent source is somewhat alien to RF designers, who usually think of RF loads as passive lumps of metal and dielectric. It is therefore necessary initially to introduce this concept. Referring to Figure 10.4, generator #1 "sees" a load resistance of R_L, if generator #2 is set to give zero current. Simple Kirchoff circuit theory shows that if generator #2 supplies a current I_2, and generator #1 supplies a current I_1, both currents flow into the load resistor so that the voltage appearing across the load resistance is

$$V_L = R_L(I_1 + I_2)$$

so the effect at the terminals of generator #1 is the same as if there were a simple passive resistor connected across it, having a value given by

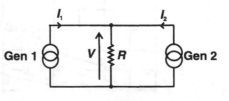

$$V = R.(I_1 + I_2)$$

Figure 10.4   Active loadpull using two signal generators.

$$R_1 = R_L \left(\frac{I_1 + I_2}{I_1} \right) \tag{10.7}$$

Simultaneously, the effect at generator #2 would be equivalent to a resistor having a value

$$R_2 = R_L \left(\frac{I_1 + I_2}{I_2} \right) \tag{10.8}$$

Equations (10.7) and (10.8) can, in the usual manner, be extended to apply for AC circuits if complex notation is used to represent the magnitude and phase of the currents and voltages, and the resistive and reactive components of the impedances. In this form, the equations show the possibility of changing, or "pulling" the impedance seen by generator #1 by controlling the magnitude and phase of the current I_2:

$$Z_1 = R_L \left(1 + \frac{I_2}{I_1} \right) \tag{10.9}$$

So, for example, Z_1 can be transformed to higher resistive values if I_2 is in phase with I_1, and to smaller resistive values if I_2 is in antiphase with I_1.

Consequently, if the two generators are now considered to be the output transconductance generators of two separate RF transistors, having co-phased input drive signals, the effective output impedance of one device can be modified by the other according to (10.9). In normal RF practice, identical devices are combined in parallel with the assumption that the impedance seen by each is the common load impedance scaled up by the number of parallel devices. This only works when the devices are identical, both in terms of parameters, bias, and drive level. We will see that there are some interesting possibilities for power amplifier design if the identity condition is relaxed, and devices of differing periphery, different bias or drive levels, are combined. In these cases, the impedance seen by each element is a function of the other elements as well as the common load, and (10.9) is the key to analyzing such configurations.

The Doherty amplifier is an example of the scope which opens up when devices are combined together in a less orthodox fashion. The basic concept is illustrated in Figure 10.5. The two devices will, for convenience, be termed "main" and "auxiliary." The final maximum RF output power is the combined power of both devices. As the input drive level is reduced, both devices contribute to the output power until a certain point is reached, typically 6 dB down from the maximum composite power, where the auxiliary amplifier shuts down and generates no more RF power; it is assumed that it will also cease to draw DC as well (either through being in Class B, or through external control circuitry). So below the 6 dB backoff point, the active device periphery has been reduced by typically 50% (in the case where the two devices have equal I_{max} values), which will significantly improve the efficiency at lower power levels. Half of the device periphery is effectively shut down except for the higher power levels where the full periphery is required. But the Doherty configuration has additional benefits, which arise through the load-pulling effects of one device on the other, as discussed above and summarized by (10.9). It can be shown

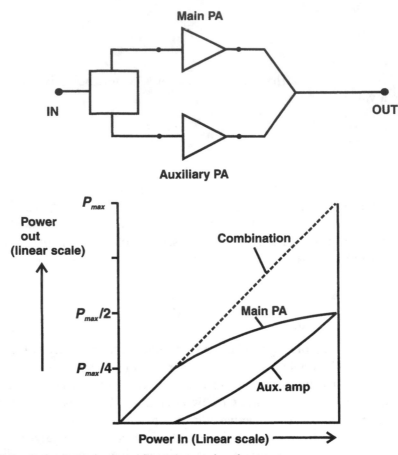

Figure 10.5   Doherty PA, basic configuration, and performance.

that under the correct impedance matching conditions, the main device can stay
close to maximum efficiency throughout the upper 6 dB of power range.

The key action of the Doherty amplifier occurs during the region where the auxiliary device is active and the main device is held in a constant maximum voltage
condition. This is achieved through the load resistance, whose effective value
decreases dynamically with increasing drive level due to the load-pulling effect of the
auxiliary amplifier, thus maintaining maximum voltage swing and efficiency. In this
regime, the output power increases in proportion to the input voltage drive level, so
that on a linear power scale, a square root characteristic is obtained. Meanwhile, the
auxiliary amplifier experiences an upward load-pull effect, so that it generates an
output power proportional to the cube of the increasing input voltage amplitude,
giving a "three-halves" power transfer characteristic. These two characteristics
combine to give a composite linear power response, as illustrated in Figure 10.5,
with close to maximum efficiency being maintained down to the 6 dB backoff point.

Figure 10.6(a) shows a schematic diagram for a Doherty amplifier, and Figure
10.6(b) shows the two device current amplitudes plotted as a function of input voltage drive. At this point, it will be assumed that these current characteristics can be
realized using the threshold characteristics of the devices, or by other suitable amplitude controls on each input. Each device, for the time being, is assumed to have the

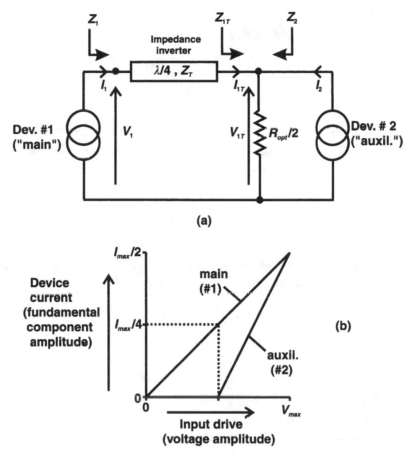

Figure 10.6   Doherty amplifier analysis: (a) schematic, and (b) main and auxiliary device RF current amplitude variation versus input drive

same maximum linear current swing of I_{max}, so that in terms of fundamental amplitude the maximum linear value for each device will be $I_{max}/2$. Note the important addition of a quarter-wave transformer between the common load resistor and the main device output. This acts as an impedance inverter, which causes the resistive impedance seen by the main device to go down as the auxiliary device current I_2 increases.

For the upper 6 dB regime, both devices are active, and their RF current amplitudes can be expressed in the form

$$I_1 = \frac{I_{max}}{4}(1+\xi)$$

$$I_2 = \frac{I_{max}}{2}\xi$$

(10.10)

where ξ has a value between 0 and 1, 0 corresponding to the 6 dB power backoff point and 1 to the maximum power point.

The load-pulling relationship of (10.9) can now be applied for the effective impedance each side of the load resistor,

$$Z_{1T} = \frac{R_{OPT}}{2}\left(1 + \frac{I_2}{I_{1T}}\right)$$

$$Z_2 = \frac{R_{OPT}}{2}\left(1 + \frac{I_{1T}}{I_2}\right)$$

(10.11)

This shows that in the maximum power condition, where $I_{1T} = I_2 (= I_{max}/2)$, then $Z_{1T} = Z_2 = R_{OPT}$, where R_{OPT} is the conventional optimum loadline impedance for the mode in use; so for an assumed Class B mode,

$$R_{OPT} = V_{dc}\left(\frac{2}{I_{max}}\right)$$

(10.12)

The quarterwave transformer has a characteristic impedance Z_T, yet to be determined, and the relationships between its input and output voltages and currents are

$$V_{1T} \cdot I_{1T} = V_1 \cdot I_1$$

$$\left(\frac{V_{1T}}{I_{1T}}\right) \cdot \left(\frac{V_1}{I_1}\right) = Z_T^2$$

(10.13)

(a 90° phaseshift between the input and output is noted, but for clarity is not included in the formulation).

Equation (10.13) can be rearranged, giving

$$I_{1T} = \frac{V_1}{Z_T}$$

and substituting for I_{1T} in the expression for Z_{1T}, (10.11), gives

$$Z_{1T} = \frac{R_{OPT}}{2}\left(1 + \frac{I_2 Z_T}{V_1}\right)$$

so the output impedance seen by the main device (#1) is

$$Z_1 = \frac{Z_T^2}{Z_{1T}}$$

$$= \frac{2Z_T^2}{R_{OPT}\left(1 + \frac{I_2 Z_T}{V_1}\right)}$$

from which the amplitude of the RF voltage swing at the output of the main device is

$$V_1 = I_1 \cdot Z_1$$

$$= \frac{2 \cdot I_1 Z_T^2}{R_{OPT}\left(1 + \dfrac{I_2 Z_T}{V_1}\right)}$$

and substituting for I_1 and I_2 in terms of the voltage drive level parameter ξ, from (10.10),

$$V_1 = \frac{Z_T^2 \left(\dfrac{I_{max}}{2}\right)(1+\xi)}{R_{OPT}\left[1 + \dfrac{\xi\left(\dfrac{I_{max}}{2}\right)Z_T}{V_1}\right]}$$

which, after some rearrangement, becomes

$$V_1 = \left(\frac{Z_T}{R_{OPT}}\right)\left(\frac{I_{max}}{2}\right)\left[Z_T + \xi(Z_T - R_{OPT})\right] \tag{10.14}$$

Examining (10.14), it appears that if the characteristic impedance of the transformer is set such that $Z_T = R_{OPT}$, then the main device voltage becomes independent of ξ, and the output voltage amplitude remains constant at

$$V_1 = R_{OPT} \cdot \left(\frac{I_{max}}{2}\right) \tag{10.15}$$

which, from (10.12), is the ideal maximum voltage swing of V_{dc}.

So the inverted load-pulling of the auxiliary device combines with the increasing current swing, to give a constant voltage amplitude swing at the output of the main device, just as originally conjectured in the previous section. The key point is that maximum efficiency will be maintained for the main device throughout this 6 dB backoff regime. The analysis in Section 10.1, however, showed that in this constant voltage regime, the power output rises in proportion to the amplitude of the input voltage drive, giving a nonlinear response. This is where the power contribution from the auxiliary device comes into play; the total power is the sum of the power generated by both devices.

Referring back to (10.9), the voltage across the load resistor, V_{1T}, can be obtained in terms of the main device current, I_1:

$$V_{1T} \cdot I_{1T} = V_1 \cdot I_1$$

$$\left(\frac{V_{1T}}{I_{1T}}\right) \cdot \left(\frac{V_1}{I_1}\right) = Z_T^2$$

giving

$$\left(\frac{V_{1T}}{I_1}\right) = \left(\frac{I_1}{V_{1T}}\right)Z_T^2$$

so that

$$V_{1T} = I_1 Z_T \tag{10.16}$$

and since Z_T will be set to equal R_{OPT}, the final composite RF voltage across the load resistor is

$$V_{1T} = I_1 \cdot R_{OPT} \tag{10.17}$$

Recalling that the main current is a linear function of the input voltage over the entire power range (Figure 10.6), (10.17) shows that the output load voltage will maintain a linear relationship with the input drive voltage over the entire range, a remarkable result. The final output power at maximum drive will be a factor of 2 higher than each device acting individually, since the voltage in (10.17) is developed across a resistor of value $R_{OPT}/2$.

Figure 10.7 shows the main and auxiliary device RF current and voltage amplitudes over the whole power range. It is clear that the main device will maintain maximum efficiency (i.e., 78.5% assuming Class B operation) over the upper 6 dB power range. The RF voltage swing at the output of the auxiliary device shows a linear

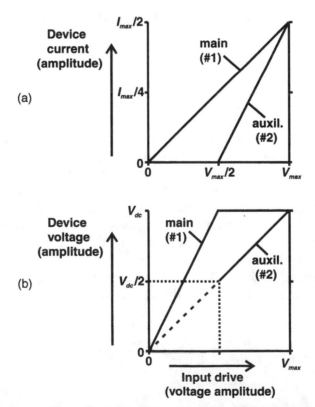

Figure 10.7 Doherty amplifier analysis: (a) device currents, and (b) device voltages, RF amplitude variations versus input drive (current amplitudes refer to the fundamental components).

decrease from the maximum supply rail swing down to half this value at the 6 dB backoff point. The auxiliary amplifier, therefore, will not have maximum efficiency over this range. But its contribution to the overall efficiency is a function of the amount of power it contributes, so in the region where the output voltage swing is lower, the power contribution is also low and the impact on overall efficiency is quite modest. An expression for overall efficiency can be determined, but it has to be recognized that the mode of operation of the auxiliary amplifier will vary according to how the input drive arrangements are implemented. Assuming that the efficiency of both amplifiers is $\pi/4$ at full rail-to-rail RF voltage swing, corresponding to Class B operation, then the efficiency of either amplifier will drop proportionately with the RF voltage swing. So for the low power regime, where only the main amplifier is active, the efficiency will be

$$\eta_{comp} = \frac{2v_{in}}{V_{max}}\left(\frac{\pi}{4}\right), 0 < v_{in} < \frac{V_{max}}{2} \tag{10.18}$$

In the upper 6 dB regime, both amplifiers are active. The composite RF output power is

$$P_{COMP} = \frac{I_1^2 \cdot R_{OPT}^2}{2} \cdot \frac{2}{R_{OPT}} = I_1^2 \cdot R_{OPT}$$

which, substituting from (10.12),

$$R_{OPT} = V_{dc}\left(\frac{2}{I_{max}}\right)$$

can be written as a function of the input voltage drive amplitude v_{in},

$$P_{COMP} = \left(\frac{I_{max}}{2}\right) \cdot \left(\frac{v_{in}}{V_{max}}\right) \cdot V_{dc} \tag{10.19}$$

The DC power consumed by the main device in this regime, assuming Class B operation, is

$$P_{DCM} = \left(\frac{v_{in}}{V_{max}}\right)\left(\frac{I_{max}}{\pi}\right) \cdot V_{dc}$$

and the corresponding DC power consumed by the auxiliary device is

$$P_{DCA} = 2 \cdot \left(\frac{v_{in}}{V_{max}} - 0.5\right)\left(\frac{I_{max}}{\pi}\right) \cdot V_{dc}$$

so the total DC power consumed by both devices is

$$P_{DC} = \left(\frac{I_{max}}{\pi}\right) \cdot \left(3\left(\frac{v_{in}}{V_{max}}\right) - 0.5\right)V_{dc} \qquad (10.20)$$

The overall efficiency, as a function of input drive level, is given by combining (10.19) and (10.20),

$$\eta = \frac{\pi}{2} \cdot \frac{\left(\dfrac{v_{in}}{V_{max}}\right)^2}{3 \cdot \left(\dfrac{v_{in}}{V_{max}}\right) - 1} \qquad (10.21)$$

It can be seen that this function equals $\pi/4$ for both the maximum power condition, $v_{in} = V_{max}$, and also the 6 dB backoff condition $v_{in} = V_{max}/2$. The low power regime efficiency ($v_{in} < V_{max}/2$) will roll off like a conventional Class B amplifier. The overall efficiency can now be plotted as a function of power backoff, in dB, as shown in Figure 10.8. The small dip in the middle of the upper power regime is due to the lower efficiency of the auxiliary amplifier where it does not display a full rail-to-rail RF voltage swing. In the original implementation of the Doherty amplifier, the auxiliary amplifier would be operating more into Class C, and the dip would be less evident, but the overall impact can be seen to be almost negligible even when Class B operation is assumed. It will be shown in the next section that a more modern implementation, using a variable attenuator to control the auxiliary amplifier, would have both devices nominally in the same mode of operation, as represented in Figure 10.8.

10.3 Realization of the Doherty Amplifier

A modern implementer of a Doherty amplifier has some benefits, but also some disadvantages, compared with its 1936 inventor. Compared to a transistor, especially a FET, a tube had considerably more flexibility in its characteristics. Tubes, in general,

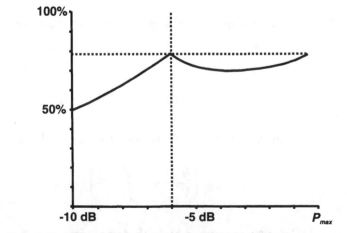

Figure 10.8 Doherty amplifier, efficiency versus input power backoff.

do not display either of the saturation mechanisms which are such important constraints in RF transistors, and they may also have extra grids which make the transconductance into a controllable parameter. In particular, the g_m and I_{max} of a tube can be varied substantially. Doherty makes use of a few such old tube tricks in order to approximate the required drive conditions for the main and auxiliary devices, and direct transistor analogies are not all that obvious. Referring back to Figure 10.6(b), obtaining the precise characteristic required from the auxiliary device is certainly a more awkward proposition when dealing with transistors rather than tubes.

The most obvious approach is, nevertheless, to follow the original implementation and use Class C bias for the auxiliary amplifier. Figure 10.9 shows how this creates an immediate problem if two identical transistors are used for the main and auxiliary PAs. The Class C device displays a useful approximation to the required hold-off behavior, but will fall well short of the $I_{max}/2$ current swing when the drive voltage reaches its maximum. The obvious solution is to scale up the periphery of the auxiliary device, although this would in principle require a scaling factor as large as 2.5. This would result in a power utilization factor ("PUF," see Chapter 3) of 2/3.5, or −2.4 dB, at maximum drive level. This is a large enough number to detract from the Doherty concept, despite its other benefits.

There are some alternatives, but these all require the use of adaptive or control devices. Such controls introduce a somewhat unwelcome baseband dimension to a concept which is appealing in its ability to operate as a self-propelled RFPA without the speed constraints that come along with video or digital processing. It is therefore worth considering the performance of a so-called "Doherty-Lite" amplifier, which uses two identical transistors, the auxiliary PA being realized by using a simple fixed Class C bias setting. Such a configuration was analyzed in [2] as part of a more general analysis of asymmetrical Doherty configurations, and some results are shown in Figure 10.10. It can be seen that careful selection of the auxiliary PA bias can still give some useful efficiency enhancement in the backed-off region. This performance can be further optimized by arranging for some extra gain in the auxiliary PA driver chain.

Given the growing use of DSP to linearize RFPAs in modern communication systems, it is reasonable to assume that accurate control signals and adaptive bias voltages can now be provided with relative ease. Figure 10.11 shows a simple adap-

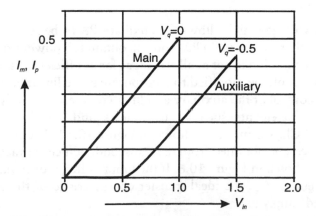

Figure 10.9 Use of Class C bias to implement the auxiliary PA in a Doherty configuration.

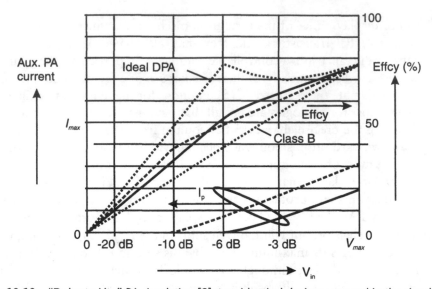

Figure 10.10 "Doherty-Lite" PA simulation [2]; two identical devices are used in the simplest possible Doherty implementation. The auxiliary device is biased into Class C and cannot be driven to its full I_{max}.

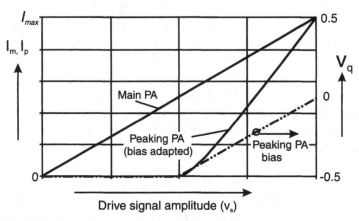

Figure 10.11 Auxiliary PA implementation using adaptive bias.

tive bias voltage which allows the auxiliary PA to be realized using a device of the same periphery as the main PA, and still obtain full power utilization. The exact profile of this adaption is not at all critical as far as efficiency is concerned, and the control signal could be modified to incorporate some linearization action.

The control of the auxiliary amplifier could alternatively be implemented using RF control elements such as attenuators and switches, rather than a heavily biased-off Class C mode, as shown in Figure 10.12. The required attenuation characteristic can be obtained from the transconductance characteristic for the auxiliary amplifier shown in Figure 10.6. If the input voltage for maximum unsaturated output current is V_{max}, the ideal transfer characteristic of the auxiliary device can be expressed simply as

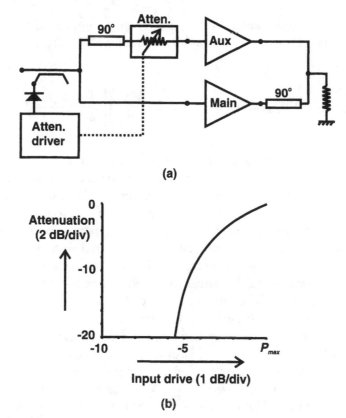

(a)

(b)

Figure 10.12   Possible Doherty amplifier input configuration: (a) block diagram, and (b) attenuator characteristic.

$$i_d = 2g_m \cdot \left(v_{in} - \frac{V_{max}}{2}\right), \frac{V_{max}}{2} < v_{in} < V_{max}$$

$$i_d = 0, 0 < v_{in} < \frac{V_{max}}{2}$$

If this characteristic is to be obtained using a similar device, with the same drive level and bias as the main device (e.g., Class B), then the attenuator has to perform all of the necessary cutoff and gain adjustment functions. In order to do this, the effective transconductance of the auxiliary device, g_a, has to have a dependency on input voltage amplitude according to

$$g_a = 2g_{in} \cdot \left(1 - \frac{V_{max}}{2v_{in}}\right), \frac{1}{2} < \frac{v_{in}}{V_{max}} < 1$$

$$= 0, 0 < \frac{v_{in}}{V_{max}} < \frac{1}{2}$$

(10.22)

The "effective" transconductance g_a gives the necessary attenuation characteristic whereby the actual transconductance of the device can be fixed at the same value as the main device, the changing characteristic being entirely handled by the input attenuator. Clearly, below the 6 dB backoff point, the attenuation has to be infinite and may be supplemented by a switch. Over the upper 6 dB of output power range, the attenuation characteristic, in dB, swings from infinity up to zero according to the law

$$A_{aux}(dB) = 20 \cdot \log\left[2\left(1 - \frac{V_{max}}{2v_{in}}\right)\right], \; \frac{1}{2} < \frac{v_{in}}{V_{max}} < 1 \tag{10.23}$$

In practice, a switch could be opened once the required attenuation drops below about 13 dB, representing less than 0.2 dB error in overall linearity. Such an attenuation characteristic (Figure 10.12) may not be too difficult to approximate using FET control elements in an RFIC implementation, although the added complexity, including input level sensing circuitry, would appear to be a substantial compromise to the basic elegance and simplicity of the original system.

The matching arrangements, including the impedance inverter and common RF load will present some challenges for the low impedance levels encountered with RF power transistors. The inverter, especially, would probably be realized using lumped elements, as discussed in Chapter 5.

Once the basic Doherty principle is fully understood, numerous variations come to mind. Throughout this description, it has been assumed that the main amplifier works in a Class B mode. The load modulation provided by the auxiliary amplifier would, however, work the same way for a Class AB, or even a Class A, main PA. There is also scope for extending the concept to more than two devices. If a suitable approach could be established for controlling the multiple auxiliary devices, it would be logical to consider a more generalized Doherty array, where devices of smaller periphery successively become active as the drive level increases. This would have the effect of greatly extending the range of maximum efficiency.

Another variation is the use of a different break point. The analysis of an asymmetrical DPA is not reproduced in detail here [2, 3] but a few highlights are worth noting. Figure 10.13 shows the device voltages and currents for an asymmetrical Doherty PA. By "asymmetrical" here we mean that the breakpoint is at a lower

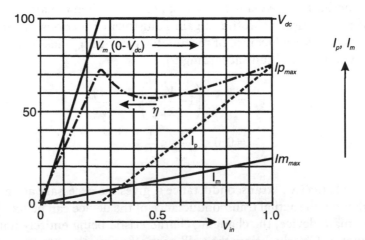

Figure 10.13 Asymmetrical Doherty PA (breakpoint at –12 dB).

point in the power backoff characteristic than the conventional 6 dB value. This variation would seem worth further investigation for signal environments where the peak-to-average ratio is significantly higher than 6 dB. However, it can be seen in Figure 10.13 that the use of a lower breakpoint requires that the auxiliary device has a correspondingly higher I_{max} value. In the prepeaking regime, the main device shows a steeply rising voltage characteristic which in practice will pose increasing difficulties for a smooth transition into the peaking regime. It is necessary to be quite sure that the efficiency benefit of such an arrangement will be worthwhile for the particular signal environment in question. We will return to the issue of making this quantification in Section 10.10.

So far little has been said about the linearity of a Doherty PA. Looking at the idealized power transfer characteristic in Figure 10.5, it is clear that if ideal linear transconductive devices are used, and the auxiliary PA control function is ideally implemented, the Doherty combination will also be ideally linear. Unfortunately, when real devices are considered, there are two factors which immediately change this optimistic viewpoint. Firstly, in the upper regime the main device voltage remains close to the clipping level, rather than backing away from it as would be the case in a conventional amplifier. This will reduce the slope of the IM backoff which would be observed in a conventional PA using the same device, where the RF voltage peaks will back away from the clipping level monotonically. Secondly, in the lower regime, a device of one-half the periphery and power capability is being used, in comparison to a conventional configuration. So for example, using a simple third degree IM3 model, the relative IM3 levels for a given output would be 6 dB higher than for the conventional PA which uses the entire periphery all of the time. So there are quite fundamental reasons why a Doherty configuration cannot be expected to give comparable linearity for equal power levels in comparison to the equivalent Class AB PA. The advantage of the efficiency characteristic, however, would seem to override linearity considerations, especially where linearization will be employed.

In conclusion, the Doherty technique appears to offer some useful possibilities for solving some of the problems which arise in both mobile and basestation amplifiers in modern wireless communications systems, and it has gradually been receiving more attention, as witnessed by many research papers in the last few years [4, 5]. But actual commercial products still seem slow to emerge and from this it is has to be concluded that some of the implementation details remain troublesome, for a technique that at the outset looks tailor-made for such applications.

10.4 Outphasing Techniques

The term "Outphasing" is somewhat ambiguous in that it can describe both a generic technique, as well as frequently being used to describe the classical high level PA modulator originally proposed by Chireix [6].

The Chireix technique dates from the 1930s, and offers a radically different approach to the PA efficiency problem. The technique was the origin of the term "LINC" (LInear amplification using Nonlinear Components) to describe linear PA systems which use highly nonlinear PAs. Indeed most LINC systems are really not amplifiers as such at all, but basically construct the required signal at the high power level. Superficially, the Chireix PA system would seem to be another strong candidate for reinvention, given the vastly improved capabilities of modern signal processing electronics technology. In fact, it seems to attract less attention than most of the other techniques described in this chapter, and has had something of a curiosity status in the context of GHz RF applications. It has to be said that Chireix's original analysis is difficult to follow, and appears to have left behind a legacy of misunderstanding and misconception in the industry as to exactly what the technique has to offer. More recent papers [7–9] have provided some helpful clarifications and underscore the real possibilities the technique presents, although successful implementations in the GHz frequency range are still quite scarce [10].

Two similar RF amplifiers are again used, but in this case both amplifiers will operate at a fixed RF drive power level and can be highly nonlinear. This is an important distinction from the Doherty technique, which requires the two constituent RF amplifiers to be linear.

The basic underlying action of the outphasing system can be simply described by a well-known trigonometric identity:

$$\cos(A) + \cos(B) = 2\cos\left(\frac{A+B}{2}\right) \cdot \cos\left(\frac{A-B}{2}\right) \tag{10.24}$$

If an amplitude modulated signal $S_{in}(t)$ is applied to a phase modulator, it is in principle possible to generate two equal, fixed amplitude signals $S_1(t)$ and $S_2(t)$, such that if

$$S_{in}(t) = A(t) \cdot \cos(\omega t)$$

then

$$S_1(t) = \cos\left(\omega t + \cos^{-1}\left(A(t)\right)\right)$$
$$S_2(t) = \cos\left(\omega t - \cos^{-1}\left(A(t)\right)\right) \tag{10.25}$$

So that if the amplifiers have a voltage gain of G, the combined output using (10.24) will be

$$S_{out}(t) = G \cdot \left(S_1(t) + S_2(t)\right)$$
$$= 2 \cdot G \cdot A(t) \cdot \cos(wt) \tag{10.26}$$

So the key element in an outphasing PA is the generation of two constant amplitude input signals which have the amplitude modulation information encoded as a differential phase shift. These signals can then be amplified by a nonlinear device and the original AM is then recovered in a summing operation. Furthermore, this procedure operates in an orthogonal manner to any phase modulation on the original signal, so signals having both AM and PM can be fully reconstituted at the output. So ideally the final distortion levels at the PA output depend on the integrity of the AM to PM conversion process in the modulator, and not on the PAs themselves, which operate at constant RF signal amplitudes.

At this stage, the outphasing technique appears certainly to qualify as a LINC scheme, but it is not yet clear how it can be described as an efficiency enhancement technique. Although it permits amplifiers of higher efficiency to be used, the overall efficiency will still scale with the envelope amplitude. This is illustrated in Figure 10.14, where the outputs of two efficient, saturated PAs are combined using a conventional power combiner. If the two inputs are offset in phase, the output power from the combiner will show the required amplitude control characteristic. However, in this configuration the amplifiers will consume the same power regardless of the outphasing action. This results in a very poor PBO-efficiency characteristic, also shown in Figure 10.14. Chireix's critical innovation was to modify the properties of the power combiner such that load modulation can take place between the two active RF devices, which leads to a much improved PBO efficiency. Before considering this next step, it is worth noting that the configuration shown in Figure 10.14 does have some advantages over a conventional amplifier having a similar PBO-efficiency characteristic. If the phase control is accurately implemented using digital look-up tables (LUTs), the resulting amplifier can be very linear, and its efficiency characteristics must be judged accordingly. It is also worth noting that the amplifiers themselves remain efficient, and do not dissipate much heat. The "wasted" power from the outphasing process could in principle be conducted away to a remote point and used as a source of energy. The rectification of RF signals to generate DC power has become a subject of increased importance in recent times, with applications in Radio Frequency Identification ("RFID"), and wireless power transmission, and will be discussed further in Section 10.9.

Figure 10.15 shows a modified schematic diagram, which is in essence the configuration proposed by Chireix [6]. The output voltages of the two nonlinear RF power amplifers are connected differentially to a common series load resistance, so the two RF generators have voltages of the form

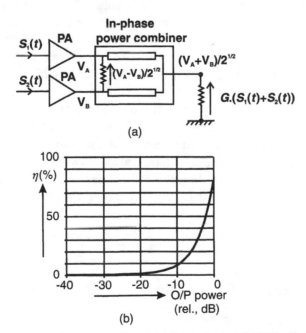

Figure 10.14   "Generic" outphasing amplifier: (a) schematic, and (b) PBO-efficiency characteristic.

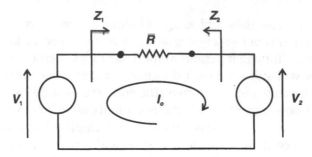

Figure 10.15   Outphasing amplifier, output schematic.

$$V_1 = V \cdot (\cos\phi + j\sin\phi)$$
$$V_2 = V \cdot (\cos\phi - j\sin\phi) \tag{10.27}$$

The differential output connection effectively forms the difference (V_1-V_2) between the two output signals across the series load resistance. This actually has little impact on the result; the RF voltage across the load resistor will now be proportional to $\sin\phi$; so that if the input modulator generates a phase shift such that

$$\phi = \sin^{-1} A(t)$$

then the final differential voltage at the load resistor will show full recovery of the original AM signal. Once again, for clarity, any phase modulation on the original signal has been omitted in the analysis, but will in principle pass through the system unmodified.

It must be noted that the original formulation used current generators, and we have switched to voltage generators in Figure 10.15. The use of voltage generators to represent the RF output of a transistor PA requires some justification. This point is not always made clear in older references, which used tubes as the RF power device. The key issue is that these amplifiers, whether tube or transistor, are assumed to be heavily saturated, so that the device will inevitably have "rail to rail" voltage swing. Under these conditions, as was discussed in Chapter 6, the device can be approximated as a fixed RF voltage generator, having essentially the rail voltage as amplitude, driving across the selected RF load. The actual RF voltage amplitude V in (10.27) can be approximated to equal the DC supply voltage in the sinusoidal case of shorted harmonics (Class B,C) or scaled up by $4/\pi$ for the square-wave case (Class D) where a selective even harmonic short has been provided.

Referring to Figure 10.15, the circulating current, I_o, is given by

$$I_O = \frac{V_1 - V_2}{R_L}$$

so that the effective RF load "seen" by the V_1 generator is

$$Z_1 = \frac{V_1}{V_1 + V_2} \cdot R_L$$

$$= \frac{\cos\phi + j\sin\phi}{2j\sin\phi} \cdot R_L \qquad (10.28)$$

$$= \frac{R_L}{2}(1 - j\cot\phi)$$

So the effect of the outphasing modulation on the RF load seen by the V_1 generator can be represented by the circuit shown in Figure 10.16(a); a resistive component equal to half the load resistance with an additional series capacitive reactance which is a function of the phase modulation angle ϕ. So in this case, the phase difference between the two generators is causing a reactive component in the device load. Clearly, $\phi = 90°$ corresponds to maximum in-phasing which is required to obtain a peak envelope output, and the reactance is zero. As ϕ decreases towards zero, the outphasing action reduces the composite output envelope amplitude, and the capacitive reactance component has a bigger impact on the RF load, causing lower efficiency operation. If the series configuration in Figure 10.16(a) is transformed to the parallel form, shown in Figure 10.16(b), it can be seen that the capacitive susceptance shunts the voltage generator and can therefore be cancelled with a shunt inductance, jX_{comp}, without affecting the AM reconstitution process. A similar argument can be applied to the V_2 generator, where a capacitive compensation of the same magnitude is indicated. So the final outphasing amplifier schematic, as proposed by Chireix, is shown in Figure 10.17.

The magnitude of the compensating susceptances requires careful selection, since the optimum value will clearly vary with the phase angle ϕ. If X_{comp} is too small, good compensation will be obtained at low outphasing phase angles (corresponding to low envelope amplitudes), but the efficiency at high phase angles (corresponding

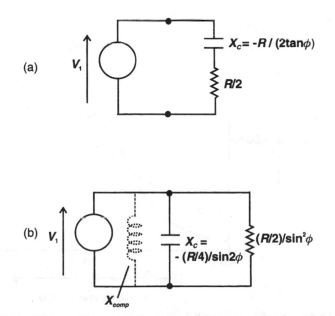

Figure 10.16   Outphasing analysis, equivalent circuit at generator #1: (a) series, and (b) parallel.

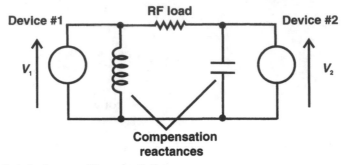

Figure 10.17 Outphasing amplifier schematic.

to higher envelope amplitudes) will be lower than in the uncompensated case. Chireix computed PBO efficiency characteristics, similar to those shown in Figure 10.18. These curves make a number of assumptions and idealizations, which can be summarized as follows:

- The outphasing angle ϕ can be converted into a resultant envelope amplitude in dB, using the mathematical relationships in (10.24) through (10.26).
- The DC supply to each amplifier will reduce as the outphasing angle increases, following a $V_{dc}/\pi|Z_1|$ relationship with the outphasing impedance Z_1 [from (10.28)].
- The final efficiency, at any particular value of the outphasing angle, will be the base efficiency (with no outphasing) multiplied by the power factor of Z_1.

These assumptions enable a direct comparison to be made between the Doherty scheme described in the last section (shown dotted in Figure 10.18), and an optimized outphasing system. On the basis of this very idealized analysis, the outphasing

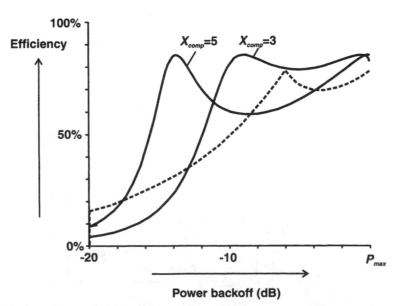

Figure 10.18 Outphasing amplifier efficiency for two values of compensation reactance (solid, X_{comp} values normalized to R/2), compared to Doherty amplifier (dotted).

system would seem to be a conspicuous winner, being apparently able to hold its efficiency close to the maximum efficiency over nearly a 10 dB output power range. Unfortunately, both simulation and experimental results appear to fall well short of the idealized analysis.

The schematic shown in Figure 10.17 is still some way from a realizable circuit at GHz frequencies. Conspicuously, the whole analysis has assumed a differential configuration, which is not favored at GHz frequencies. This situation is particularly troublesome in that the action of the outphasing does not preserve a virtual ground at the midpoint of the load resistor. This means that most of the balanced-to-unbalanced transformers (baluns) commonly used at low GHz frequencies may not work properly. An alternative configuration was reported by Raab [8], Figure 10.19, whereby quarter-wave transmission line transformers are used at the output of each device to transform the assumed voltage generators into current generators. Thus the two outputs can be summed using a common grounded load connection. Simulation of this circuit seems to indicate that the classical behavior of a Chireix outphasing system is preserved, although the action of the compensating reactances becomes more difficult to realize [2].

The overhead requirements, in terms of generating the drive signals and inevitable DSP corrections, have been something of a detraction from the Chireix outphasing system, in comparison with other techniques described in this chapter. Simulations of practical configurations show some significant deviations from the idealized theory as presented in this section, especially in the action of the compensating reactances. It seems however that useful performance can be obtained using a basic uncompensated combining structure [10].

10.5 Envelope Elimination and Restoration (EER)

Both the Doherty and the Chireix techniques use concepts which are somewhat unfamiliar to RF designers and require some effort to understand on a first pass. Envelope elimination and restoration (EER) is, by comparison, quite easy to understand. EER is a logical alternative to high level amplitude modulation, which was for many years the standard method for modulating AM transmitters. Kahn [11] originally proposed the EER technique as a more efficient alternative to linear Class AB RF power amplification for single sideband (SSB) transmitters. The SSB modulation technique was, in some respects, a forerunner of modern digital modulation

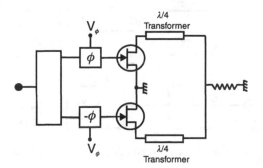

Figure 10.19 Practical configuration for Chireix outphasing system.

systems in that some information was carried in the phase, as well as the amplitude of the modulated RF carrier. This meant that the final RF power amplifier could not be modulated by simple control of the DC supply as was common practice in tube PAs, and the obvious alternative was to amplify the composite low level signal using the same device operating in a linear Class AB mode, rather than the more efficient Class C.

Kahn demonstrated a system, shown in Figure 10.20, where the phase modulation of the input signal was preserved by passing it through a limiter. The limiter, ideally, eliminates the possibility of AM-PM distortion in the nonlinear RFPA and so the output of the PA still retains the original, undistorted, phase characteristics of the input signal. The envelope amplitude can be "restored" at the output by using a conventional high level voltage supply modulation, but in this case the modulating signal is derived from an envelope detector, followed by suitable conditioning and linearization functions. In principle, the high level AM modulation process is much more efficient than a linear amplification approach; firstly because the amplifier itself can be saturated and nonlinear, just as in the outphasing approach, but the efficiency of such an amplifier remains essentially constant as the supply voltage (and consequently the output power) is reduced. Recall once again, as discussed in the last section and more extensively in Chapter 6, a well saturated amplifier can be modeled as an RF voltage generator whose amplitude is equal, or proportional to, the "DC" supply voltage. So to a first order of approximation, the RF envelope amplitude will be proportional to the modulating supply voltage, and no functionality needs to be applied to the modulating signal to achieve the desired result.

In practice, things are not quite so straightforward. In particular, the process of amplifying the detected envelope signal up to the necessary voltage and current capacity to modulate the PA device will consume a significant amount of power. If this conditioning process can be done with 100% efficiency, then high PA efficiency could be maintained over a substantial range of envelope amplitude. This would make the EER technique a clear winner among the three efficiency enhancement techniques considered so far, and probably explains the level of resurgent interest in EER [12–15]. Notwithstanding the challenge of implementing a suitable power converter, it should be further noted that the EER technique is potentially immune to RF bandwidth and matching issues, which become severe limitations of the Doherty and Chireix techniques for bandwidths greater than 10%.

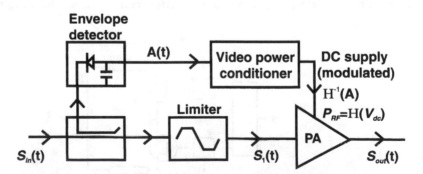

Figure 10.20 Envelope Elimination and Restoration ("EER") system.

Modern approaches to envelope restoration usually dispense with the limiter as a means of generating the necessary constant amplitude,[1] phase modulated signal. In a modern system, it is now the current trend to assume that the necessary phase and amplitude modulation drive signals can be generated directly in the system digital processor, giving a basic system block diagram as shown in Figure 10.21.[2] Confusingly, such modern derivatives of the Kahn EER technique have become known as "Polar" PA systems, having some generic similarity to the "Polar Loop" feedback linearization system (see Chapter 14), but with the notable distinction that in some cases they remain open-loop configurations.

ER systems present a number of challenges, in addition to the obvious one of realizing a high efficiency power converter. The assumption that the phase modulation on the input to the nonlinear RFPA will be preserved is quickly invalidated when the supply is varied, and there is a serious issue of dynamic range when attempting to implement ER for a zero-crossing signal. The latter problem has potential solutions, such as the use of tracking bias controls on the driver stages. Linearity problems, as always, can in principle be handled using DSP techniques, however this introduces some serious overheads to the cost and complexity of the system. In particular, the need to calibrate each individual RFPA device characteristics is a detraction from the ER and Polar techniques.

10.6   Envelope Tracking

In the EER configurations considered in the previous section, the amplitude modulation was created by modulating the supply to a well-saturated RFPA stage. This requires great accuracy in the generation of a suitable supply voltage and may have dynamic range limitations. There are some interesting potential benefits in applying

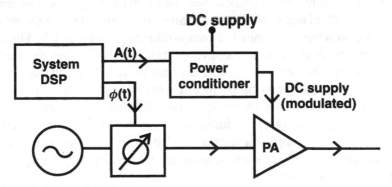

Figure 10.21   "Polar" modulation system.

1.   Indeed, it is not clear as to how Kahn realized this function for zero-crossing modulation systems; his much quoted publications gave no details on this critical item.
2.   It is not clear why such systems are still given the generic label of "EER," since there is neither elimination nor restoration of the envelope, it is simply "constructed" by suitable amplitude modulation of the supply to the RFPA. Replacing "EER" with "C" does not appear to have much support at present, and "ER" would seem to be a good compromise in terminology.

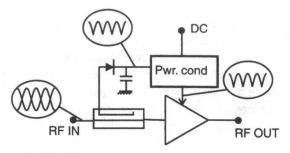

Figure 10.22   Envelope Tracking ("ET") RF PA system.

a similar envelope-derived modulation to the supply voltage of a conventional linear RF amplifier. Such a technique is generally termed "Envelope Tracking," or "ET," and is shown schematically in Figure 10.22.

Recalling the analysis of a conventional Class B amplifier (Figure 10.1), we have already seen that if the RF load resistor is decreased in inverse proportion to increasing drive voltage amplitude, the efficiency remains constant at its maximum value while the power increases as the square root of the drive power; this is an important mode of dynamic behavior in the Doherty amplifier (Section 10.2). An alternative, and considerably simpler, scenario is that the load resistor remains fixed and the supply voltage is increased in proportion to the increasing drive voltage. In this case, maximum efficiency is maintained (due to full rail-to-rail voltage swing) and the output power increases linearly with input drive power. This process can be continued down to a selected point on the lower side, and up to a point where the RF swing reaches breakdown level. In this manner, maximum efficiency can be maintained over a wide linear power range.

One of the appealing aspects of this technique is that the modulation control voltage does not have to replicate the signal envelope with great accuracy, as is the case in the ER process. For example, the supply voltage could be tracked for just the upper few dB of the signal envelope range; this would then show an overall PBO efficiency characteristic somewhat comparable to a Doherty PA. The still-present challenge of the tracking power converter can also be further reduced by the use of a power supply having two or more discrete switched output voltage levels. Problems of dynamic range, particularly in zero-crossing signal environments, are essentially eliminated by allowing for normal linear operation at a reduced supply voltage in the small signal regime. An additional attraction of ET is that the efficiency enhancement process is completely decoupled[3] from the RF matching. This is in contrast to the Doherty and Chireix configurations, which depend heavily for their operation on resonant RF circuit elements.

It is instructive to consider a simple case, where two supply voltages are available, as shown in Figure 10.23. The supply switch can, in principle, be realized using low cost and readily available semiconductors, which for envelope speeds in the MHz range do not have to be exceptionally fast, by modern standards. The drive signal to the switch would conventionally be derived by using an envelope detector

---

3.   Both literally and metaphorically; the physical integration of power converter and RFPA circuitry is one of the less publicized challenges in implementing ER systems.

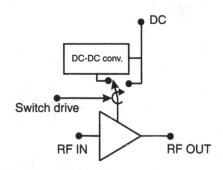

Figure 10.23 ET System using a two-level switched supply.

and a threshold circuit which sets the level at which the switch was activated. A modern implementation would probably use a drive signal generated by the system DSP, which could conveniently be time-synchronized with appropriate pre-distortion of the input RF signal. This will include necessary compensation for the small changes in gain and phase in the RFPA as the supply is switched. The "break-point," or envelope level at which the supply is switched, is clearly an important design choice. This choice will have an optimum value, dependent on the statistics of the signal environment. As with the selection of the breakpoint in a Doherty PA, the intuitive choice would be to set the lower voltage to switch in at the mean power level, the higher supply being used up to the peak power. The efficiency characteristic shown in Figure 10.24 assumes that the PA follows the ideal Class B PBO/efficiency characteristic for each supply voltage setting, reaching a maximum value of 70% at the onset of envelope clipping in each case.

In order to evaluate the impact of such a characteristic in a specific signal environment, it is necessary to generate a representative signal burst and evaluate the average efficiency. This process can be repeated for various breakpoint and PEP backoff settings. For example, Figure 10.25 shows the results from performing this exercise for a two-level supply voltage on an EDGE signal. The average efficiency

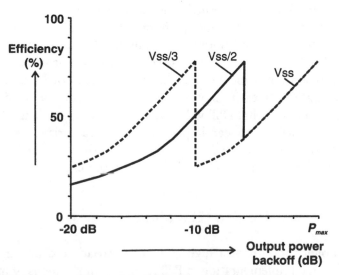

Figure 10.24 PBO-Efficiency characteristic for two-level switched ET system.

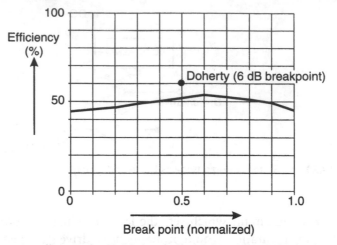

Figure 10.25 Average efficiency for EDGE signal, Class B PA with switched voltage supply; (switch break point plotted as x-axis). Doherty PA efficiency shown for standard 6 dB break point.

for the two-level supply switch is plotted for a range of breakpoint values. It is clear that a simple two-level switch, in this signal environment, can give a useful increase in efficiency (53.5% at optimum break point) in comparison to a standard Class B PA (45.3%), but falls significantly short of the Doherty PA result for the same signal environment (60.3%). In the case of an EDGE signal, it will in practice be necessary to back the PEP level down from the clipping point in order to obtain acceptable ACP performance, this will reduce all of the efficiency results by a similar factor. The efficiency gains in Figure 10.25 seem to deliver somewhat less than the PBO efficiency curves might suggest. The returns are, however, quite critically dependent on the signal environment. Some further details of the method used to perform the efficiency computations contained in Figure 10.25, and some more simulation results, will be presented in Section 10.10.

As is customary when promoting efficiency enhancement techniques, little has been said about the possible effects of the technique on the PA linearity. As far as envelope tracking is concerned, this has to be raised as a potentially serious issue. The next chapter in this book, Chapter 11, is largely devoted to minimizing the harmful effects of small amounts of unintended modulation on the supply voltage to an RFPA. Clearly, when it is proposed that this same supply voltage be intentionally yanked around over most of its allowable range in order to improve the efficiency, questions about linearity have to be addressed. The standard answer, as always in the modern era, is that DSP will save the day. The gain and phase variations which the tracking will create can in principle be characterized and compensated by suitable adjustment (DPD) to the input signal. But there can be little doubt that the large supply variations will make the DPD task more challenging.

10.7 Power Converters for EER and ET

The implementation of a high efficiency, broadband power converter design for EER and ET implementation in RFPAs has been the focus of much research activity for many years. In older times, the converter only had an audio signal to track, and the use of a modulated supply voltage and a Class C RFPA was normal practice in the tube era for the generation of an amplitude modulated RF power. It is still instructive to look at a typical tube AM transmitter schematic from the 1960s era, such as that shown in Figure 10.26(a). The key components are a Class C RFPA stage, probably operating in the region of 85% efficiency, a necessarily linear audio power amplifier, and a "modulation transformer," whose secondary winding is connected between the decoupled RFPA supply point and the prime DC supply. The audio power amplifier had to supply sufficient voltage swing, through the coupling of the transformer, to modulate the RF output.

Glassware notwithstanding, this system has some features which are conspicuously absent in most modern ER configurations. The use of a linear Class B push-pull audio amplifier is of interest, but most noteworthy is the division of the supply power between the PA "B+," which is an unmodulated DC supply point, and the entirely AC audio power supplied by the modulator amplifier. This configuration is able to conserve the audio power requirement from the modulator through the continued use of a simultaneous DC supply to the RFPA.

This is worth a brief quantification. If the mean DC supply (B+) to the RF tube plate is V_{dc}, then the action of the audio amplifier and modulation transformer is to provide a swing of $\pm V_{dc}$ around this mean value. Assuming that the RF tube presents a high impedance R_T to both the B+ supply and the modulating AC, the DC component of power supplied to the RF tube will be

$$P_{dc} = \frac{V_{dc}^2}{R_T}$$

and the AC audio power will be

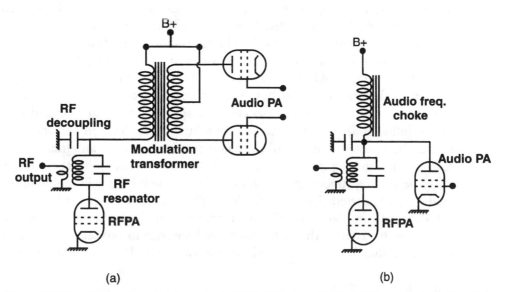

(a)   (b)

Figure 10.26   High level modulation of tube RFPA (EER circa 1960): (a) plate supply modulation using transformer, and (b) "choke" modulator.

$$P_{AC} = \frac{V_{dc}^2}{2R_T}$$

for the case of a sinusoidal modulating signal, or "tone."

Thus the audio component of the power supply to the RF tube is only one third of the total power, 85% of which is duly converted into RF energy. So even if the audio amplifier average efficiency is 33%, the overall efficiency of the modulator is still a respectable 60%, assuming that the B+ voltage is a prime supply.

The recognition that a high percentage of the power supplied to the RFPA is still at DC can be a major efficiency savior, so long as it can be separated from the AC component. This concept becomes clearer in an alternative implementation, of considerably greater antiquity,[4] shown in Figure 10.26(b). This arrangement was superseded by the use of a modulation transformer, which gave the additional feature of impedance matching between the audio stage and the RFPA supply. But there remains an important lesson here, which is the value of removing the DC component from the required envelope modulation drive signal and using a DC supply to provide it.

This arrangement is quite distinct from a typical modern EER configuration where the "modulator," or power converter, is the sole source of supply for the RFPA. Modern communications systems, of course, greatly extend the modulation bandwidth requirement from audio frequencies into tens of MHz, which along with important differences between tube and semiconductor RF power devices, has resulted in a significantly different approach to ER implementation. Unfortunately much of the extensive development work in this area has been wrapped in corporate secrecy, and is only accessible through a mountain of patents. Techniques which have already been mentioned, such as the use of multiple switched levels for envelope tracking, have been an especially active area for patent seekers [16–18]. Echoing Figure 10.26(b), perhaps, various schemes which make use of an inductor to provide selective routing between different supplies have also been patented [19–21]. More conventional approaches, which use so-called "Class S" switching power converters, are older in genesis and can be adapted to work at higher frequencies if faster switching transistor technology is employed. The logical choice is to use the same device technology for power converter and RFPA [22], but this raises cost issues.

Two techniques appear to have been around long enough to qualify for safe-harbor status in this patented minefield. The split band modulator and the humble source follower are therefore worthy of a little more detailed consideration. It has already been noted that one of the challenges in designing an efficient power converter for use in ER systems is the bandwidth required to replicate the envelope and phase describing functions, which are usually much higher in bandwidth than the final fully formatted signal itself. This is illustrated in the case of a single channel EDGE signal, in Figure 10.27, which shows that the bandwidth of a power converter used to generate the high power envelope tracking voltage would need to be at least three times higher than the signal bandwidth itself. This requirement is moder-

---

4.   Such circuits can be found in many older radio textbooks which can still be found on library shelves; in this case the source was the *U.K. Admiralty Handbook of Wireless Telegraphy, Volume 2,* U.K.: H.M.S.O., 1938.

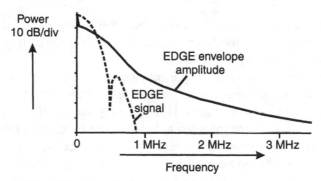

Figure 10.27   Bandwidth of EDGE envelope, compared to composite baseband signal.

ated by the observation that a high proportion of the energy resides in a spectral range significantly smaller than the signal channel.[5]

This will arguably be the case for the kinds of signals used for digital communications [23]. In any event, it can be asserted that the DC component alone will still constitute a high percentage of the envelope drive energy. For example, using the data in a spreadsheet such as **EDGE_effy**, it can be quickly shown that for an EDGE envelope, the DC component is 85% of the total power.[6]

It has been proposed [12, 24, 25] that a split-band technique could therefore be used for an ER power converter, as shown in Figure 10.28. Conventional high efficiency switching power converters can be realized readily for frequency bands up to about 20 kHz, and the upper band can use a linear broadband amplifier having lower efficiency. The challenge in this approach appears to be the design of suitable diplexing networks, coupled closely with the accuracy of the underlying assumptions in the selection of the break point frequency. Recalling the above discussion on

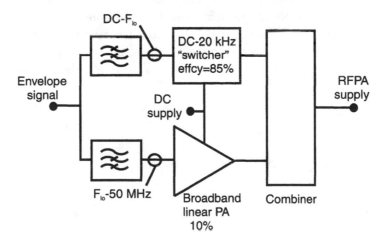

Figure 10.28   Split band ER modulator concept.

5.   This may seem a surprising fact when looking at conventional logarithmic spectrum display, such as in Figure 10.27, but displaying the same spectrum on a linear scale gives stronger support to this statement.
6.   This result gives a somewhat optimistic view on the requirements for generating the AC component, due to the fact that the converter has to supply a peak, rather than a mean, AC component.

tube modulation, it has to be speculated whether such systems are in effect reinventing the function of a modulation transformer, thus enabling the DC component to be generated with high efficiency.

The most obvious and simple approach to the generation of a high current tracking signal, whose voltage tracks a low-level drive signal, is the source follower; a configuration shown in Figure 10.29. It may appear that such a configuration will be of little value, since the power that is saved by tracking the PA supply voltage is dissipated in the "pass" transistor. Closer inspection however suggests that some significant benefit can be obtained from this simplest of solutions, so long as the control voltage can be specified with reasonable accuracy. Figure 10.30 shows how the supply to the PA can be tracked along a line of constant gain, taking account of the typical variations of gain and gain compression of an RF power transistor. Such a device operates in a twilight zone between ER and ET, and the possible outcome is shown in Figure 10.31. A conventional Class AB PA operating with fixed bias will show significant gain compression at, say, the 70% efficiency point. In most applications this will mean that the peak power level of the signal would have to be backed off, as shown in Figure 10.31. The source-follower tracked device could possibly be kept at a higher efficiency level, maybe 80%, in the supply-tracking regime. Thus the two PBO-efficiency curves will be well separated, and for a signal with a PAR in the 6 dB region this could translate into an average efficiency difference of 10%.

Some additional comments on envelope-tracking converters will be made in Section 10.9, under the heading of RF to DC conversion.

10.8   Pulse Width Modulation (PWM)

A more radical approach to solving the power converter problem is to dispense with the analog drive signal to the PA altogether, and use pulse width modulation (PWM) to encode the amplitude information. For example, in Figure 10.32, the RFPA is pulsed either on or off by a suitably buffered 1-bit digital drive signal. This switching can be implemented either as modulation on the input signal or (as shown in Figure 10.32) by switching the supply to the RFPA. As the pulse on/off duty cycle is varied,

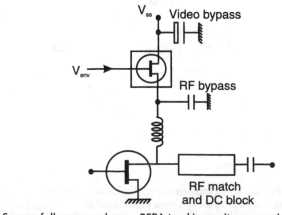

Figure 10.29   Source-follower used as an RFPA tracking voltage supply.

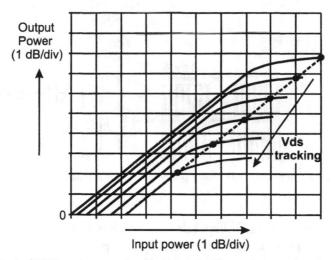

Figure 10.30 Typical RFPA power transfer characteristics for varying supply voltage.

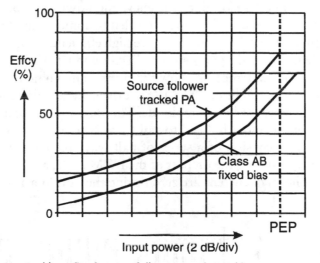

Figure 10.31 Conceptual benefit of source-follower supply tracking, versus conventional Class AB PA having fixed bias.

this will generate an output spectrum from the RFPA, which consists of a central carrier now having amplitude modulation, as well as some modulation sidebands which can in principle be removed using an output bandpass filter. Such techniques have been widely implemented in the audio power amplifier industry, and the possibility of applying the same concepts to high efficiency RFPAs has attracted some commercial interest from this distant industry sector. It is, inevitably, another patented minefield.

Whether these techniques justify the claims made for them still remains to be seen. The manifestations and motivations for realizing high efficiency amplifiers are quite different in the audio and RF sectors, and the role and requirements of the output filter are very much at the center of this burgeoning debate. A point which seems frequently to be missed is that the output filtering process will always reduce the RF

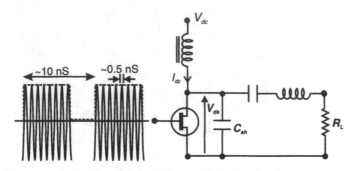

Figure 10.32 RFPA supplied with a pulse-width modulated RF signal.

voltage swing at the RF output device, which will in turn reduce the efficiency as the PWM duty cycle is reduced. Indeed, this process is basically the same as that observed in a conventional linear RFPA, and can be shown to give the same familiar PBO characteristic considered in earlier chapters in this book. This troublesome subtlety often seems to be overlooked, when at first sight it appears that the RFPA is either off, or in a state of high efficiency, and nowhere in between. This misreading of the true operation of the system has led to some wildly misleading claims about the potential of such PA systems to give high efficiencies, in the 90% region, even for amplitude modulated signals such as EDGE and CDMA.

Figure 10.33 attempts to show the current and voltage waveforms[7] in a reduced envelope power condition. The PA is driven by a pulse-width modulated signal, which has a constant "on" amplitude. This amplitude is assumed to be sufficient to drive the device into a full rail-to-rail Class B condition. The device current waveform therefore consists of a series of half-wave rectified sinusoidal pulses, having amplitude I_{max}, during the on part of the PWM cycle, and of course both the RF and DC components of current are zero during the "off" part of the cycle.

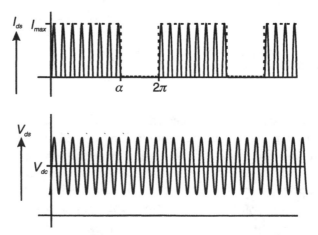

Figure 10.33 Waveforms for PWM RFPA.

---

7.   As always with such waveforms, in order to show both the envelope and the RF waveforms it is necessary to "slow down" the RF carrier cycles.

The output circuit is familiar enough, and is much the same in schematic terms as a normal Class B amplifier, the only difference being some added stipulations about the passband of the fundamental matching resonator. The shunt capacitor, C_{sb}, is assumed to be a large capacitor, whose reactance is very low compared to R_L, the conventional loadline resistance for the device. Thus at all relevant frequencies above, or *in the vicinity* of the fundamental, this capacitor terminates any nonfundamental components of current with a short circuit so that no voltage is generated. The series resonator cancels out this capacitive reactance at the carrier frequency, and presents a resistive impedance R_L to the device. The key point here is that for the purposes of this analysis, the first sideband frequencies of the PWM modulation, which will be separated from the RF carrier frequency ω by the PWM frequency Ω, are assumed to fall *outside* the fundamental resonator passband, and thus also shorted by C_{sb}.

Thus the voltage waveform at the device output is a sinewave at the carrier frequency, whose amplitude is the product of the fundamental component of the modulated carrier, with the load resistor R_L. The current waveform can be written in the form

$$I_d(t) = I_{max}\big[p(t)\big]\sin \omega t \quad \sin \omega t > 0$$
$$= 0, \quad \sin \omega t < 0$$

where

$$p(t) = 1,\, 0 < t < \alpha,$$
$$= 0,\, \alpha < t < 2\pi/\Omega \ (\Omega \ll \omega)$$

The PWM amplitude function, $p(t)$, can be expanded as a Fourier series,

$$p(t) = p_0 + p_1 \cos \Omega t + p_2 \cos 2\Omega t + p_3 \cos 3\Omega t + \dots$$

and by inspection, clearly we have $p_0 = \dfrac{\alpha}{2\pi}$ so that the fundamental component of the current waveform is

$$I_1(\alpha) = \frac{\alpha}{2\pi}\frac{I_{max}}{2} \tag{10.29}$$

It is important to recognize that this component, if filtered out from the pulsed spectrum, is a *continuous cw carrier*, just in the same way that the filtered carrier of an amplitude modulated signal is continuous; the modulation appears only in the sidebands[8]

---

8.   It is worth noting that the physical existence of cw carrier and modulation sidebands as separate entities in the frequency band was hotly debated in the early days of radio, before receivers could be made selective enough to prove that the mathematical analysis was correct. It seems that there is a residual reluctance in some circles even today, in accepting that the voltage in Figure 10.33 will indeed be continuous across the "off" condition of the PWM, due to the action of the filter.

So the device output voltage waveform shown in Figure 10.33 is a continuous sinewave at the fundamental frequency, whose amplitude will vary in direct proportion to the duty cycle, $\alpha/2\pi$, and will reach its maximum rail-to-rail value of V_{dc} when $\alpha = 2\pi$. Thus RF power at PWM angle α is

$$P_{rf} = \left[\frac{\alpha}{2\pi} V_{dc}\right]^2 \bigg/ 2R_L$$

where

$$R_L = \frac{2V_{dc}}{I_{max}}$$

The DC power is

$$P_{dc} = V_{dc} \cdot \frac{I_{max}}{\pi} \cdot \frac{\alpha}{2\pi}$$

so the efficiency is

$$\frac{P_{rf}}{P_{dc}} = \frac{\alpha}{2\pi} \cdot \frac{\alpha}{2\pi} V_{dc}^2 \frac{I_{max}}{4V_{dc}} \frac{2\pi}{\alpha} \frac{\pi}{I_{max}} \frac{1}{V_{dc}}$$

which reduces to

$$\frac{P_{rf}}{P_{dc}} = \frac{\pi}{4} \frac{\alpha}{2\pi} \tag{10.30}$$

This shows that both the efficiency and the output RF voltage magnitude are inversely proportional to the PWM duty cycle. Thus the PBO efficiency behavior of this system is the same as an ideal, conventional Class B PA, a highly disappointing result.

The PWM approach to improving PA efficiency has thus run aground[9]; in terms of efficiency this system does not offer a comparable alternative to an ER or ET system which uses an efficient power converter. On the other hand, the PWM PA does have some interesting features. The use of an RFPA at only two distinct RF drive levels has some attractions. The power and linearity become essentially decoupled from the detailed quirks of the active device in question, and almost entirely controlled by the single-bit drive signal, which in turn can be easily generated by the system DSP and contain appropriate corrections for linear operation. Memory effects can likely be minimized by ensuring that the power dissipation in the device is maintained constant at the two drive levels. But such advantages are heavily offset by the need for an output filter to clean up the spectrum. More detailed analysis on the requirements

9.   Protagonists for PWM techniques will argue that this conclusion is unfairly based on a specific circuit configuration which short-circuits all the PWM harmonic components. It is conceivable that configurations may be proposed that yield better results.

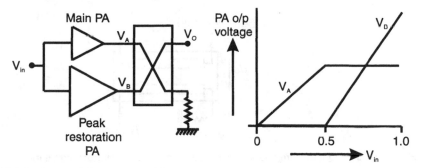

Figure 10.34 Sequential Power Amplifier.

of this filter in a typical case indicates that it would likely be a substantial component, in terms of cost and size.

The use of PWM techniques in RFPAs continues to be a very active area for R&D groups both in industry and academia, and it remains to be seen whether these efforts will be justified.

10.9   Other Efficiency Enhancement Techniques

As the wireless communications revolution progresses, new modulation systems seem to become ever more unfriendly, with escalating peak-to-average ratios and widening signal bandwidths. The devising of yet more alternatives to the PA power backoff efficiency problem continues to occupy the minds of PA designers, and this section attempts to summarize some of these ideas.

10.9.1   The Sequential Power Amplifier

The basic requirement that the active periphery of the RF power transistor needs to be reduced at lower power levels immediately suggests a configuration such as that shown in Figure 10.34. A lower power device ("main" PA) handles the signal up to a certain level, when a higher power device ("peak restoration" PA) is activated. The lower power device will start to saturate, but the higher power amplifier can generate the necessary additional power and maintain linear performance up to a higher level. This configuration offers a comparable, but more pragmatic, solution compared to the Doherty PA, but its viability depends critically on the means chosen for combining the RF output power from the two devices. A conventional directional coupler can be considered, but presents problems in terms of the power loss when one of the PA devices is shut down. A 3 dB hybrid coupler, for example, would present 3 dB of loss in the path of the main amplifier, when the peak restoration amplifier is shut down at lower power levels. Use of a higher coupling factor, say, 10 dB, would lead to wastage of power from the peak restoration amplifier in the higher power regions, due to the high coupling factor.

It is nevertheless worthwhile to consider the possibilities when a coupler in the 6 dB range is used in such a configuration. Figure 10.35 shows the voltages at the two output ports of a conventional quadrature coupler, given two inputs, V_1 and V_2 at

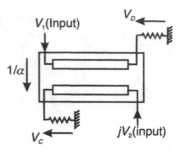

Figure 10.35 Quadrature coupler with two co-phased quadrature signal inputs.

the input and isolated ports; V_1 and V_2 represent the amplitudes of two co-phased sinusoidal signals, with a 90° phase difference. The output from the direct port of the coupler will have a magnitude V_D, given by

$$V_D = V_1 \sqrt{1 - \frac{1}{\alpha}} + V_2 \frac{1}{\sqrt{\alpha}}$$

where α is the coupling factor expressed as a power ratio; for example, for a 6 dB coupler, $\alpha = 4$. The output voltage at the coupled port will be

$$V_C = V_2 \sqrt{1 - \frac{1}{\alpha}} - V_1 \frac{1}{\sqrt{\alpha}}$$

so that for any given coupler there is a ratio between V_1 and V_2 which causes the coupled port voltage to vanish, this ratio is given by

$$\frac{V_1}{V_2} = \sqrt{\alpha - 1}$$

so if the two signal magnitudes happen to have this ratio, the power at the direct port will be equal to the sum of the two input powers. So for a 6 dB coupler, if the power input at port 1 is 3W, and the power input at port 2 is 1W, the output taken from the direct port will be 4W. So for this specific power ratio, the coupler behaves as a lossless power combiner. Even for input power ratios which are significantly displaced from the ideal value, the power delivered to the output port remains surprisingly close to the total input power. This is quantified in Figure 10.36, which plots

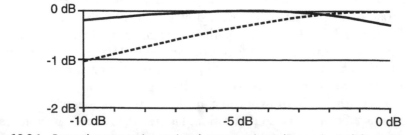

Figure 10.36 Power loss versus input signal power ratio: 6 dB coupler (solid), and 3 dB coupler (dotted).

the power loss in dB for a range of port 2 input powers around the optimum value, relative to the port 1 input.

Applying these results to the system in Figure 10.34, the peak restoration PA can supply an equal power to the main PA, and only 0.3 dB will be lost in the coupler termination. With the peak restoration PA shut down, the main PA output will be attenuated, but only by 1.25 dB. Using the standard inverse square law PBO relationship for the efficiency of the two amplifiers, and assuming the peak restoration PA is activated at the point where the main PA starts to saturate, it is possible to construct a PBO efficiency relationship for this system, shown in Figure 10.37. Although this may at first appear to be a poor relation to the equivalent Doherty response, this system has some potential advantages. In particular, it is broadband in nature, couplers being inherently very broadband devices. The two amplifiers are self-contained and are connected to well-isolated ports of the coupler. Although some further thinking is required in order to profile the necessary turn-on characteristic of the peak restoration PA, much the same considerations apply to the implementation of the auxiliary PA in a Doherty configuration. Unlike the Doherty PA, however, it may be advisable to limit the signal to the main PA in the upper regime, to prevent excessive limiting behavior. Although potentially improving the efficiency characteristics shown in Figure 10.37, the onset of AM-PM effects may be worth restricting.

10.9.2   Pulse Position Modulation

With the availability of logic families that can generate binary signals having clock rates in the tens of Gb/s range, it has become relevant to consider the possibility of generating a fully formatted digital communications signal in the low GHz range, directly from a digital source. A typical starting point would be a string of binary pulses whose repetition rate corresponded to the original carrier frequency. Phase modulation of the fundamental component can be applied by suitable variations in the phasing of the bit stream, and amplitude modulation can be realized by drop-

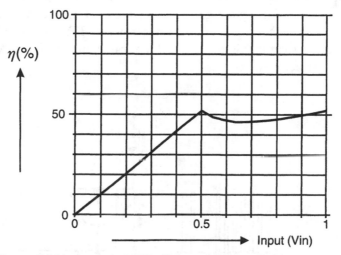

Figure 10.37   Sequential PA, theoretical PBO-efficiency (6 dB coupler).

ping pulses strategically in order to reduce the magnitude of the fundamental component. From an RFPA viewpoint, such schemes run into much the same problem as described in the above section on PWM techniques. If the stream of pulses is fed into an RF transistor, which is biased somewhat beyond its cutoff point, something similar to a Class C PA will be created, as indicated in Figure 10.38.

Once again, it is the action of the output filter that rains on the parade. As the pulse stream is modified to change the fundamental component, the sinusoidal voltage swing in the output of the RF power transistor will also modulate in magnitude, thus reducing the efficiency in the case of reduced amplitude. Much like the baseband PWM system analyzed in Section 10.8, the resulting PBO-efficiency characteristic will be much the same as a regular Class C amplifier.

This system nevertheless does have some interesting possibilities, so long as overzealous claims about its potential for 99% efficiency are quietly dropped. As with the baseband PWM system, it has a major advantage over linear, or analog, RF amplifiers in that the RF power device is only being used at a single "on" operating point. Thus the characteristics of the final system can be almost entirely decoupled from the quirky and nonreproduceable properties of the RF device. This is in contrast to the thousands, or even hundreds of thousands, of data points which are required in the look-up tables of digital correction schemes which linearize and control conventional analog RF amplifiers.

10.9.3   RF to DC Conversion

When dealing with high efficiency techniques for RFPAs, there are often times when it appears that RF power is being wasted, usually in a termination, which could be put to better use. Rightly or wrongly, in a competitive commercial environment, big orders can be won or lost on the basis of one or two percentage points of efficiency. It is therefore quite justifiable that no stone should be left unturned in the quest for higher efficiency, even when the return appears to be marginal. Two obvious examples of wasted RF power are the basic outphasing technique, described in Section 10.3 in this chapter, and the uncoupled power from the error amplifier in a feedforward loop, described in Chapter 14. A more striking application, perhaps, is the possibility of using a high efficiency PA in combination with an RF to DC converter to implement a power converter in ET or ER systems. Using the RF signal to hand, rather than introducing switching transients at kHz or MHz frequencies, has obvious attractions.

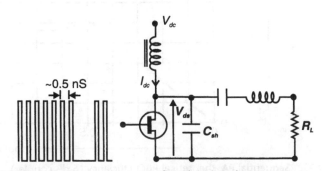

Figure 10.38   RF Power amplifier configuration for Gb/s digital input.

Rectification and smoothing of an alternating, sinusoidal, voltage into a "DC" supply lies at the heart of most electrical and electronic power systems. It is therefore logical to proceed in the same manner to convert an RF signal into a DC power source. One of the immediate differences between a 60 Hz and 2 GHz source is the impedance level at which the alternating voltage presents itself. Line supplies are "stiff," and present a voltage source with very low impedance, whereas an RF voltage source will always have a significant impedance associated with it. If we have a 1 watt source of power at 2 GHz, in the RF world what we mean is that this is the power that will be dissipated in a matched load. There is also the usual culture conflict in the use of balanced voltages at low frequencies (through the use of transformers) versus the strong preference for unbalanced circuits at GHz frequencies.

A simple half-wave rectifier circuit, shown in Figure 10.39, can in principle be used as a very efficient RF to DC converter. Such circuits are usually associated with small signal detection, and efficiency is not considered to be as important as in the case of a power converter. The key to efficient rectification is to recognize the need for a harmonic short across the sinusoidal RF source, to provide a path for the harmonic components of the nonsinusoidal rectified current. Another important design issue is to determine what matching is required in order to obtain maximum power transfer from the RF source into the rectification circuit.

Referring to Figure 10.39, the detector is initially assumed to be perfect, which means it turns on with zero resistance at zero terminal voltage, and has no parasitics. It is also assumed that we have contrived to present a fundamental impedance match to the generator, so that the voltage at the diode input is

$$V_d = V \cos \omega t$$

or half of the input signal amplitude, indicating a matched condition.

In the steady state condition, the shunt capacitor will charge up to a voltage marginally lower than V, so that the diode conducts pulses of current at the peaks of the input voltage. With no energy being dissipated in the diode, and all harmonic voltages short-circuited, it can be stipulated that the DC power in the load resistance R_v will be equal to the available power from the RF source,

$$\frac{V^2}{R_v} = \frac{V^2}{2R_o}$$

giving the somewhat unexpected result that the matched RF condition occurs when the DC load resistance has twice the value of the RF source impedance.

Clearly, higher DC output voltages can be obtained, for any given RF input level, by suitable scaling of R_v. Appropriate impedance matching would then be required on the input, so that the system interface impedance is transformed, proba-

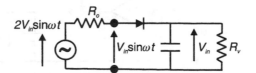

Figure 10.39   RF to DC conversion using a saturated RFPA and efficient detector circuit.

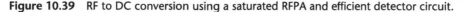

bly upwards, to $R_v/2$. This would appear to have the makings of a transformer-less DC-to-DC converter technique, using a well saturated and highly efficient PA to drive the rectifier circuit, as shown in Figure 10.40. Such a converter could be realized using the same semiconductor technology as the main PA itself, and provide a means of generating a higher voltage supply for use on amplitude peaks. The actual RF power required from the converter PA could be estimated as being equal to the system peak power, scaled down by the duty cycle of peak usage of the higher voltage. Thus it would seem that the converter PA transistor could be significantly smaller than the main RFPA device.

Unlike the design of detectors for receiving applications, the choice of bypass capacitance is uncritical, since in this application the detected output is being used to establish a reservoir of charge at a fixed voltage.

In practice, the non-ideal characteristics of the diode will reduce the efficiency, but measured values in excess of 80% have been reported [26, 27]. One of the problems associated with trying to design such circuits for higher power levels is the lack of microwave detector diodes which are capable of handling currents higher than a few tens of milliamps. It would seem quite feasible, in principle, to manufacture multiple cell diodes, in much the same way that RF power transistors are designed. It has to be assumed that diode manufacturers do not perceive this to be a profitable area at the present time.

10.9.4   RF Switching Techniques

RF switching technology has advanced in recent times, driven mainly by the need for switched frequency band coverage in mobile phones. Usually implemented using GaAs PHEMT technology, multiple pole, multiway RF switches are available which can pass several watts of RF power with low loss and introduce minimal extra distortion. Although not typically fast enough to be able to implement configurational changes at envelope speeds, it would be quite feasible to shut down an output stage and connect the driver to the antenna when the operating conditions require low power. Changes could also be made to the matching network, by switching to alternate paths between the PA output transistor and the antenna. Different matching in each path could effect a form of slow-tracking load modulation, maintaining a high RF voltage swing at low power levels. These options are illustrated in Figure 10.41. Such slow-tracking techniques, which adapt the system to changes in average power, can have as much impact in long-term PA power consumption than the more exotic techniques which seek to operate at the much higher envelope speeds.

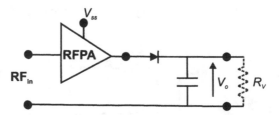

Figure 10.40   DC-to-DC converter using an RFPA and RF to DC converter.

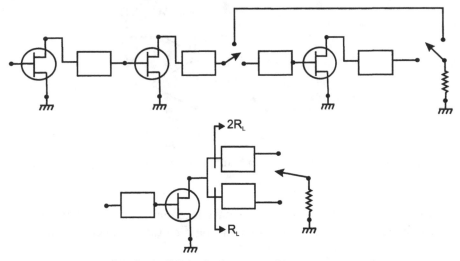

Figure 10.41   "Slow-tracking" possibilities for long-term PA power conservation.

10.9.5   Smart Antennas

The use of so-called smart antenna systems is a growth area, especially in WLAN and WiFi applications. It is therefore becoming a subject in its own right, but it is worthwhile to consider one aspect of the subject, viewed from an RFPA perspective: the ability of a pair of antennas to deliver a power-combined signal at a distant receiver, regardless of the individual power levels. This is a function which, as we have already seen in Section 10.9.1, can only be performed, frustratingly, at circuit level for specific power ratios. The most common example of this is the 3 dB hybrid combiner which will combine equal, co-phased signals without loss of power. Any change in this power ratio will result in power being lost in the fourth port of the hybrid.

Given that in the case of a radio transmitter circuit the final output will be radiated from an antenna (i.e., the actual power is never used directly, other than to measure the output with a power meter), it is reasonable to ask whether "space combining" using two antennas, may be a better option. If this comparison is done at the receiver end of the link, the power received will be proportional to the sum of the two powers radiated from each antenna, regardless of the power ratio, provided that the signals from each antenna are in phase with each other. For example, recalling the configuration in Figure 10.34, where a larger PA output device is combined with a smaller one, with the intent of using the larger device only on amplitude peaks, we have seen that the circuit combining techniques available will always result in some wastage of power. In particular, the smaller main device will have to work through an attenuation anywhere between 1 and 2 dB before reaching the antenna, when the peak restoration PA is inactive.

Figure 10.42 shows a plausible alternative, which from the viewpoint of a distant receiver, will perform the power addition in a lossless fashion. It has even been suggested that this same configuration could be used to combine the main and error signals in a feedforward loop (see Chapter 14) without incurring the coupling losses inherent in a conventional implementation. Phasing issues and multipath effects will of course complicate, and potentially exterminate, this possibility. In fairness,

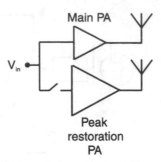

Figure 10.42 "Space combining" of transmitter PAs for different signal envelope ranges.

however, the more conventional smart antenna techniques that are being proposed, which are mainly aimed at implementing directional selectivity, will have to combat similar problems.

10.10   Case Studies in Efficiency Enhancement

It was already shown, in Section 10.6 (Figure 10.25), that some care needs to be taken when assessing the value of a particular efficiency enhancement technique. The value will be closely related to the modulation environment, taking due account of linearity requirements. Based on experience to date with mobile phone and WiFi PA designs, the regular Class AB PA still seems able to hold its own against heavily promoted "smarter" designs, using one or another of the efficiency enhancement techniques described in this chapter. One of the key issues, in making quantitative comparisons between different approaches, is the assumption that the peak power has to be lined up with the onset of clipping. Some modulation systems can tolerate a certain amount of peak clipping, and this means that the efficiency roll-off starts at a lower power level. This can result in quite respectable average efficiency performance, say, greater than 40%, from conventional Class AB PAs.

For example, Figure 10.43 shows several choices for efficiency backoff characteristics, using an OFDM (802.11a) signal. A straight Class AB PA with fixed bias will give an efficiency characteristic which backs off from the peak value as the inverse square root of the power. If the PEP is set to coincide with a peak efficiency of 70%, the average power level will be approximately 10 dB lower, where the cw efficiency will be $70/\sqrt{10}$, or 22%. It can however be shown that the statistical probability of power levels above the 7 dB PAR is sufficiently low that these occasional peaks can be allowed to clip, without any significant effect on the ACP or EVM. Thus the second efficiency PBO characteristic in Figure 10.43 shows the upper 3 dB being hard-clipped and the efficiency is assumed to remain at 70% in this region. Then the efficiency backs off in the usual way, but is now 31% at the average power level. In order to check how close these numbers are to the average efficiency in the presence of the signal in question, it is necessary to generate a representative number of envelope points, perform the efficiency calculation, and average the results. Although some commercial RF system simulators will perform this task, it is useful and instructive to perform the necessary computations using a simpler, and much cheaper, spreadsheet.

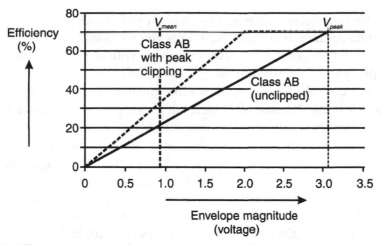

Figure 10.43   Efficiency versus envelope magnitude (voltage); mean and peak levels for 802.11a signal shown. Dotted trace shows clipping at 7 dB above mean level.

Referring to the Excel spreadsheet "OFDM_effy" on the CD-ROM, columns 3 and 5 are the I-Q components of an 802.11a OFDM signal. There are 4096 points, sampled at a rate which gives a representative signal burst. Column 7 performs a simple envelope amplitude calculation,

$$V_m = \sqrt{\left(V_I^2 + V_Q^2 \right)}$$

and at the head of this column, Excel informs us that the maximum and average values in this burst are 3.06 and 0.906, respectively, corresponding to a PAR of 10.6 dB.

Various efficiency options are computed in the subsequent columns. Column 11 uses the basic Class AB efficiency relationship, taking 70% as the peak efficiency value:

$$\eta = 70 \cdot \frac{V_m}{V_{pk}}$$

where the normalized envelope amplitude, V_m/V_{pk}, has been computed in column 9.

This gives an average efficiency result of 20.7%, quite close to the approximate result obtained using the average power efficiency of 22%.

Moving over to column 27, a more general case can be defined, which simulates a two-level supply ET system, for a stipulated clipping level (V_{bk}, r2c29) and peak clipping level (V_{clp}, r3c29). For convenience, this column equation uses the absolute values of the envelope amplitude, in column 7. The column function, r5c27, has the clumsy but reasonably self-explanatory form

$$\eta = \left(\text{if } V_m > V_{clp}, 1, 0 \right) * 70 + \left(\text{if } V_m > V_{bk}, 1, 0 \right)\left(\text{if } V_m < V_{clp}, 1, 0 \right) * 70$$

$$* \frac{V_m}{V_{clp}} + \left(\text{if } V_m < V_{bk}, 1, 0 \right) * 70$$

This expression has some elements of optimism that could be easily changed; the formula assumes that the lower supply voltage is switched in at the precise moment that the RF voltage swing reaches the lower voltage supply value, and that the lower supply is available without using a DC to DC converter.

The case $V_{clp} = 2$ corresponds approximately to clipping at 7 dB above the mean power level, and the use of a regular single supply can be represented by setting $V_{bk} = 2$. The efficiency result for this case is 31.6%, a sizeable increase from the unclipped value. If a voltage supply switch is now included as well, some experimentation reveals an optimum level around $V_{bk} = 1$, giving an average efficiency of just over 45%. The spreadsheet also does the efficiency calculation for a classical Doherty PA, giving a result of 40%. In this case, no clipping has been allowed; it may be argued that this is an unfair comparison with the two-level supply switch which also allows peak clipping. The ruling which has been applied here is that the behavior of the Doherty PA in the clipping region is not as straightforward as for a regular Class AB device, and in any case the ideal efficiency characteristic is not likely to be realized in practice. Figure 10.44 shows a plot obtained using the spreadsheet data. It plots a sample of the data points for the three cases discussed, which additionally gives some indication of the statistical distribution of the envelope amplitudes, through the density of the points in different regions.

The spreadsheet can also be used in a more abstract fashion, to propose possible efficiency/PBO characteristics that are not as yet linked to physical PA configurations. For example, an efficiency characteristic having the form

$$\eta = 70 * \frac{V_{bk}}{V_m}, \quad V_m > V_{bk}$$

$$= 70 * \frac{V_m}{V_{bk}}, \quad V_m < V_{bk}$$

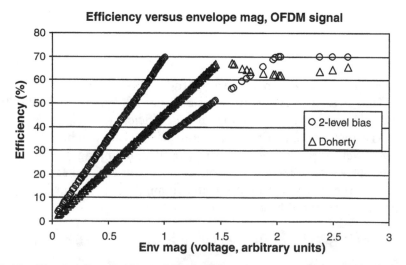

Efficiency versus envelope mag, OFDM signal

Figure 10.44 "Dynamic" spreadsheet plots of efficiency versus envelope magnitude for two cases: Doherty (triangles) and two-level supply switch (circles). Density of points gives some indication of signal statistics.

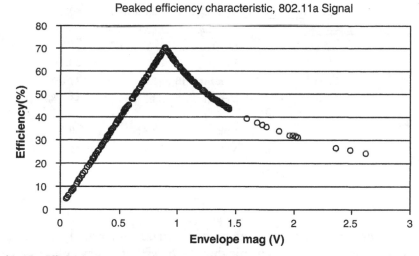

Figure 10.45 Efficiency versus envelope magnitude for postulated system having efficiency peak at mean level; points plotted for 802.11a signal.

has a peak efficiency deliberately set at the V_{bk} level, tailing off at higher and lower levels. Intuitively, it seems logical to start with the average level for setting the peak efficiency. In the OFDM case considered here, the efficiency at the peak power level will have dropped to about 20% (see Figure 10.45). The average efficiency for this characteristic is 49%, the highest yet obtained in this exercise for an OFDM signal.

Of course, this efficiency characteristic has been stipulated without any clear idea as to how it may be implemented, but the result is clearly relevant to the design of power converters, both for fully tracked ET and switched supply systems. It has to be speculated that an optimum efficiency-power characteristic can always be determined for a particular signal environment. This result shows that there may be some value in taking a more *a priori* approach to the design of ET systems. The spreadsheet OFDM_effcy can be used with different modulation systems, all that is necessary is to paste an appropriate set of data points into columns 3 and 5. EDGE system efficiency analysis, **EDGE_effy**, is already available on the accompanying CD-ROM. As described in Appendix B, companion spreadsheets are also available which provide rudimentary, but useful, ACP analysis for stipulated PA nonlinearity characteristics. The results in this section are summarized below (Table 10.1).

10.11 Conclusions

This chapter has considered many techniques, mainly old but some with newer twists, that can be considered for efficiency enhancement of RFPAs in complex signal environments. It is a matter of experience and observation that although attractive on paper, actual implementation of these techniques seems to present substantial problems, especially in high volume products. The interaction of any efficiency enhancement technique with the system linearity always seems to be a negative one, to the extent that there seems to be no single approach which does not

Table 10.1 Summary of Efficiency Simulations for 802.11a Signal

System	Clipping Level	Efficiency at Mean Level	Mean Efficiency
Class AB	Vpk	22%	20.7%
Class AB	Vmean+7 dB	31.9%	31.6%
Class AB, two-level switch, V_{bk}=1.0	Vmean+7 dB	63.7%	45.3%
Doherty	Vpk	40.3%	40.3%
Peaked efficiency characteristic (Figure 10.45)	Vpk	70%	49.1%

carry with it the excess baggage of linearization, usually in the form of digital predistortion. It is indeed ironic that the oldest efficiency enhancement technique in the book, that of reduced conduction angle PA operation, still remains very much the default approach in the industry.

References[10]

[1] Doherty, W. H., "A New High Efficiency Power Amplifier for Modulated Waves," *Proc. IRE*, Vol. 24, No. 9, September 1936, pp. 1163–1182.

[2] Cripps, S. C., *Advanced Techniques in RF Power Amplifier Design*, Norwood, MA: Artech House, 2002.

[3] Raab, F. H., "Efficiency of Doherty RF Power Systems," *IEEE Trans. on Broadcasting*, Vol. BC-33, No. 3, Sept. 1985, pp. 1094–1099.

[4] Takenaka, I., et al., "A 240W Doherty GaAs Power FET Amplifier with High Efficiency and Low Distortion for W-CDMA Base Stations," *2004 IEEE Intl. Microw. Symp.*, Fort Worth, TX, WE-5A-2.

[5] Gajadharsing, J. R., et al.,"Analysis and Design of a 200W LDMOS Based Doherty Amplifier for 3G Base Stations," *2004 IEEE Intl. Microw. Symp.*, Fort Worth, TX, WE-5A-2.

[6] Chireix, H., "High Power Outphasing Modulation," *Proc. IRE*, Vol. 23, No. 11, November 1935, pp. 1370–1392.

[7] Stengel, R., and W. Eisenstadt, "High Efficiency LINC Power Amplifier," *Proc. Wireless & Microwave Technology Conf.*, Chantilly, VA, 1997.

[8] Raab, F. H., "Efficiency of Outphasing RF Power Amplifier Systems," *IEEE Trans.Commun.*, Vol. 33, No. 10, October 1985, pp. 1094–1099.

[9] Hetzel, S. A., A. Bateman, and J. McGeehan, "A LINC Transmitter," *Proc. 42nd IEEE Vehicular Tech. Conf.*, Denver, CO, May 1992, pp. 759–763.

[10] Grundlingh, J., K. Parker, and G. Rabjohn, "A High Efficiency Outphasing Power Amplifier for 5GHz WLAN Applications," *2004 IEEE Int. Microw. Symp.*, Fort Worth, TX, TH3B, pp. 1535–1538.

[11] Kahn, L. R., "Single Sideband Transmission by Envelope Elimination and Restoration," *Proc. IRE*, Vol. 40, July 1952, pp. 803–806.

[12] Asbeck, P., et al., "High Efficiency Techniques for Wide Dynamic Range Handset Applications," *2004 IEEE Intnl. Microw. Symp.*, WFD-6.

[13] Hietala, A., "A Quad-Band 8PSK/GMSK Polar Transceiver," *2005 Radio Frequ. Integrated Circ. Symp.*, Long Beach, CA, RMO1A-1, pp. 9–13.

---

10. Patent references are given for bibliographic use. In citing specific patents, the author is not indicating any view on priority issues.

[14] Sowlati, T., et al., "Polar Loop Transmitter for GSM/GPRS/EDGE," *2005 Radio Frequ. Integrated Circ. Symp.*, Long Beach, CA, RMO1A-2, pp. 13–16.

[15] Nesimoglu, T., et al., "Improved EER Transmitters for WLAN," *2006 Radio and Wireless Symp.*, San Diego, CA, WE2B-3.

[16] Domokos, J., "A Power Amplifier System," Patent WO2004019486, 2004.

[17] Wilson, M. P., "High Efficiency Amplification," Patent WO2005057769, 2005.

[18] Akihisa, S., and K. Kunihiko, "Power Amplifier for Radio Transmitter," Patent JP9064757, 1997.

[19] Daniele, R., and A. Abbaiti, "Linear Microwave Power Amplifier with Supply Power Controlled by Modulation Envelope," Patent WO9534128, 1995.

[20] Bar-David, I., "Method and Apparatus for Improving the Efficiency of Power Amplifiers Operating Under a Large Peak-to-Average Ratio," Patent US20006437641 (Paragon), 2001.

[21] Festoe, A., and K. Onarheim, "Efficient Power Supply for Rapidly Changing Power Requirements," Patent WO2005041404, 2005.

[22] Hanington, G., P. F. Chen, and P. M. Asbeck, "A 10 MHz DC to DC Converter for Microwave Power Amplifier Efficiency Improvement," *IEEE MTT Symp.*, Baltimore, MD, 1998.

[23] Price, R., "A Note on the Envelope and Phase-Modulated Components of Narrow-Band Gaussian Noise," *IRE Trans. Info. Theory*, Vol. IT 1, September 1955, pp. 9–13.

[24] Meinzer, K., "A Linear Transponder for Amateur Radio Satellite," *VHF Communications*, Vol. 7, January 1975, pp. 42–57.

[25] Raab, F. H., "Split Band Modulator for Kahn-Technique Transmitters," *2004 IEEE Intl. Microw. Symp.*, Fort Worth, TX, WE2D-1, pp. 887–890.

[26] Brown, W. C., "Progress in the Design of Rectennas," *J. Microw. Power*, Vol. 4, No. 3, 1969, pp. 168–175.

[27] Zhang, X., et al., "Analysis of Power Recycling Techniques for RF and Microwave Outphasing Power Amplifiers," *IEEE Trans. Cir. and Sys.*, Vol. 49, No. 5, May 2002, pp. 312–320.

CHAPTER 11
Power Amplifier Bias Circuit Design

11.1 Introduction

The oscillatory habits of RF power transistors are an infamous aspect of RF power amplifier design. This much-dreaded skeleton jumps out of its closet at the most inconvenient times, frequently causing destruction and panic in its wake. Its effects are most prevalent at lower frequencies in the MHz to VHF range, where the terminating impedances of an RF power device are mainly defined by the bias insertion networks. The design of bias networks for any RF power amplifier therefore plays an important part in establishing stable operation.

The subject has a curious history, which takes us into the murky waters of cultural divides—not just the different world-views of low frequency analog and microwave designers, but also the overlapping roles of circuit designer and lab technician. While the RF designer uses an array of advanced analytical tools and techniques to design an RF matching circuit for a 2 GHz power amplifier, an oscillation at 60 MHz is passed off as a minor irritation (smoke notwithstanding) that can be removed using a shotgun-style artillery of capacitors and ferrite beads. The plot (not necessarily the smoke) then thickens; having established a combination of randomly soldered components that seem to stabilize a particular device, the same *pot pourri* becomes enshrined as a standard for others to duplicate. Data sheets, application notes, demonstration test boards, all start to recommend or use a bias network configuration which is largely based on empirical design methods, without any real appreciation of what the problem was in the first place. Needless to say, hand-waving explanations for the necessary functions of the various stabilizing elements are readily available.

The irony of all this is that bias networks can be analyzed quite easily using the same tools as are extensively used to do the RF design. Perhaps one of the reasons this is still such a minority activity is that RF CAD programs like to use RF units: GHz, pF, and nH, whereas bias network design requires to deal with MHz, nF, and mH. RF designers are perhaps reluctant to type in so many zeros and wonder whether the software will crash. But the analytical approach is the main theme in this chapter, both in dealing with stability problems and establishing networks which can, in principle, eliminate the problem.

More recently, the wireless communications revolution has reemphasized another aspect of RFPA bias network design, which is the need to prevent vestiges of the signal modulation from appearing on the supply rails to the PA device. Such "remodulation" is almost certainly harmful, causing additional distortion of the conventional kind, with an unwelcome variation in the form of asymmetry in the intermodulation sidebands. Although this is hardly a new problem in PA design, it is worth noting that not so many years ago most mobile radio transmitters were required to handle only a single analog voice channel, and in any case used mainly constant envelope frequency modulation. This placed few restrictions on the kinds of networks which could be used successfully to stabilize large RF power transistors. Older manufacturers' catalogs, still to be seen on shelves around engineering offices, were often full of applications notes showing bias networks littered with large inductors wound on ferrite rings, which were quite acceptable so long as the modulation was either constant envelope, or had a very low signal bandwidth. Modern wireless and satcom systems not only use amplitude modulated signals, but the bandwidth can be well into the MHz, or tens of MHz, regions.

The very term "DC," as applied to the supply requirements of a typical Class AB RFPA becomes inappropriate. Any RFPA stage which is operating in a Class AB mode will draw an alternating current through its supply circuitry, the variation following the peaks and dips of the amplitude-modulated signal envelope. Ideally, the on-board biasing circuitry of any RFPA should smooth out these variations in supply current such that under all operating conditions the user, or system interface, is not required to supply any alternating components. Unfortunately, physical limitations on capacitor technology can make this a near-impossible task for higher power RFPAs, and as a result problems of supply modulation can arise from the manner in which an RFPA is connected to the primary power source, as well as the on-board bias network itself.

In this chapter, stability issues will be treated first, in Section 11.2. Subsequent sections will focus on various aspects of bias supply design, the deleterious effects of unintentional bias supply modulation, and practical techniques for alleviating these effects.

11.2 Stability of RF Power Transistors

The stability performance of electronic amplifiers is a vast subject, and the focus in this chapter is on the specific problem of instability of an RF power device at low out-of-band frequencies. As described in the introduction above, this is a particularly troublesome issue, and the cause of much frustration, among PA designers at GHz frequencies. In presenting a simplistic view of the theory behind such effects, some liberties will be taken in terms of compliance with the very rigorous dissertations on stability which have appeared in the literature over many years. The goal here is to identify the principal mechanism by which large RF transistors are caused to oscillate at low frequencies, when typical biasing and RF matching networks are used. This is a pragmatic approach which in no way competes with, or refutes, the intellectual stability edifice built by the likes of Rollett, Mason, and Nyquist, and well summarized by Gupta [1].

Figure 11.1 shows a model for an RF power amplifier circuit, suitably stripped down for low frequency analysis. The only elements which survive the cut in the device model are the output transconductance generator and the feedback capacitance. For simplicity, the input and output capacitances are assumed to be negligible

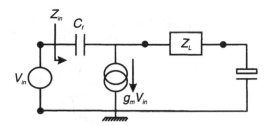

Figure 11.1   RF device model for low frequency stability analysis.

at the frequencies of interest here. The RF matching networks are assumed to be open-circuited through the use of low value blocking capacitors, and the bias networks are represented as large outboard decoupling capacitors with series impedances connecting the bias points to the active device. It is the design of the series impedance networks which determines the stability of the device, since the use of large bypass capacitors at the bias terminals is mandatory for system interface purposes. The stability analysis technique used here will be to unhook the input or output bypass capacitor and to analyze the impedance seen with the opposite port fully decoupled. The stability criterion, or assumption, is that any negative resistance component at any frequency over the validity range of the analysis constitutes an oscillation hazard when the decoupling capacitor at the port in question is replaced. Of course, oscillation will in fact be restricted to the frequencies at which the reflection back from the decoupling capacitor has the appropriate phasing. In many practical cases such a frequency may not exist, and the device will operate without oscillation despite the existence of negative resistance over certain frequency ranges. The safest design approach, however, is to seek to eliminate the negative resistance as much as possible, notwithstanding the probability that in many cases we would get away with it.

The circuit of Figure 11.1 can be analyzed using simple linear circuit theory. Taking initially the case where the output is fully decoupled, the impedance seen at the input is

$$Z_{in} = \frac{Z_L + \dfrac{1}{j\omega C_f}}{1 + g_m Z_L} \tag{11.1}$$

Taking the two extreme but relevant cases of open and short circuit terminations,

$$Z_{in} = \frac{1}{g_m}, Z_L = \infty$$
$$\tag{11.2}$$
$$Z_{in} = \frac{1}{j\omega C_f}, Z_L = 0$$

neither of which gives immediate cause for concern. Unfortunately, there will always be some significant inductance which connects the output of the device to the decoupling capacitor, not to mention the parasitic inductance of the capacitor itself. So rather than an ideal short, a more realistic case to consider is $Z_L = j\omega L$, when

$$Z_{in} = \frac{j\left(\omega L - \dfrac{1}{\omega C_f}\right)}{1 + jg_m\omega L}$$

and making the (very likely) assumption that

$$\frac{1}{\omega C_f} \gg \omega L$$

$$Z_{in} = \frac{1}{(1 + g_m\omega L)^2}\left(\frac{1}{j\omega C_f} - \frac{g_m L}{C_f}\right) \tag{11.3}$$

which has a *global* negative resistance component. Here, in symbolic form, is the "skeleton" of low frequency PA instability.

Extracting the real part, and making the further approximation

$$g_{in}\omega L \gg 1 \tag{11.4}$$

$$\mathrm{Re}(Z_{in}) = -\left(\frac{1}{\omega C_f}\right)\left(\frac{1}{g_m}\right)\left(\frac{1}{\omega L}\right) \tag{11.5}$$

Before putting some typical values into this equation, it is necessary to complete this analysis by considering the reciprocal situation, where the input is decoupled through a series load Z_L. Somewhat remarkably, we obtain the same expression for the impedance, Z_{out}, measured at the decoupling point,

$$Z_{out} = \frac{Z_L + \dfrac{1}{j\omega C_f}}{1 + g_m Z_L} \tag{11.6}$$

so that there will be a similar negative resistance component at the output, for an inductive input termination, as given in (11.5). The choice of the input value of Z_L will however in practice be much less restricted than the equivalent output value, due to the large and usually modulated current which has to flow into the device output. For example, it is clear from (11.5) that making the series bias inductance L value arbitrarily large will reduce the negative resistance component to an acceptably low value for all relevant frequencies. This may be a viable strategy for the input bias network, but would have some disastrous consequences if used on the output, were amplitude modulation present. Nevertheless, it would seem that the use of large series bias inductors, suitably wrapped around lossy ferrite to eliminate resonances, was quite widely practiced in older RFPA designs.

Returning to the key expression for negative resistance (11.5), some care needs to be exercised in making generalizations about the size of the negative resistance

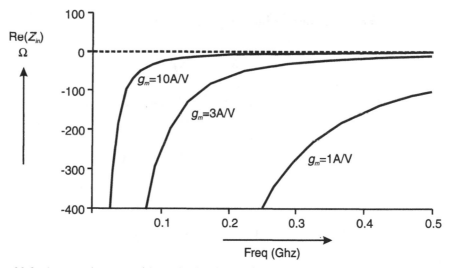

Figure 11.2 Input resistance at bias point for three values of device transconductance; inductive output termination of 10 nH, and feedback capacitance of 0.1 pF/S assumed.

component in specific practical situations. It should be noted that the terms in the first and second brackets will both change in inverse proportion to the peripheral scaling factor of a particular device type. The key design parameter is the series bias inductor L, which is the only parameter in (11.5) under design control, for a particular device. This inductor, on the output side, is also a critical design parameter for minimizing bias modulation effects, as will be discussed in subsequent sections of this chapter. So in any application using AM, and where the modulation bandwidth is of the order of a few MHz, the output bias inductance will need to be as low as practically possible, and preferably in the range of 10 nH. Figure 11.2 shows an impedance plot of the input bias point impedance with this value of L for three values of device scaling, corresponding to transconductance values ranging between 1s (1A/V) and 10s. All three cases show wide frequency ranges where there is a high possibility of oscillation. In a formal stability analysis, the existence of a negative resistance component is merely a necessary and not a sufficient condition for oscillation. The input termination has to present a load having a specific magnitude and phase for oscillation to occur, and the analysis can be extended to consider this. But in general the goal should be to eliminate the negative resistance region as much as possible, especially at higher frequencies where the phase of the load will rotate much more quickly as a function of frequency.

A simple cure for cases where the output bias inductor is being kept low is to place a series resistor between the bias point and the decoupling capacitor, to "quench" the negative resistance. For FET devices, this is quite satisfactory since the gate draws little or no current. This solution is clearly not appropriate on the output side, however, where the current will drop too much voltage across any series resistance. Fortunately, the inclusion of some series resistance on the input port will greatly alleviate the stabilization process on the output. If the input series impedance in (11.6) is now reformulated to include a series resistance,

$$Z = R + j\omega L$$

the real part of the output bias port impedance becomes

$$Z_{out} = \frac{R(1 + g_m R) - \dfrac{g_m L}{C_f}}{(1 + g_m R)^2 + (g_m \omega L)^2} \qquad (11.7)$$

which, assuming $g_m R > 1$, will be positive if

$$R^2 > \frac{L}{C} \qquad (11.8)$$

Taking some median values for L and C,

$$L = 10 \text{ nH}, \; C = 1 \text{ pF}$$

gives $R > 100\Omega$.

This is quite a high value for a series gate resistance, even for a FET device. In practice, a lower value of R than that specified by (11.8) can be used with adequate effect, and simulation of specific cases will determine a suitable value.

The overall conclusions from this analysis are quite surprising and are worth summarizing:

- Almost any RF power transistor (e.g., devices with $g_m > 1S$) will show a negative resistance at the input bias port, over a wide range of baseband and VHF frequencies, if the output bias is fed through an inductance anywhere between 10 nH and 1 mH.
- This is the primary culprit for low frequency oscillations in RFPAs.
- The same, precisely, applies to the output impedance in the reverse situation.
- A series resistor can be used to cancel the negative resistance on the input port.
- This in turn greatly reduces the magnitude of the output negative resistance.

The main purpose of this section has been to show how the oscillation problem can be attacked using a systematic analytical approach. The models are somewhat idealized but the whole process can easily be repeated using any linear circuit simulator, which can include more of the parasitic elements that may have some impact on the final answers. The simple cure of using a series input resistance between the RF bias point and the decoupling capacitor works well in many cases, but there will be some more challenging situations where alternative or additional networks will be required to achieve complete stability. One such addition is the use of a feedback resistor which only comes into play at the lower frequency bands, as shown schematically in Figure 11.3. Such a network is easily realizable using good quality SMT components which will not introduce undesirable resonances in the RF band. Examples of specific networks will be considered further in Section 11.4.

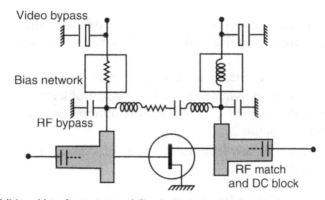

Figure 11.3   Additional low frequency stabilization using a feedback network (typical values for the feedback network components: $L = 20$ nH, $C = 1$ μF, $R = 500\Omega$).

11.3   Bias Supply Modulation Effects

Class AB amplifiers draw a current from the main output bias supply which varies from a low value, maybe 10% of the peak value, as the amplitude modulation on the input signal swings the envelope amplitude over its full dynamic range. Any impedance which is placed in the bias supply path will cause voltage modulation to appear at the output terminal of the device; this will in turn modulate the gain and phase of the amplifier and cause additional distortion products. The designer therefore has to ensure that this series impedance is sufficiently low that the resulting voltage modulation has an acceptably low amplitude.

A simple numerical example is appropriate. A 100 watt PA stage, running from a bias voltage of 28V draws a peak current of 7 amps from the bias supply. Suppose that the modulation is a sinusoidal variation from zero to peak power with a frequency of 1 MHz. If the bias supply has a total series inductance of 50 nH between the device output tab and the main bias supply, the resulting voltage modulation at the device output solder tab (drain/collector) will be 175 mV. If the modulation frequency drops to 100 kHz, the voltage ripple will be 17.5 mV, and at 10 MHz it will be 1.75V.

This very simple but representative case raises a multitude of questions:

- How much ripple on the supply can be tolerated?
- Is 50 nH a realistic number, given that it has to include the routing through the on-board RF matching networks, the PC tracks and wiring to the nearest bypass capacitor, and the internal inductance associated with the bypass capacitor?
- How big does the bypass capacitor have to be in order not to contribute more ripple itself, regardless of the additional effects of its internal series inductance and resistance ("ESR")?
- Where do voltage regulators fit in to this picture?

Possibly the most difficult of these questions to answer is the first, since it is both application and device dependent. Different RF transistor types will show considerable variation in their sensitivity to bias voltage ripple, and to the actual effects such ripple will cause, in terms of increased or additional distortion products. For the present purposes we will assume the following scale of impact on voltage ripple:

- 17.5 mV will be taken as "tolerable" in all but the most stringent applications.
- 1.75V will be assumed to cause major problems with PA distortion, almost certainly including asymmetrical intermodulation sidebands and poor predistortability.
- 175 mV may be tolerable in some cases, depending especially on the particular device being used.

Thus we will take 100 mV maximum power supply ripple to represent a useful design goal for a wide range of PA applications. Clearly, it is the stipulation of a 1 MHz baseband frequency that really swings the bias supply ripple issue. As discussed in the introduction to this chapter, when radio transmitters only had to handle voice channels of a few kHz bandwidth, the results of such calculations were very different and bias supply ripple was not a serious consideration. There is another factor; in the GHz frequency range it is a fairly recent development that solid state devices have become available to generate power in the 100 watt sector. The ongoing need to use much lower supply voltage rails than used by tube predecessors has made this issue, to some extent, uncharted territory which still catches PA designers unaware. Most multicarrier PAs will require signal bandwidths well into the MHz region, and even 2G and 3G mobile phone handsets, respectively, straddle the 1 MHz bandwidth requirement.

In order to address the second and third questions above, some more detailed analysis is required. Figure 11.4 shows a realistic model for the bias supply to an RFPA. The PA itself is operating in a deep Class AB mode, so the supply current will vary over about a 3:1 range as the signal amplitude modulation varies from peak to minimum. This current is supplied to the output transistor from a bias network consisting of a series inductance, a series resistance, and a bypass reservoir capacitor. The inductance, as noted above, includes several physical components: the inductance of the RF matching (microstrip) elements through which the bias is routed to the device, additional PC board tracks which connect the RF subassembly to the bypass capacitor, and the internal parasitic inductance and resistance associated with the bypass capacitor itself.

For simplicity, a sinusoidal (two-carrier) modulation is assumed, and the effect of each individual bias element on the device voltage is shown in Figure 11.5. For analysis purposes, it will be assumed that the gain of the amplifier will decrease with lowering bias supply. This may not always be the case, but is very typical. Looking first at the effect of a series resistor, the voltage shows a flattening in the peak region, very reminiscent of the effect of gain compression on a sinusoidal enve-

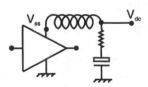

Figure 11.4 PA Bias supply schematic.

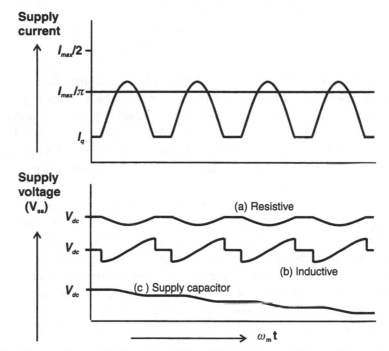

Figure 11.5   Bias supply modulation: (a) resistive, (b) Inductive, and (c) supply capacitor.

lope (see Figure 9.7). This is a case where the effect of bias modulation will be very similar to gain compression, and it is possible to obtain a quantitative estimate of this effect at the peak power condition. The flattening effect can be modeled as a third degree envelope component, whose amplitude is equal to the voltage drop caused by the series resistance at the envelope peaks. So if the undistorted output envelope has the form

$$e(t) = V_{dc} \cos \Omega t \cos \omega t$$

where Ω is the modulation frequency and ω the RF carrier frequency, the peak flattening can be modeled by adding a third degree term in the modulation function,

$$e'(t) = V_{dc} \left(\cos \Omega t - \left(\frac{v_b}{V_{dc}} \right) \cos 3\Omega t \right) \cos \omega t$$

where v_b is the amplitude of the bias modulation caused by the series resistance. This can be expanded to obtain the spectral components,

$$\frac{2 \cdot e'(t)}{V_{dc}} = \cos\{(\omega + \Omega)t\} + \cos\{(\omega - \Omega)t\} - \frac{v_b}{V_{dc}} \cos\{(\omega + 3\Omega)t\} -$$

$$\frac{v_b}{V_{dc}} \cos\{(\omega - 3\Omega)t\}$$

which shows that the relative level of the IM3 products caused by the bias modulation is

$$20\log\left(\frac{v_b}{V_{dc}}\right)$$

So for a 100 mV ripple on a 10V supply, an estimate for the resulting IM3 level would be −40 dBc. This is a nontrivial level for IM3s in most applications, and is sufficient to trigger some thoughts on whether inherent device nonlinearities have been unfairly blamed as the sole source of spectral distortion. The 100 mV of ripple in question here can be generated by a bias current swing of 5A and a capacitor with an ESR of 20 mΩ. We will see that this represents a low value for readily available capacitor technology.

Figure 11.5(b) shows the corresponding effect of a series inductance. The interesting aspect here is that the voltage ripple passes through zero at the peak of the envelope and through opposed positive and negative maxima on the falling and rising portions of envelope. So the envelope distortion is asymmetric, again very similar to the asymmetrical envelope distortion postulated in Chapter 9. This case would require more information on the PA gain/supply voltage characteristic in order to make a quantitative estimate of the envelope distortion, but suffice it to say that this is a clear physical mechanism by which asymmetrical spectral distortion can be generated.

Figure 11.5(c) shows the possible effect of using a reservoir capacitor of insufficient size. This is again a difficult effect to quantify, since the ripple function is also dependent on the time constant of the recharging supply from the prime power source which will be located at a yet more remote point, with yet more board traces and wiring. It is however important that the first reservoir capacitor in the chain should be able to supply the current with minimal ripple for the upper portion of the modulation bandwidth. In this case, the capacitance required can be estimated by integrating the charge for a single modulation cycle.

$$Q = I_{pk} \int_0^{\pi/\Omega} \sin \Omega t \cdot dt$$

where for estimation purposes Class B operation has been assumed, and I_{pk} represents the *supply* current at the envelope peak. Putting $I_{pk} = 5A$, $\Omega = 2\pi \times 1$ MHz, $Q \approx 10^{-6}$ Coulombs, a capacitor which can supply this charge and show a voltage reduction of only 100 mV will have a capacitance of 10 μF ($\delta Q = C\delta V$). Such a value is available as a small SMT component and can be placed very close to the point where bias is inserted into the RF matching circuitry surrounding the transistor. It is necessary however to examine what kind of series resistance values can be expected from readily available capacitor types. Rather than relying on manufacturers' data, which can be misleading if not measured at the frequency and current density in question, Figures 11.6 and 11.7 show details of some actual measurements on three SMT capacitor types: regular tantalum, "Low ESR" tantalum, and a parallel pair of X7R ceramic parts.

These measurements were performed using an RFPA under equal two-carrier excitation, which drew a supply current of 3A peak (10% quiescent). Figure 11.6 shows the test board configuration. In these measurements it is important to note there are two measurement points: the bias point, which is decoupled using the

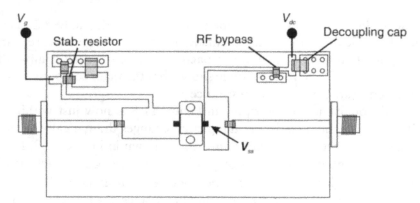

Figure 11.6   Test board for supply modulation measurements.

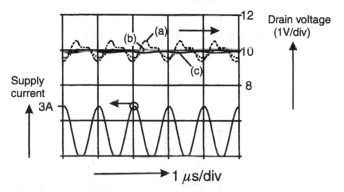

Figure 11.7   Decoupling capacitor ripple caused by 1 MHz PA supply current variation: (a) 10 μF regular tantalum, (b) 10 μF low ESR tantalum, and (c) 4.7 μF X7R 1206 SMT.

capacitor in question, and the actual drain of the RF transistor itself.[1] Figure 11.7 shows the voltage measured at the bias point for the three capacitor types. The voltage ripple can be assumed to be caused primarily by the internal parasitics of the on-board bypass capacitor. Trace (a), the regular tantalum capacitor, shows a large inductive component as well as a resistive component. Trace (b), the low-ESR tantalum capacitor, shows mainly a resistive drop which can be estimated to be due to an ESR about 0.2Ω; this actually ties in quite well with the specified ESR value for the component in question. Clearly, the only case which comes anywhere close to meeting the design goal of 100 mV maximum ripple is the SMT ceramic part, case (c). This however needs some additional qualification. Packing a capacitance of 4.7 μF into a 1206 SMT package outline stretches conventional[2] capacitor technology to its limits. On the other hand, the small size allows two, or even four, parts to be

---

1.   Clearly, some care is needed in the choice of a suitable probing technique to measure the baseband bias voltage variation in the presence of a high power RF voltage.
2.   Recently, there have been some technological breakthroughs in capacitor technology, resulting in so-called "super-" or "ultra-" capacitors which have higher storage capacity per unit volume than common batteries. These devices usually have low voltage breakdown, and are currently quite expensive. They are not therefore included in the present discussion.

stacked in parallel, resulting in corresponding reduction in ESR and inductance. Figure 11.8 shows the same measurement, on a finer voltage scale, repeated for a parallel pair of such parts, giving a capacitance which is now comparable to the two previous cases. The voltage ripple shows that the main component is the recharging transient, rather than series resistance.

Although the voltage ripple in Figure 11.8 is now just within the established 100mV maximum guideline, the picture changes dramatically when the voltage is measured at the actual drain terminal, as shown in Figure 11.9. In this design, the DC drain supply is inserted through a $\lambda/4$ SCSS, having a characteristic impedance of 90Ω. Although this bias stub has been located as near as possible to the device drain pad in order to minimize the baseband inductance, the inductance of the stub itself is enough to cause an additional ripple of 300 mV. This head-on collision between the traditionally disparate worlds of baseband and microwave circuit effects comes as quite a shock when first encountered. Once again, some simple rational analysis can help the shock recovery process. The input impedance of a short-circuited stub, characteristic impedance Z_0 and electrical length θ is

$$X_s = Z_0 \tan \theta$$
$$\approx Z_0 \theta$$

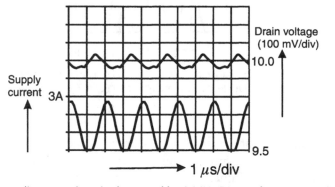

Figure 11.8 Decoupling capacitor ripple caused by 1 MHz PA supply current variation; parallel pair of 4.7 μF X7R 1206 SMT parts (note x10 voltage scale change from Figure 11.7).

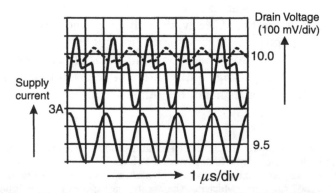

Figure 11.9 Additional ripple caused by 2 GHz $\lambda/4$ bias SCSS for 1 MHz PA bias supply current variation. (Same scale as Figure 11.8, shown dotted.)

at baseband frequencies where the line is obviously very short in comparison to a wavelength. So at baseband frequency ω, the equivalent inductance is given by

$$\omega L_s = Z_0 l \frac{\omega}{c}$$

where l is the physical length, so that

$$L_s = Z_0 \frac{l}{c}$$

In the above example, $Z_0 = 90\Omega$, $l = 37.5$ mm ($\lambda/4$ at 2 GHz), giving

$$L_s = \frac{90 \times 37.5 \times 10^{-3}}{3 \times 10^8} = 11.2\text{nH}$$

So at 1 MHz, a 1.5A amplitude sinusoidal current will show a peak-to-peak ripple amplitude of 200 mV due to the SCSS inductance alone. This is a minimal estimate, as shown in Figure 11.9, where the ripple is substantially higher due to the higher harmonics present in the current waveform. So the much-used SCSS for bias insertion into microwave circuits has to be carefully reconsidered in higher power applications. This issue, and some potential alternatives, will be considered further in Section 11.5.

The final question, posed at the start of this section, asks about the role of voltage regulators in reducing ripple. The previous analysis of the effects of even a simple SCSS at 2 GHz on adding voltage ripple between the bias point and the RF device partially answers this question; clearly, it is not feasible to introduce a voltage regulator within the RF matching itself. But even at the bias point, the effectiveness of an electronic regulator decreases rapidly as the speed of the ripple becomes comparable to the time delay of the regulator action. This problem is entirely comparable to the limitations of envelope feedback techniques for PA linearization [2], and becomes formidable once the signal bandwidth enters the MHz domain. Open loop compensation, however, would appear to be viable but not yet widely practiced. Just as it is possible to predistort the input signal to a PA in order to compensate for the PA distortion, it would seem equally possible to predistort the supply to a PA in order to remove the impedance effects of known, fixed, passive elements in the bias supply chain. This possibility will also be further considered in Section 11.4.

This section has attempted to quantify the possible effects of unintentional bias supply modulation of the supply voltage to an RF power transistor. These effects multiply rapidly as supply current and signal frequency step up through orders of magnitude to a point where a device running in deep Class AB, drawing in excess of 10A peak current, with an envelope amplitude modulation having components above 10 MHz may present problems of supply modulation that are essentially intractable using conventional passive bias networks.

11.4   Bias Network Design

Based on the results and analysis of the previous sections, this section presents some design guidelines, and physical network options, for RFPA design. It is a difficult subject to generalize, since the problems of bias network design and stability escalate with device periphery, which may not be quite the same thing as RF power level. As discussed in the last section, the problems also escalate with signal bandwidth.

Figure 11.10 shows a schematic for an RFPA, showing the demarcation between the RF matching and video biasing networks. The interface point will usually take the form of an RF bypass capacitor, which provides a high quality short circuit, or sufficiently low reactance, path to ground for the RF current components. On the output side, there will be an intermediate biasing network which connects between the RF decoupling point and a local voltage supply in the form of a suitably large reservoir capacitor. On the input side, there is not such a requirement for a current reservoir, but for system interface purposes a bypass capacitor which appears as a near-short circuit over the whole signal bandwidth is desirable.

In its simplest form, the output bias network in Figure 11.10 could be simply a short length of wire, or PC board track, which shunts the RF bypass capacitor with the much larger reservoir capacitor. As discussed in the previous section, even a few nH of inductance between these two capacitors can cause significant extra supply ripple in higher power applications using larger signal bandwidths. However, such an arrangement is quite satisfactory for smaller RF power devices and/or lower signal bandwidth applications. But in some cases there will be an increasing conflict between the requirements for a larger reservoir capacitor, and a lower series inductance in the supply path. As discussed in the previous section, capacitor technology continues to advance, and the range of applications where satisfactory performance from a simple shunt bypass capacitor, such as shown in Figure 11.11(a), is expanding.

A common solution to this conflict is to use a multiple bypass network, such as that shown in Figure 11.11(b). The basic reasoning behind such networks is that the higher frequency components of the device current can be supplied by lower value reservoir capacitors, which have smaller physical size, and thus lower series inductance. Lower frequency components require higher value reservoir capacitors but a higher series inductance can be tolerated. The impedance plots in Figure 11.11 show the impedance presented to the device (excluding the RF circuit elements) assuming

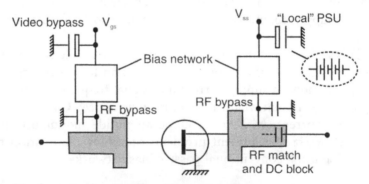

Figure 11.10   PA schematic, showing RF and bias networks.

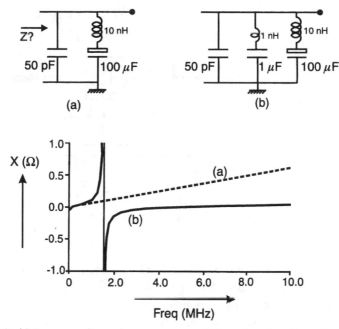

Figure 11.11   (a, b) Frequency/impedance response of bias network options. (Impedance measurement assumes back-end open circuit.)

that the external prime power source appears as a highly reactive source due to its remote location. It can be seen that the "graduated" bypass network has admirable properties except for a resonance at 2 MHz. The impact of such a narrowband resonance may not be too serious in broadband signal applications, and is in any case somewhat exacerbated by the rather draconian assumption that the prime power supply appears as an open circuit at baseband frequencies. In most cases, suitable permutations of component values may be possible to move the resonance to a point which is effectively above the signal bandwidth. Once again, the virtue of using CAD tools to design a suitable bias network is apparent, rather than reliance on *ad hoc* methods based on a combination of hearsay and possibly unrepresentative past experiences. The use of a network analyzer to measure the bias network impedance characteristic is also a readily available, but seldom used, option.

It has already been discussed that in more extreme cases of high peak supply current and MHz signal bandwidths, the ripple problem may still persist. Given the technology which has been developed for the digital predistortion of RF signals to linearize RFPAs, it would seem reasonable to consider the much less challenging concept of predistortion of the supply voltage, as indicated in Figure 11.12. Unlike RF predistorters, which have to compensate for a nonlinear PA property, the compensation of supply ripple would appear to be an essentially linear process. Given the luxury of a supply voltage which can be controlled, on the fly, by a DSP driver, it would however be logical to attempt some PA linearization as well. The objection to such an approach will always be the power loss in the pass transistor, but it is not clear that this should be a fatal objection in some applications. Such a configuration would be distinct from two other techniques described in this book; envelope restoration (EER, Kahn) which actually creates the AM by modulation of the PA transis-

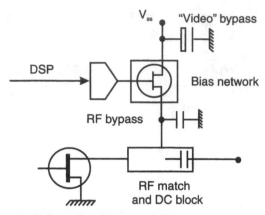

Figure 11.12   "Predistortion" of supply voltage modulation.

tor supply, and envelope tracking ("ET") which drops the supply voltage at low envelope amplitudes in order to conserve efficiency. Both of these familiar techniques require much larger swings of supply voltage in order to accomplish their desired function, and the power consumption of the control element is indeed a major consideration (see Chapter 10).

On the input side, the results of the analysis in Section 11.2 come into play, and a series resistor is usually required between the separate RF and baseband decoupling capacitors. Many low frequency oscillation problems are caused by misguided attempts to combine these two functions with a single capacitor. The value of the series resistor has to be chosen such that stability conditions are met, as described in Section 11.2, but this can in some cases lead to a conflict if bias current needs to be supplied. This can be an issue with higher power bipolar transistors, whose base current is a scaled-down version of the output current in a Class AB amplifier. In such extreme cases, it is usually possible to devise a bypass branch, as indicated in Figure 11.13, such that the resistive termination is still provided at frequencies above the signal baseband, but is bypassed within the baseband down to DC. This is one application where an antiparasitic ferrite bead can serve as a useful bypass inductor, as indicated in the measured impedance plot shown in Figure 11.13.

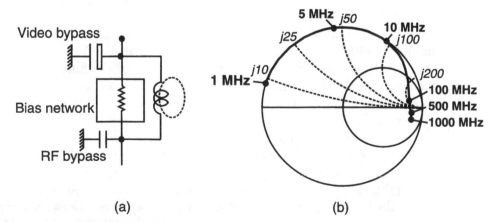

(a)   (b)

Figure 11.13   (a) Inductive bypassing of bias stabilization resistor, and (b) measured impedance of 3-turn ferrite bead RF choke.

11.5   Bias Insertion Networks

Section 11.3 showed some surprising results concerning the inductance of microwave networks at baseband frequencies. In more extreme cases of peak current
swing and high signal bandwidth, traditional biasing techniques for microwave circuits such as short-circuited stubs become problematic. In the example quoted, a 2
GHz SCSS with a 90Ω characteristic impedance was shown to present a series
inductance of 11 nH at low frequencies. An obvious way to reduce this is to reduce
the value of Z_0, in particular through the use of a symmetrical parallel bias feed. The
physical width of lower impedance stubs can however limit the range of this
approach.

Another approach is to reduce the length of the stub. This becomes an admissi-
ble strategy when dealing with larger RF power devices, due to the low impedance
environment at the transistor output. For example a $\lambda/8$ SCSS has an RF impedance
of jZ_0, which will be almost a negligible reactive component for higher power
devices. Even a $\lambda/20$ SCSS will present a shunt reactance of approximately $j\,Z_0/3$,
which will still be a small reactive perturbation in many cases.

An alternative view, in the use of short SCSS elements to supply output bias, is
that they can be absorbed advantageously into the matching network. For example,
Figure 11.14 recalls a result from earlier chapters (see Figure 4.9), where it has been
noted that the device output capacitance has the effect of transforming the device
loadline impedance in the wrong direction, increasing the impedance transformation requirement from the output matching network. The use of a shunt inductance
to resonate the output capacitance, as shown in Figure 11.14, removes this undesirable feature and creates a convenient, and optimally positioned, biasing point. Such
an element could also be used as part of a broader-band matching network.

It is appropriate to include some comments here about the use of commercially
manufactured external biasing devices, traditionally called "bias tees" in RF labs.
Such components are frequently used when evaluating and characterizing devices,
for example, in load-pull measurements. These devices have acquired some mystical
connotations, rather like ferrite isolators, of allowing devices to be tested without

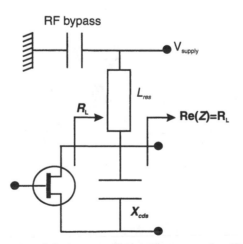

Figure 11.14   Shunt resonance of device output capacitance, creating bias insertion point and
eliminating downward impedance transformation.

having to confront the dreaded oscillation skeleton. The usefulness and cost of such devices act as a formidable restraint on the desire to crack one open to discover its inner secrets, a desire which just about any microwave designer will have experienced at one time or another. The secret, such that it remains, is usually the use of a large bias inductance which is cleverly designed to eliminate resonances over a very wide band of frequencies, perhaps 10 MHz to 20 GHz.

This can be achieved, for example, by winding the coil on a conical former using slightly resistive wire. So in higher power applications, the DC voltage drop can be quite significant, and a separate bias sensing connection is often provided to allow the correct voltage to be set. Such devices are therefore fundamentally unsuited for use with modulated signals, since they use inductances which can typically be in the mH range. As noted in Section 11.2, this is a viable and quite effective approach in preventing oscillations, but is usually inadmissible in RFPA applications where the output current is an AC signal of possibly several MHz bandwidth. Another hazard is to place the bias tee too far, electrically, from the device under test; this applies to any form of bias network in high current applications using modulated test signals. One meter of 50Ω transmission line will add 167 nH effective series inductance in the baseband bias current path.

11.6 Prime Power Supply Issues

Most RF designers treat power supplies as closed boxes, and satisfy themselves with learning how to push the right buttons. It would be inappropriate to attempt to cover what is in effect an entire branch of the electronics industry in a book such as this one, given that many books are available which are entirely devoted to the subject of power supply design. There is however one important aspect of connecting an RFPA to a source of prime power, which is the amount of "AC" that the RFPA still requires, from a supply that has usually only been specified for supplying "DC."

This is illustrated in Figure 11.15, which gives a more system-level view. It is important to recognize that in any system, the RFPA will probably be the biggest user

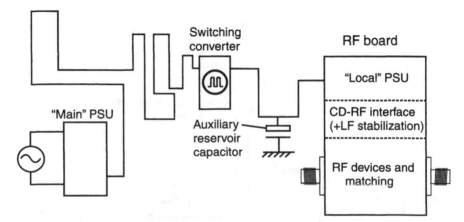

Figure 11.15   System level supply of power to RFPA subassembly.

of prime power. Other than in the special case of mobiles, prime power will usually come in the form of a line alternating voltage. There will usually be an intermediate step between the rectification and filtering of the prime power and the power connection to the PA circuit board. This will usually take the form of a high efficiency switching DC to DC converter, which converts the prime power to the PA supply requirement. This converter will also have an electronic regulator which stabilizes the output voltage between specified limits for a wide range of DC current output.

Note the emphasis on "DC." Here the term is used in its strictest and most inflexible sense; the manufacturers of DC supplies will not usually make any guarantees that the specified voltage and current regulation will apply for currents having an alternating component. In practice, of course, some specifications will be given for the regulation performance over a low frequency range, perhaps from DC up to a few kHz. Quite what will happen when a user attempts to extract a current which has an alternating component in the MHz region will depend on the design of the converter. In particular, it is unlikely that the electronic regulation will respond to such "fast" variations in current drain. Suitable placement of a large auxiliary reservoir capacitor, possibly in the 10,000–100,000 μF range may be necessary in some cases, as shown in Figure 11.15. Schematically speaking, this represents the addition of a third branch to the overall bias network shown in Figure 11.11, and extra resonances are a probable result. Such components are very large and are not usually accommodated inside RFPA housings. But as a result of this, users of RFPAs should probably consider placing specifications on the AC components of current which the device will require under conditions of full drive, and expected signal modulation environments, so that appropriate power supply provisions can be made.

11.7   Bias Control Circuits

This chapter has been concerned primarily with the routing of suitable bias voltages to RF power transistors, in a manner that results in stable operation. The various RF and video networks considered will be used regardless of the nature of the bias voltages themselves, which in some cases may be generated by additional control circuitry. Such controls are implemented in practical PA designs for numerous reasons:

- Maintaining constant current bias conditions over a population of devices having variable DC IV characteristics;
- Power control;
- Linearization.

Power control and linearization are covered as topics elsewhere in this book (mainly in Chapters 10 and 14). The first item is worthy of further comment, since it is an issue very much at the heart of modern electronic design methodology, and can expose some culture conflicts between the RFPA designer and low frequency analog and digital IC design principles.

A high power RFPA is still very much a board-level subsystem. Even if some integrated components (RFICs) are being used in driver or gain stages, the whole assembly can easily accommodate a couple of potentiometers for the purpose of setting critical bias voltages. Manual adjustments of this kind have become a universal anathema in highly integrated electronic design, and many techniques have been developed for sensing and accommodating process variations in the design of VLSI chips, which may contain thousands up to millions of individual transistors. For example, the circuit shown in Figure 11.16(a) can set the quiescent current in a FET device as a fraction of its saturated level, whereby the accuracy of the setting becomes a function of the accuracy of the ratio of the fixed resistance elements rather than the device IV characteristics. Such a ratio will be determined geometrically, and can be designed such that it will be virtually process independent. Figure 11.16(b) shows another technique, widely used in bipolar analog IC design, known as a "current mirror." The bias control device is connected up as a diode, so that the base-emitter voltage can be applied to a device of similar type and the current will "mirror" the control device current. This technique has been adopted by manufacturers of larger packaged bipolar power transistors who sometimes give external access to a small segment of the RF transistor for bias purposes. The current mirror circuit can then give a low impedance, temperature-regulated bias voltage for the main transistor.

There is however a big difference between controlling the current in a Class AB RF power amplifier, as compared to most analog or digital integrated circuit requirements. As has been extensively discussed in this chapter, the current in a Class AB amplifier changes substantially as the input signal amplitude varies. In a sense, the Class AB configuration uses the RF signal itself as the control of current through the RF device. Any attempt to use external bias controls will likely conflict with the intended operation of the Class AB mode. Other than some circuitry to set the quiescent level, perhaps using a small but otherwise identical transistor element that does not have an RF signal applied, the control of transistor bias is still mainly done using individual voltage controls. These controls may still take the form of physical potentiometers, although it is becoming more common to use firmware settings and DACs in modern PA system designs.

11.8   Conclusions

The use of broadband and multicarrier modulation systems in modern wireless systems has had a major impact on the design of biasing networks for RFPAs. This

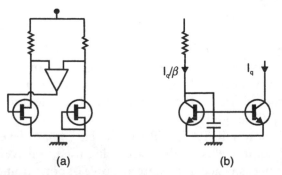

Figure 11.16   (a) Circuit for setting FET quiescent current, and (b) bipolar current mirror.

chapter has mainly emphasized the need for designers to use CAD circuit analysis tools to design suitable bias networks for the required signal bandwidths and also to ensure stable operation. Instability should not be regarded as a counter-serendipitous effect, and can be explained and analyzed with considerably more ease, using the same tools and techniques as the in-band stability considerations with which RF designers are more familiar.

References

[1]   Gupta, M. S., "Power Gain in Feedback Amplifiers, A Classic Revisited," *IEEE Trans. Microwave Theory & Tech.*, MTT-40, May 1992, pp. 864–879.

[2]   Cripps, S. C., *Advanced Techniques in RF Power Amplifier Design*, Norwood, MA: Artech House, 2002.

CHAPTER 12

Load-Pull Techniques

"Load-pull" is something of a euphemism when applied to PA design. The term originated in the world of oscillators, where a key specification is the amount of frequency change, or "pulling" the oscillator displays as the output load is tuned from its nominal value. In PA design, the term is used to describe an empirical process by which the matching requirements of a PA device are determined, using some form of variable impedance tuning device.

It might be thought that load-pull techniques, along with the slide rule and the slotted line, would be teetering on the brink of extinction due to the availability and almost universal intrusion of the CAD approach into RF circuit design. Judging, however, by the continuing evolution and availability of more complex and versatile commercial load-pull hardware, it would seem that this is a preserve which has successfully survived the CAD revolution. This chapter is therefore intended as an update on current load-pull techniques and capabilities, rather than an extended justification for its continuing use.

We pick up the story, in some respects, from Chapter 2, which describes the basic loadline approach to Class A RFPA design. As long ago as 1983, this author (Chapter 2, [1]) showed that the oval-shaped load-pull power contours which were frequently measured using fundamental load-pull equipment could be predicted using loadline principles. The use of a simple fundamental tuner was appropriate for Class A design, where up to the onset of gain compression the harmonic impedances can be expected not to play much part in the performance of the device. Basic tuner design will be covered in Section 12.1. Class AB operation, which has been the main focus in the intervening chapters, poses some more challenges for the load-pull technique in that harmonic impedances will need to be adjustable as well. In pursuing this goal, mechanical tuners become more complicated and the option of active load-pull has to be considered. These techniques will be compared and contrasted in Sections 12.2 and 12.3, along with some actual experimental data.

12.1 Tuner Design for Fundamental Load-Pull

Most microwave tuners are configured, and judged, based on their ability to "cover the Smith chart." To this end, a reactive discontinuity is introduced to a length of 50Ω transmission line such that the magnitude of the discontinuity and its position along the line can be adjusted with suitably high precision. Such a tuner, even without direct calibration, can give an intuitive feel for the impedance path which is being followed, unlike the multiple stub tuners that can still be found lurking in the drawers of microwave labs. Two classical tuner designs dominate load-pull applications and are illustrated schematically in Figures 12.1 and 12.2.

The "slide-screw" tuner, Figure 12.1, has a metal plunger whose end can be gradually moved closer to the inner conductor of a 50 ohm airline; the end of the plunger can be profiled to surround, but never touch, the inner conductor. The plunger is mounted on a carriage which can be moved along the line to control the phase of the reflection introduced by the plunger. This form of tuner is intuitive in its operation, and has the advantage that in any position of the carriage the discontinuity can be moved out to return to a reasonable 50 ohm termination. The reflection coefficient presented by the plunger will be frequency dependent; the reflection will increase with frequency until a point is reached where the plunger length exceeds a quarter wavelength, when the reflection will peak and start to decrease. For this reason, commercial tuners of this type sometimes provide two separate plungers to extend the useful frequency range.

The twin-slug tuner, Figure 12.2, is of simpler construction although less intuitive in its operation. The "slugs" are essentially lengths of transmission line, having much lower characteristic impedance than the main 50Ω coaxial line, within which they can be moved. Considerations of loss, and the realization of maximum impedance tuning range suggest the use of metal tuning slugs, which slide along the inner conductor and make contact with it. This however raises the thorny issue of engineering a sliding metal-to-metal contact. Dielectric slugs remove this requirement, but it becomes necessary to trade the dielectric properties of various materials. At any particular frequency, the maximum reflection can be generated using two $\lambda/4$

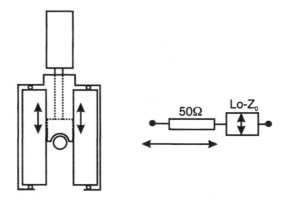

Figure 12.1 "Slide-screw" tuner.

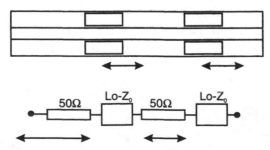

Figure 12.2 Twin slug tuner.

slugs, which are spaced $\lambda/4$ apart. Multiple application of the well-known transmission line transformer relationship quickly reveals that the minimum impedance will be $50/\varepsilon_r^2$, so that if Teflon ($\varepsilon_r =2.2$) is used, this impedance will be only about 10Ω. Various materials having relative dielectric constants in the 3–4 range are available, whereby the maximum reflection can be increased to around 0.9, but with increased loss at higher GHz frequencies. So despite the engineering challenges, most commercial computer controlled tuner systems use metal slugs, which are moved using worm drives.

Such tuners are still widely used to determine the optimum load for an acceptable tradeoff in the power, efficiency, and linearity of an RF power transistor. Despite the extensive analysis and commentary in this book and others, on the importance of tuning at harmonic as well as fundamental frequencies, it is a surprising fact that many successful products appear to be based on essentially a fundamental tuning characterization. Manufacturers of such tuning test systems will admit that sales of fundamental tuners greatly exceed the now available systems that allow for independent harmonic tuning as well. Harmonic tuning will be considered in the next section; however it is worthwhile to take a somewhat more quantitative look at the potential hazards of fundamental-only load-pull characterization.

Figure 12.3 shows a Smith chart plot of two typical tuner topologies. The key point about the two impedance plots is that although in each case the fundamental has been matched to the desired low resistance value, the individual tuners show major divergences in the harmonic impedances which they present. Although the two cases in Figure 12.3 represent the two standard tuner types shown in Figures 12.1 and 12.2, it becomes apparent that even tuners having the same basic geometry, but some differences in dimensions, can give wildly different responses at harmonic frequencies despite being tuned to the same fundamental impedance. Another manifestation of this effect is that the harmonic behavior from a given tuner can fortuitously enhance, or degrade, the power performance at different measurement frequencies.

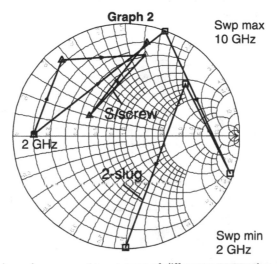

Figure 12.3 Harmonic performance of two tuners of different construction, set to give the same fundamental impedance

Figure 12.4 shows the kind of unsatisfactory results that this effect can cause. A device, biased for Class A operation, has been tuned for maximum P_{1dB} over a frequency range of 1 to 6 GHz. Assuming that the measurements have been performed with due care and attention, it appears that the power performance varies somewhat randomly over this broad bandwidth and does not show the kind of decreasing monotonic characteristic that would be expected. Figure 12.4 is derived from some actual measurements on a GaAs MESFET device, and the variations probably represent the fortuitous peaking of the second harmonic impedance, giving some Class J performance (see Chapter 4).

Such variations will be less apparent when measuring devices whose output capacitance is large enough to provide a reasonable short circuit at harmonic frequencies. This subject has been discussed and analyzed in much detail in previous chapters and once again returns to prominence when using simple fundamental tuners to optimize power performance. Most LDMOS transistors, for example, operating above 2 GHz will show low sensitivity to external harmonic impedance variations due to their larger relative output capacitance.

Despite these caveats, it is still a fact that the high-density data packages that come from lengthy computer controlled load-pull evaluations will always contain some vestiges of the harmonic signature of the tuner. Recent commercial tuner systems typically provide calibration data for harmonic as well as fundamental tuning states. When selecting specific tuning conditions as the basis for a design, designers can therefore at least make some attempt to reproduce the harmonic conditions, both in the subsequent simulations and also in the network realization.

12.2 Harmonic Load-Pull

In introducing the subject of harmonic load-pull, it is necessary to make a distinction. As discussed in the previous section, even the simplest form of tuner will always have some measurable impedance characteristic at harmonic frequencies, even if in some cases this is dominated by such things as cheap low frequency connectors, lossy FR-4 board, or resonant tuning elements. The key point is that the user has no

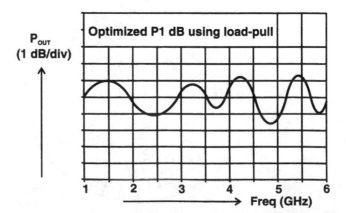

Figure 12.4 Typical result using fundamental-only tuning to optimize P_{1dB} over a broad bandwidth.

control of the harmonic impedances, and may in some cases be driven away from a particular fundamental impedance setting due to undesirable, but essentially unseen, harmonic effects in the tuner impedance characteristic. This scenario becomes more likely when using a device at a frequency where the X_{Cds}/R_L ratio (see Chapter 5, Figure 5.25) is well above unity. It is also more likely as the device is biased into deeper Class AB operation.

For some years several manufacturers have attempted to provide tuning systems which allow the user to select one or two harmonic impedances (usually second and third) independently from the fundamental. Such systems are necessarily more complex, more expensive, and generate extra dimensions in experimental space that can demand considerable skill on the part of the user to reap the stipulated rich rewards.

Figure 12.5 shows a possible configuration in schematic form. Following the twin-slug fundamental tuning section, and placed on the same low loss 50Ω airline section, are moveable elements which have bandstop characteristics at the selected harmonic frequencies. In their simplest form, these elements could be λ/4 open circuit stubs at the respective harmonic; the use of a high characteristic impedance to realize the OCSS will reduce the reactive loading at the fundamental. A commercial manufacturer has to do a little better than this, in order to increase the useable bandwidth of the system and further reduce the interaction between the harmonic and fundamental tuning [1]. This becomes an ever-increasing challenge, given the need to move the harmonic elements in order to shift the phase of the harmonic reflections, and the need to keep losses to a minimum. In any event, such an arrangement only allows for a high reflection coefficient, albeit with full phase control, at the harmonic frequencies.

An alternative system is based on the properties of a microwave multiplexer. Such devices have been staple items in ECM systems for decades, and their design has been well covered in the literature. Figure 12.6 shows the swept frequency performance of a typical multiplexer. Basically, each specified band transmits only from the input port to a single designated output port, and is highly reflected from the other ports, regardless of their terminations. In their more customary role of

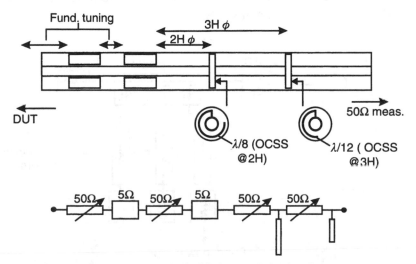

Figure 12.5   Schematic representation of possible harmonic load-pull tuner.

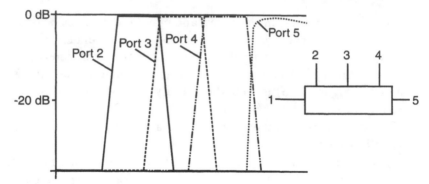

Figure 12.6   Typical microwave multiplexer response.

bandsplitting, the various transmission bands will intersect in a carefully prescribed manner; the response of the "thru" port (port 1-5 response in Figure 12.6) will also vary according to specification. It is clear that such a component can be used to configure a harmonic tuning system, as shown in Figure 12.7. The isolation properties of the multiplexer can be used to give independent tuning at however many harmonic bands are deemed necessary. In particular, separate tuners on each port can set impedances at any point on the Smith chart, rather than just a phase controlled-reflection.

One disadvantage of this system is that the multiplexer will usually have significant in-band insertion loss which will restrict the tuning range, especially when dealing with high power devices that require very low impedance to match the fundamental. Even though the band-pass attenuation may look reasonable in a 50Ω environment, say, 1 dB, this can escalate alarmingly when the device is placed in a high VSWR situation. Allowing for these losses becomes quite a challenge when both magnitude and phase of the reflection presented to a high power device will have a major impact on the loss correction on the measured power. This problem is illustrated in Figure 12.8, where a lossy 50Ω device is modeled as a matched "L" attenuator. Clearly, if the leading series resistance of the model equals the impedance of an external device, 6 dB attenuation will occur due to the apparently "reasonable" loss as measured in a 50Ω system.

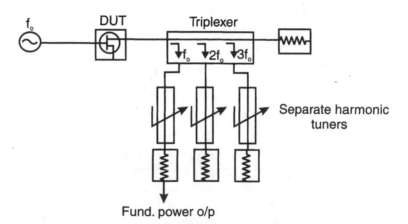

Figure 12.7   Harmonic load-pull system based on a multiplexer.

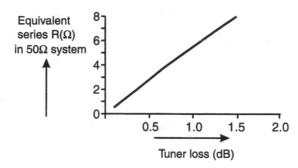

Figure 12.8   Input series resistance for matched attenuator having same loss as tuner.

The potentially catastrophic effect of even just a few tenths of a dB loss in any 50Ω lines, connectors, and filtering elements between an RF power device and the first physical tuning element is highlighted in Figure 12.8. This shows that when the desired device matching impedance drops to the vicinity of 1Ω, even two or three tenths of a dB of tuner loss will have a major effect on the observed performance, reducing the gain by several dB as well as limiting the matching capability of the tuner to a higher resistance value. This effect has been the cause of much confusion and frustration to the unwary in the evaluation of larger RF power transistor performance using external tuners. Although the values in Figure 12.8 represent a somewhat worst-case characterization of the problem, it is clear that the fundamental matching device has to be placed physically very close to the device, with a minimum of connecting cables, no matter how low loss the cables may be. It can be further stated that any device which requires an impedance lower than about 5Ω will cause serious loss issues in load-pull testing and should be prematched in the immediate vicinity of the device chip.

12.3   Active Harmonic Load-Pull

The multiplexer system shown in Figure 12.7, and the associated problems of losses, highlighted in Figure 12.8, form a good starting point to consider an alternative approach to load-pull known as "active" load-pull. The various reflections which are created by the tuners in Figure 12.7 could, from the viewpoint at the device plane, be created by replacing the passive tuners by phase-locked signal generators, and the tuner settings simulated by adjusting the amplitudes and phases of the individual harmonic signals. Such a system is shown in basic form in Figure 12.9. The simple argument already presented is convincing enough when the system is viewed from the device plane. Things get a little more complicated when questions are then asked about power flow. We now have a system in which multiple sources of power are being fed into the device, variously reflected out again, back through the multiplexer, and finally put to rest in a very necessary isolation device at each generator output. Although such a system, intuitively, would appear able to simulate the effects of selected fundamental and harmonic loads at the device plane, how can the equivalent of the device output power be measured in such a system?

As with so many old chestnuts in the RFPA business, this issue has been the subject of much debate over decades, rather than years. By far the most satisfactory

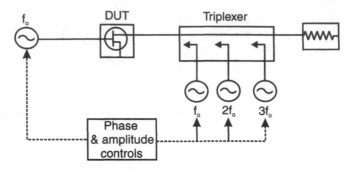

Figure 12.9 Concept of active load-pull system.

approach appears to be to stay at the device plane and measure the voltage and current waveforms at this point. In fact, any reference point in the 50Ω section between the device and the multiplexer can be selected in order to establish the voltage and current time domain waveforms using the forward and reflected waves which can be measured directly using directional couplers. Fourier analysis of the waveforms yields not just the effective power being generated by the device, but also the harmonic and fundamental impedances. Such a system, shown in Figure 12.10, has the added value that provided the device package and fixture are sufficiently well characterized, the reference plane can be set to the actual device output plane, which enables the current and voltage to be monitored. This can give valuable information about the mode of operation, as well as details about the device operation, such as dynamic IV characteristics at the actual frequency of use [2]. A key point about an active LP system is that the detrimental effects of losses (for example, the additional directional couplers that are now required) can be removed by making suitable adjustments to the generator power outputs. Figure 12.11 shows a typical measurement from such a system, showing remarkable detail in the current and voltage waveforms under controlled harmonic loading.

Such a system can, in principle, set any number of harmonic impedances by suitable settings of the generators. Harmonic source terminations can also be simulated by using a similar multiplexer and generators on the input, as shown in Figure 12.10. In practice there will be considerable interaction between the settings when measuring a device in a nonlinear condition, and computerized search algorithms

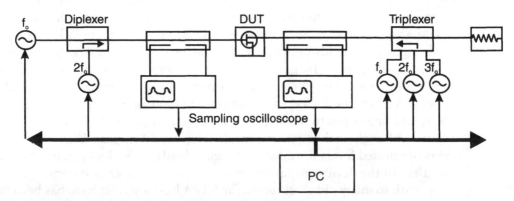

Figure 12.10 Practical active load-pull system (see [2]).

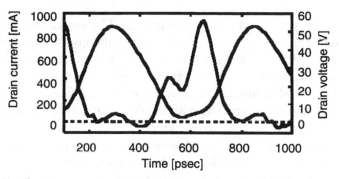

Figure 12.11 Waveform measurements using system of Figure 12.10. (*Courtesy* High Frequency Measurements Group, Cardiff University, United Kingdom.)

become almost essential in a practical system. Impressive results from such active LP systems have been reported in the literature [2, 3], but the overall complexity, along with challenging requirements for driver and calibration software, appears so far to have deterred commercial manufacturers.

12.4 Variations, Results, Conclusions

The current generation of commercial LP systems can trace their origin to much more basic manual tuning tests performed using simple, and in some cases crude, tuning devices. The engineer or technician who "tweaks" the output of a PA by moving a capacitor along a section of transmission line on a preproduction PC board is engaging in a form of load-pull activity. At higher frequencies the capacitor transmogrifies into a small piece of metal foil. Although much mocked as a now-unnecessary activity by the CAD merchant community, it is still alive and well in the back rooms and dark corners of RF development labs. It is therefore quite relevant to consider load-pull systems that in cost and complexity lie between the two extremes of a commercial harmonic system costing a significant fraction of $1 million, and a small piece of foil.

One variation on a low-cost system is to design a test fixture which includes some harmonic matching. The rest of the experimental space can then be explored using a tuner at the fundamental only. For example, the provision of a λ/4 SCSS connected as close to the device output as possible, can in principle take care of the all-important second harmonic for Class AB designs, while also providing a convenient bias insertion point. Although this technique will become less useful, and less effective, for larger devices at higher frequencies, in these cases it may be that the device capacitance itself becomes large enough to act as an internal harmonic short. These two possibilities however leave an intermediate zone where harmonic terminations can still have an important effect, and the choice of tuner can give misleading results, as described in Section 12.1.

A configuration which seems to have been overlooked, both by the user and tuner manufacturers, is to design a fundamental tuner which is "transparent" at one or more harmonic frequencies. Such a tuner can be used to ensure that only the fundamental is being tuned, and suitable harmonic terminations can be placed at any point in the system. For example, Figure 12.12 shows a system where the fundamental tuner uses two slugs which are exactly one quarter wavelength long at the frequency of interest. Such a tuner will be "transparent" at the second and fourth harmonics. If the tuner is terminated by a broadband attenuator, as would normally be the case, the harmonics are at least presenting a fixed invariant termination to the device. This is in contrast to possible wild gyrations of harmonic impedances which can be exhibited by some commercial tuners. An additional feature could be the placement of a highly reflective element at the second harmonic, with a line stretcher to vary the phase of the reflection. The key feature of such a system is that the effect of the second harmonic phase rotation can be examined with a simple measurement.

Figure 12.13 shows the result of such a test on a deep Class AB GaAs MESFET device at 2 GHz. The cyclic variations of power and efficiency are a reminder of the importance of harmonic matching in amplifier designs of this type, and the value of being able to characterize these effects using fairly simple experimental test equipment.

In conclusion, it would be fair to state that the ever-increasing demands being placed on RF power amplifiers in communications systems creates a situation where models and simulators seem to be forever playing catch-up. The ability of CAD simulators to predict detailed PA performance, especially spectral distortion, has to be regarded as a still improving, but as yet not totally adequate, area. Load-pull tech-

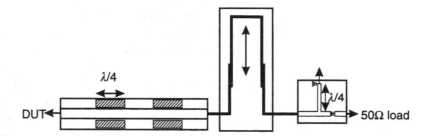

Figure 12.12 Harmonic tuning using fundamental tuner having "transparent" tuning elements.

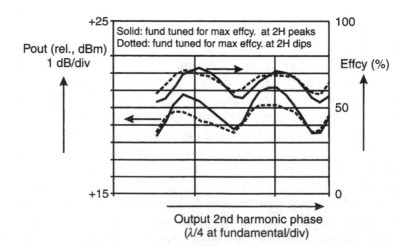

Figure 12.13 Power and efficiency tuning results obtained using second harmonic tuner shown in Figure 12.12.

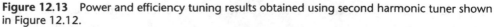

niques are still widely used by PA designers, especially in higher power applications, and show no signs of imminent retirement.

References

[1] Tsironis, C. (Focus Microwave), "Multi-Purpose, Multi-Probe Tuner for On-Wafer Harmonic Tuning," *IEEE MTTS Intl. Microw. Symposium*, Long Beach, CA, 2005.

[2] Williams, D. J., J. Leckey, and P. J. Tasker, "A Study of the Effect of Envelope Impedance on Intermodulation Asymmetry Using a Two-Tone Time Domain Measurement System," *2002 MTT-S Intl. Microw. Symposium*, Seattle, WA, June 2–7, 2002.

[3] Aboush, Z., et al., "Active Harmonic Load-Pull System for Characterization of Highly Mismatched High Power Transistors," *IEEE MTT-S Intl. Microw. Symposium Digest*, Long Beach, CA, June 12–17, 2005.

CHAPTER 13

Power Amplifier Architecture

Introduction

Most of this book has been concerned with the design of a single power amplifier gain stage, using a single transistor. Any practical power amplifier will be a subassembly consisting of several driver and gain stages, and the final power stage itself will probably use some form of power-combining network. This chapter addresses several of the specific topics which arise in designing complete amplifier assemblies. These topics include balanced and push-pull configurations, multistage design, power combining, biasing and stability considerations.

The previous focus on single transistor PA stages is easily justified. A PA stage with 10 dB of gain is most likely to consume around 80% of the DC power of the transmitter RF chain and dissipate about the same high percentage of heat. The transistor itself will be by far the most costly individual item, and may be over 50% of the total material cost of the final assembly. Due to its thermal requirements, the final stage will demand the heaviest design and manufacturing overheads, with physical size, mechanical integrity, and heat management requiring attention from a wide range of engineering and manufacturing services.

Indeed, with all these justifications in mind, the design of even the PA driver stage seems a relatively noncritical task. The contribution of the driver stages to the overall efficiency of the assembly is reduced by the factor of the final PA stage gain, and it is quite justifiable to design all the driver stages as simple Class A, or light AB, amplifiers. This has great benefits in terms of linearity and AM-PM issues, and avoids the demanding harmonic termination requirements of higher efficiency modes. Ironically, the "walk-in-the-park" attitude toward driver stage design can often lead to the biggest problems when the assembly is tested. Stability problems, and driver linearity issues in particular, can ruin much good work on the final PA stage design. The key problem in designing a PA driver chain is to limit the distortion, or gain compression, of the drivers so that most of the distortion takes place in the PA stage itself.

Power combination using circuit techniques is an important topic which will be covered in this chapter. In particular, the status of the circuit combiner as a last resort to achieve a specific power level, when all other options run out, will be called into question. What can be achieved with circuit board costing $5 per square foot is well worth leveraging in comparison to the continual invention of ever more exotic semiconductor processes to meet the needs of modern communication systems.

Initially, two important techniques, push-pull and balanced amplifiers, will be discussed. Both of these techniques feature prominently in almost any commercial

PA assembly. There is a tendency, especially in the lower frequency PA regime, to regard the push-pull configuration as mandatory, and to ignore the quadrature balanced approach. Conversely, the microwave ECM amplifier community use quadrature balanced modules almost exclusively, and claim little or no benefit from push-pull operation. The wireless communication bands occupy an interesting middle-ground position, and both techniques should be examined and well understood before making any decisions on the architecture of a new PA assembly.

13.1   Push-Pull Amplifiers

Most of the RF power amplifiers built for operation below about 500 MHz are of the push-pull type. Here again, an old technique which was invented in the early days of tubes has survived well into the semiconductor era with its benefits largely intact. Differential techniques are recognized throughout the entire electronics arena as advantageous, and RF power amplification is no exception. There are important differences, however, between a low frequency, or audio, push-pull amplifier and one at GHz frequencies. Most notably, RF communications signals are fundamentally sinusoidal in nature, the information being carried as much slower amplitude and phase variations. In this case, it is quite acceptable to distort, or rectify, the RF carrier provided the modulation is kept intact. The audio case is quite different, in that the input signal consists of a highly complex time-varying signal whose form has to be accurately preserved during the amplification process. A simple single-ended Class B amplifier would be quite unacceptable for this application, since the negative excursions of the signal would be wiped out. This was the original motivation for the development of push-pull amplification.

Figure 13.1 shows the basic push-pull amplifier as seen in electronics textbooks. The two devices are driven differentially so that the equivalent circuit shows the two devices being driven in antiphase, both devices being biased to a Class B, or zero quiescent current. The output load resistor is connected differentially between the two drains (or collectors), through a center-tapped transformer. Due to the differential excitation, the positive excursions of current are linearly amplified by one device and the negative excursions are amplified by the other. Each device is driven beyond its cutoff point whenever the other is conducting, as shown in Figure 13.2. These amplified upper and lower portions will be added back together again by the action of the center-tapped transformer, and so the amplified signal is reconstructed at the output load resistor. The most important benefit of this configuration at audio frequencies is that some of the benefits of Class B operation can be leveraged while maintaining broadband linear operation.

There are important differences between this kind of amplifier and the Class B RF mode discussed in previous chapters. It is essentially a broadband signal amplifier which requires a resistive output termination. The load is maintained constant over the entire frequency range of the input signal; there are no harmonic shorts since these would obviously corrupt the signal waveform. This means that the efficiency will not be as high as 78%, but there are still considerable benefits from making amplifiers in this way, which are mainly associated with power management. Unlike a single-ended Class A amplifier, the DC consumption will rise and fall in

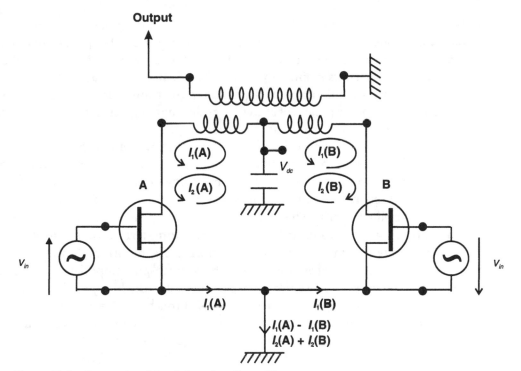

Figure 13.1 Conventional (audio) push-pull amplifier.

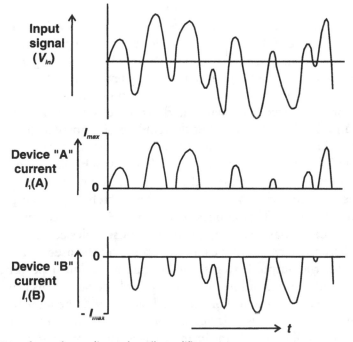

Figure 13.2 Waveforms for audio push-pull amplifier.

synchronism with the varying input waveform amplitude, and the overall efficiency, most conspicuously manifested as lower heat dissipation, will be much improved.

Because of the essentially random nature of the signal it is not possible to derive simple expressions for the efficiency improvement, but it can be quite dramatic due to the inherent power management capability of the Class B mode. There will also be possible improvements in linearity over a Class A amplifier, due to the inherent cancellation of any second harmonic distortion in the transistor characteristics. Unfortunately, as any audiophile will know, an additional new distortion process know as "crossover" distortion, makes an unwelcome appearance.

Such an amplifier can still have applications at RF and microwave frequencies, particularly for broadband amplifiers having an octave or more bandwidth. In these applications, amplifiers have broadband matching circuits which cannot, for example, filter out second harmonics such as would be the case in narrower band communications applications. The benefits of power management for variable envelope, or multicarrier, signals would still result in lower long-term power consumption and heat-sinking requirements than a Class A design. It is important, however, to make a distinction between the traditional broadband push-pull amplifier of this kind, and the simple push-pull connection of a pair of Class AB, B, D, or F stages designed according to the principles in earlier chapters of this book. Such an amplifier, shown in Figure 13.3, retains the even harmonic short circuit on each device in order to constrain the individual drain-to-source voltages to a sine or squared-up sinewave required by Class B or Class F operation. This kind of amplifier has the same efficiency as each individual single-ended component, but has two important advantages over single-ended operation: reduced common lead effects and impedance doubling.

One of the key benefits of push-pull operation of this kind is that the fundamental frequency components of the two device currents will be equal and opposite (see Figure 13.3). So if there is a common lead connecting the sources to ground, cancellation will occur and no feedback voltage will be developed across the lead inductance. This of course assumes that the two devices and the differential excitation are exactly in balance. Thus in the ideal case, common lead effects are eliminated. Such common lead effects are particularly troublesome to the RFPA designer, causing substantial gain reduction at each doubling of transistor periphery in single-ended designs. The second benefit is probably the most important in RFPA design. If it is assumed, for the time being, that differential RF excitation is available, then the composite impedance presented by the push-pull pair at both the input and the output, is a factor of two higher than that which would be presented by one single-ended device. This applies to the input impedance, where the two transistor gate-source (or base-emitter) junctions appear series connected, and also at the output where the usual loadline considerations can be used to come up with a value of $2.R_{OPT}$ for the output load resistor for maximum linear power performance (where R_{OPT} would be the loadline match for a single device). This is in somewhat dramatic contrast to the impedance of a parallel pair, where the impedances would be halved in comparison to a single device. So there appears to be a 4:1 impedance benefit for a push-pull combining scheme in an equal-power comparison to a simple parallel connection. When device impedances are down in the 1Ω range, this becomes a highly significant benefit.

This analysis would seem to indicate that push-pull connection offers the RF designer enough benefits that its use would be widespread. This is very much the

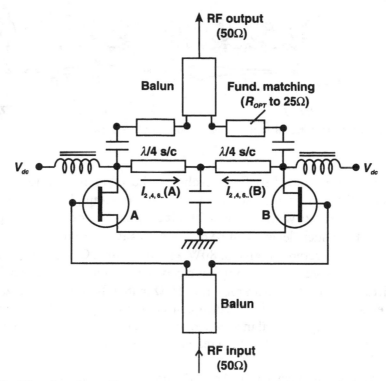

Figure 13.3   RF push-pull amplifier.

case at frequencies up into the UHF range. At higher frequencies, into the GHz region, some problems start to appear. First and foremost, the assumption that a differential signal is readily available becomes more questionable, and baluns must be provided. This is not quite such a straightforward task for a power amplifier as it is for other applications requiring baluns, such as mixers and antennas. There is another issue which concerns one of the key benefits of push-pull operation, that of the cancellation of common lead effects. Figure 13.4 shows that at higher frequencies, the distributed nature of a typical microwave power transistor starts to dilute this benefit. A typical device will have multicell construction, and the common lead ground currents are distributed using multiple ground bonds around the periphery of the chip. At higher microwave frequencies, the ground bonds can be replaced by plated-through vias through the die itself. In this kind of layout, equal and opposite currents from the two main antiphase devices will not flow in each and every inductance.

Without attempting a more detailed quantitative analysis, it can be reasonably concluded that differential operation will not necessarily result in higher gain at frequencies where the on-chip interconnections have significant inductance. Another way of looking at this issue is to note that any microwave power transistor has already been designed, using technological aids such as via hole grounds, to minimize common-lead inductance effects. So a push-pull connection seeks only to reduce what may already be a small effect; microwave devices are always designed, as a matter of necessity, to have low source inductance. It remains a moot point whether such attention to grounding could be relaxed in a device designed specifically for push-pull operation, with possible cost and yield savings.

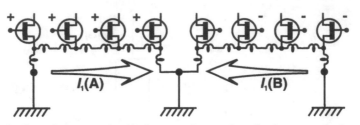

Figure 13.4 Distributed common lead inductance in a push-pull microwave device.

The impedance benefit is probably the main motivation to persevere with push-pull design at microwave frequencies. For example, it has been seen in earlier chapters that a 20W transistor at 2 GHz may have an optimum output power match of 1Ω. Four such devices can be connected as two parallel push-pull pairs, giving a composite structure capable of 80W with the same 1Ω impedance. This would compare very favorably to a simple four-way parallel connection, which would result in a virtually unusable device requiring 1/4Ω match. In order to realize this benefit, it is necessary to design baluns which can handle the power level, have the necessary phase and amplitude balance, and have negligible insertion loss. The loss, in particular, becomes a critical factor in reaping the benefits of push-pull operation at higher frequencies.

At lower frequencies, in the VHF and down into the HF range, RF balun design is dominated by the use of multifilar-wound transformers on ferrite cores. These transformers are typically ingenious applications of low frequency transformer concepts harnessed at much higher frequencies through the properties of ferrite materials and the transmission line coupling effects of multiple twisted wires. Much practical and theoretical work has been published to assist in the design of such components [1, 2], although their useful frequency range for PA applications would seem to extend no higher than UHF. For higher frequency applications, particularly above 2 GHz, other options must be considered.

Figure 13.5 shows a balun which is commonly used for higher frequency RF push-pull amplifiers. It has its origins in antenna applications, where balanced dipoles need to be fed with unbalanced coaxial feeder. The key requirement of any balun structure is to float the ground connection such that the impedance back to ground, as measured from either of the balanced outputs, is suitably high compared to the currents flowing in the balanced termination. A length of 50Ω semi-rigid cable, with a conventional unbalanced input termination, is suspended above the ground plane to form a short-circuited stub between the outer sheath of the coaxial cable and the groundplane. Viewed from the balanced outputs, this places the

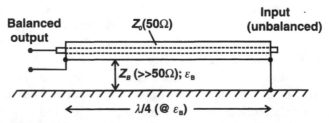

Figure 13.5 Coaxial balun structure.

required open-circuit to ground at the resonant frequency of the stub. Some analysis is required to determine how the characteristic impedance of the stub, and its electrical length, affect the performance of the balun over a usable frequency range.

Although the problem lends itself to analysis using transmission line equations, the performance is most easily evaluated using a linear microwave circuit simulator. In order to analyze the circuit, shown schematically in Figure 13.6, it is necessary to have a simulator which recognizes a transmission line as a 4-node element. The analysis, also shown in Figure 13.6, computes the power division into two balanced 25Ω loads, which together comprise a single balanced 50Ω termination. For convenience, each 25Ω load has been transformed up to 50Ω using an ideal 2:1 transformer. This would not be done in practice since the 25Ω termination in part represents the advantageous impedance transformation property of push-pull operation. The power split is perfect at the resonant quarterwave frequency, and the rolloff either side is largely a function of the characteristic impedance of the outer of the co-ax to ground, higher values giving broader band performance. At lower frequencies, such baluns can be made into very broadband devices by wrapping the cable around a ferrite core. Figure 13.6 shows that the phase split, ideally 180°, has a similar acceptable bandwidth to the amplitude response. The key feature to note in the design of such a balun is that the center frequency is determined by the quarterwave resonant frequency of the short circuit stub formed between the outer of the co-ax and ground. This will usually be an airspaced line, or may include some dielectric in PC board realizations. In either case, the length of the cable will be substantially greater than an electrical quarter wavelength of the cable itself, as indicated in the parameters shown in Figure 13.6.

These simulations would seem to indicate that balun design is not much of a problem. In practice, the manufacturing issues raised by this structure at higher fre-

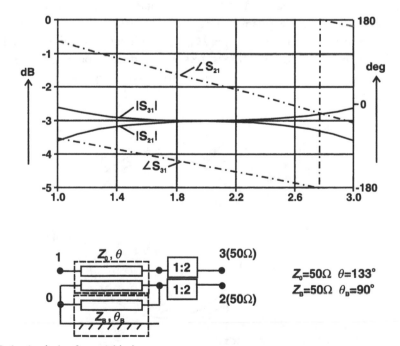

Figure 13.6   Analysis of coaxial balun.

quencies are considerable, particularly regarding the accuracy required in the cutting, stripping, and soldering of the cable. The search for a truly planar balun structure which is compatible with conventional microstrip board manufacturing has been the subject of many papers and patents, stretching over numerous decades [3]. The venerable magic tee, or rat race as it is called in microstrip format, has much to commend it, as shown in Figure 13.7. Although occupying more board space than some 3-dimensional structures, it has precision, repeatability, and zero assembly time in its favor. It should however be noted that the terminations of this structure are all Z_0 (50Ω) to ground and in this respect the rat race is not a "true" balun, since the balanced ports present a $2Z_0$ impedance level. This could be considered as a significant dilution of some of the impedance transformation benefits of push-pull operation. The bandwidth, shown in the analysis in Figure 13.8, is substantially lower than the coaxial balun, but would seem to be adequate for communications band applications.

With the improved quality and low cost of surface-mount RF components now available, the possibility of using a lumped element balun structure which incorporates some matching becomes realistic. Figure 13.9 shows a possible network, which was originally used as an antenna balun [4]. If each arm of the bridge type network has a reactance X_B (either capacitive or inductive), then the balanced impedance Z_{BAL} is given, in terms of the unbalanced termination Z_0, by

$$Z_{BAL} = \frac{X_B^2}{Z_0} \tag{13.1}$$

Such a circuit could be designed to combine the functions of matching and phase inversion required for push-pull operation, as shown in Figure 13.10. In this form, it can be seen that the circuit consists of complementary low-pass and high-pass sections on respective push-pull ports, each giving the same impedance transformation but generating the required differential phase inversion. As a specific design example, consider again a push-pull pair of unit devices, each requiring a 1Ω output power match. Substituting $Z_{BAL} = 2Ω$, $Z_0 = 50Ω$ in (13.1), gives $X_B = 10Ω$. So for a 1.9 GHz design, the inductor value is 0.84 nH, and the capacitor value is 8.4 pF. The circuit is analyzed in Figure 13.11, where for analysis purposes the 1Ω loads have been

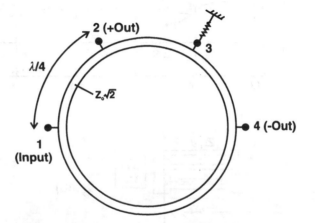

Figure 13.7 Rat race 180° power splitter.

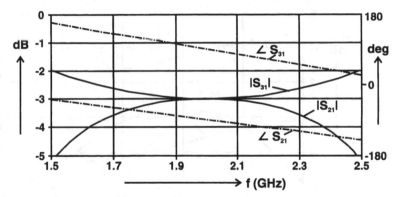

Figure 13.8 Rat-race simulation.

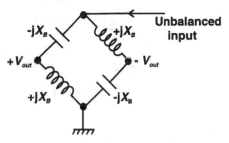

Figure 13.9 Lumped-element balun. (*From:* [4]. © 1992 Peter Peregrinus. Reprinted with permission.)

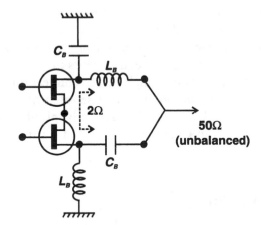

Figure 13.10 Lumped-element balun with integral match form 2Ω (balanced) to 50Ω (unbalanced). Values for 1.9 GHz: $L_B = 0.84$ nH; $C_B = 8.4$ pF.

transformed back up to 50Ω using ideal transformers. The power split is maintained within 0.5 dB over a bandwidth of 100 MHz, which is adequate for some communications band applications, and the 180° phase split is very flat over a much wider band. Clearly, some care would be required in allowing for parasitic effects in realizing the ideal performance shown in Figure 13.11, but the advantages in board space and assembly costs would appear attractive.

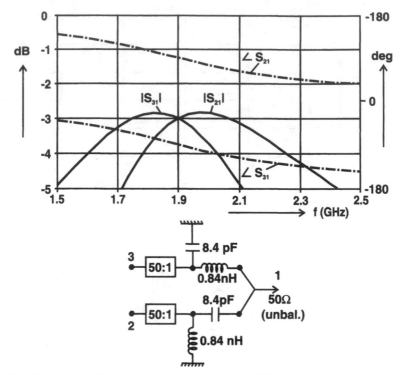

Figure 13.11 Simulation of lumped-element balun/match (Figure 13.10).

This brief survey of possibilities for the realization of suitable baluns at higher frequencies would seem to indicate that push-pull operation is still worthwhile, but requires some preliminary development effort to ensure that the required power and phase characteristics are being obtained from whatever structure is chosen. It should also be noted that the matching requirements of each device in a push-pull configuration are still the same as in a single-ended connection, and this includes the appropriate harmonic termination in the case of amplifiers operating with shorter conduction angles such as Class B. Essentially, all of the techniques and circuit configurations discussed in Chapters 3, 4, 5, and 6 can be used with some benefits in push-pull mode.

13.2 Balanced Amplifiers

Although the push-pull configuration may be considered to be balanced in the more generic sense of the term, in the microwave amplifier world a "balanced" amplifier is something quite different, and is shown schematically in Figure 13.12. Two identical amplifiers are fed from an input power splitter which produces two signals in phase quadrature, the outputs being recombined using a similar device connected in reverse. The principal advantage of this configuration is that any mismatch reflections from the amplifiers pass back through the couplers and appear in antiphase and therefore cancel at the RF input (or output) port. Reflected energy is diverted to the terminated coupler ports. Provided the two amplifiers are identical in their characteristics, and the couplers have ideally flat amplitude and quadrature differential

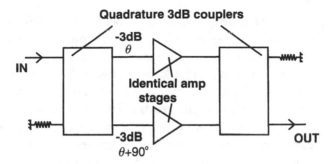

Figure 13.12 Balanced amplifier configuration.

phase responses, the match seen at the input and output will be essentially perfect, regardless of the reflective nature of the individual amplifier stages. Thus the functional effect can be compared to the use of ferrite isolators on the input and output ports of a single-ended amplifier having twice the transistor periphery.

The benefits of the balanced amplifier configuration were first recognized in 1965 in a classic paper [5], which in effect founded a new microwave amplifier industry. It was an enabling technique which made octave and multioctave microwave amplifier designs straightforward and manufacturable. The ability to design gain "modules" which had flat, but imperfect VSWR, characteristics over wide bandwidths and then by the emplacement of couplers create almost perfect cascadability was a luxury which seemed almost to defy the normally unforgiving laws of nature. This approach, teamed up with thin film hybrid technology, quickly became the mainstay of a burgeoning ECM amplifier industry. Not only was the balanced module a technical breakthrough for the microwave circuit designer, it also greatly increased efficacy in manufacturing and production control. Different amplifier specifications could be easily accommodated by using permutations of a few standard balanced modules. These modules could be pretested and inventoried in reasonable quantities. Intrepid marketeers could respond to customers' changing requirements through the use of simple spreadsheet cascading routines, which until the advent of the laptop PC often took the form of a pencil and a used envelope.[1]

Given the major impact of the balanced amplifier on the microwave business, its more detailed technical ramifications seems to get less than their fair share of treatment in the literature. This is perhaps due in part to the unpalatability of a simple pragmatic technique that instantly downgrades, although by no means invalidates, some basic textbook results and teachings. For example, balanced amplifiers are much more stable than their single-ended equivalents. So much so, in fact, that in the balanced world some of the dire predictions of k-factor analysis can be quietly set aside. This statement will be quantified in due course, but it is an issue which is often missed by the RF communications amplifier industry, who tend to view the balanced amplifier as an exclusively broadband technique.

Another related issue that has the greatest significance to power amplifier design, whatever the bandwidth, is interstage matching. If amplifier stages are all balanced, interstage matching becomes a redundant issue. Essentially, all balanced

---

1. Cigarette packets were also popular for this application, until the available workspace became filled with health warnings.

modules have a 50Ω impedance, and have performance designed and measured with 50Ω terminations. These terminations can be input and output 50Ω ports, or other balanced modules. This is important in developing the architecture of a power amplifier. A power module, with known power characteristics, can be confidently expected to produce the same performance when used as a driver for a higher power output stage if both modules are realized in balanced form. This is far from being the case when a driver transistor output is matched directly into the PA device in a single-ended configuration. Indeed, many such designs fail due to inadequate performance of the driver device, and such a situation does not lend itself to easy troubleshooting.

On the downside, it is commonly stated that the balanced approach is expensive, consuming twice the number of components and higher current. For PA design, this is not a fair assessment, because due account must be taken of transistor periphery. Quadrature couplers are very effective power combiners, and a single-ended design for a given power level will require a transistor having twice the periphery of each device in a balanced equivalent. There is an additional issue of unfair comparison here; the matching Q-factors and consequent gain-bandwidth product of the devices in the balanced module will be much more favorable than the single-ended version. This may impact the cost of the matching and bypass capacitors which are, for sure, twice in number in the balanced version.

Just as with the push-pull configuration, these benefits cannot be realized unless a suitable power splitter and combiner can be implemented. In the push-pull case, a balun is required, and for balanced operation a quadrature 3 dB coupler is needed. Fortunately, the 3 dB quadrature coupler is a well-established component in the microwave toolbox, and detailed design information is spread throughout the literature and CAD software libraries. Essentially, a pair of coupled quarter-wave transmission lines does the job, providing certain well-documented physical criteria are correctly met, see Figure 13.13. A pair of coupled lines are analyzed in terms of two characteristic impedances, called the odd and even mode impedances. The odd mode impedance, Z_{0o}, physically represents the characteristic impedance of the two

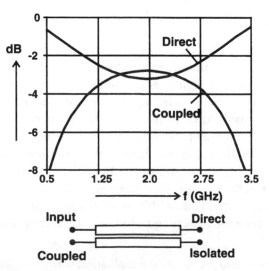

Figure 13.13   Quadrature coupler response.

strips under differential excitation, and the even mode impedance, Z_{0e}, represents the characteristic impedance of the two strips in conventional unbalanced parallel connection. (More precisely, Z_{0o} is defined as half of the physical value of Zo under odd-mode excitation and Z_{0e} is defined as twice the even mode impedance of the two strips in parallel.) Two key relationships enable values of Z_{0o} and Z_{0e} to be determined for a required mid-band coupling value, K:

$$Z_0^2 = \sqrt{Z_{0e} \cdot Z_{0o}}; \ K = \frac{Z_{0e} - Z_{0o}}{Z_{0e} + Z_{0o}}$$

For example, values of $Z_{0e} = 125\Omega$, $Z_{0o} = 20\Omega$, for $Z_0 = 50\Omega$, give a coupling factor $K = 0.724$, or -2.8 dB.

The 20Ω value for Z_{0o} represents especially tight coupling between the lines, such as cannot be obtained by simple edge-coupling, as shown in Figure 13.14(a). Several solutions to the coupling problem are shown in subsequent frames in Figure 13.14. A symmetrical multilayer structure, Figure 13.14(b) is by far the most satisfactory in terms of RF coupling and directivity performance, and is the structure used for most commercially available "drop-in" coupler products. The widespread availability of multilayer PC board and thru-via processing technology would appear to make this structure compatible with board layouts used for modern wire-

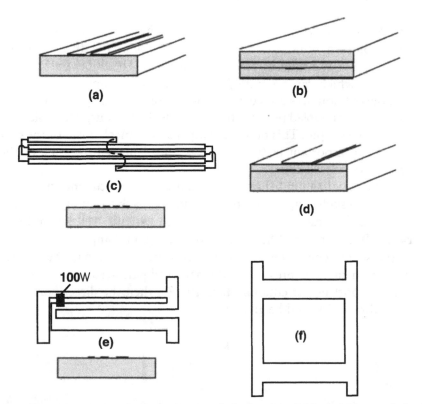

Figure 13.14 Quadrature couplers: (a) edge-coupled microstrip; (b) broadside coupled stripline; (c) interdigitated Lange coupler; (d) microstrip with capacitive overlay; (e) Wilkinson with 90° phase "jog"; and (f) branchline coupler.

less systems, although low cost surface mount couplers are now commercially available as separate components.

The microwave hybrid industry had a requirement for a single-layer solution, which was admirably provided by the interdigitated Lange coupler (Figure 13.14(c), [6]), where the two coupled lines are broken up into a number of strips in order to multiply the number of edges available for coupling purposes. Such a structure still has fairly narrow gaps and is quite a challenge to realize using PC board etching techniques, and there is the added issue of making interconnections between alternate strips. Figure 10.14(d) shows a compromise approach, where a simple overlay can provide the necessary increase in coupling factor. This would appear to be more appropriate for manufacture using modern multilayer board techniques if the added cost of drop-in couplers is a problem, although this structure will tend to radiate more than any of the other structures shown. A 3 dB quadrature coupling can be achieved in other ways. Figure 13.14(e) shows a simple in-phase Wilkinson combiner with a 90° "jog" on one port. This lends itself to a convenient and compact layout, as shown. There is also the so-called branchline coupler, shown in Figure 13.14(f). This has the disadvantages of fairly large size and narrow bandwidth. It will be shown, shortly, that some of the side benefits of balanced operation will be eroded if narrow band couplers are used, and both of the structures shown in Figure 13.14(e, f) may suffer in this respect.

Returning to the basic block diagram of the balanced amplifier, Figure 13.12, the effects of nonideal coupling response on the gain and power needs to be analyzed in order to determine the coupler specifications for a particular application. Figure 13.13 shows a typical coupler response. The midband coupling is usually set at a value somewhat lower, or tighter, than the nominal 3 dB so that the 3 dB crossover points occur at a suitable bandwidth spacing. The direct port has a complementary response which can accurately compensate the coupled port response if the input and output couplers are correctly orientated. The upper amplifier, for example, is fed from the coupled port on the input side, but its output couples through the direct port to the output. This built-in compensation mechanism of the coupled line structure is an important feature which is not displayed by some of the other coupler options discussed above.

The equalization behavior works fine when both amplifiers are completely linear. If two such couplers are placed back-to-back to simulate a unity gain amplifier, a broadband response is obtained, which extends well beyond the 3 dB coupling range. When the amplifiers are driven into gain compression, the unequal coupling response will cause one amplifier to compress more than the other, and the overall impact on gain compression of the balanced pair needs more detailed analysis. If the coupled port power coupling factor is K_C, then the direct port coupling, assuming good directivity, will be given by

$$K_D = \frac{1}{1-\left(\dfrac{1}{K_C}\right)} \qquad (13.2)$$

So an input signal of voltage amplitude V_{IN} will split into two incident amplifier signals, V_{IC} and V_{ID}, at the respective amplifier inputs, according to

$$V_{IC} = \frac{V_{IN}}{\sqrt{K_C}}; \; V_{ID} = \frac{V_{IN}}{\sqrt{K_D}}$$

The measured transfer characteristic of a single amplifier, $V_O = T(V_I)$ can now be applied to each respective amplifier, and then the output coupler will perform the complementary voltage division and superposition, giving a final output voltage,

$$V_{OUT} = \frac{1}{\sqrt{K_D}} \, T\!\left(\frac{V_{IN}}{\sqrt{K_C}}\right) + \frac{1}{\sqrt{K_C}} \, T\!\left(\frac{V_{IN}}{\sqrt{K_D}}\right) \tag{13.3}$$

Clearly, if $T(V_I)$ is a simple linear function, so that $V_{OUT} = A.V_{IN}$, then

$$V_{OUT} = \frac{2AV_{IN}}{\sqrt{K_D K_C}}$$

which simplifies to the gain of a single amplifier, A, if $K_D - K_C = 2$, and shows a rather insensitive dependency on the coupling factors so long as they follow the coupler relationship of (13.2). Figure 13.15 shows the power transfer characteristic represented by (13.3) plotted for a few representative values of coupling factor and a typical compression characteristic, as discussed in Chapter 9. It can be seen that even the power characteristic is remarkably unchanged from the ideal 3 dB coupling characteristic for coupling factors down to 1.5 dB. A coupling factor of 2 dB appears to soften the characteristic by only a matter of 0.2 dB. Essentially the same self-compensation mechanism is at work; the overcoupled input amplifier contributes correspondingly less to the output power than the undercoupled (direct) input amplifier which, in turn, operates more linearly.

The above analysis confirms that the quadrature coupler makes an excellent power combiner. For narrowband operation, it is usual to select a low amount of overcoupling, maybe 2.8 dB as in the above example. Wider bandwidths are easily

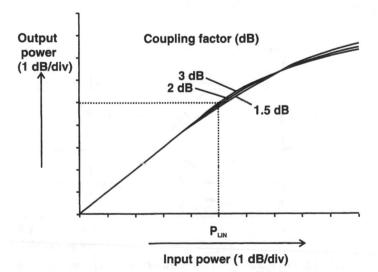

Figure 13.15   Balanced power amplifier compression characteristics.

achieved with higher amounts of overcoupling. Successful designs up to 4:1 band-width have been demonstrated with coupling factors in the region of 1.5 dB.

It has already been stated that the balanced configuration provides greater sta-bility. This can be demonstrated quite dramatically by recalling the Class A ampli-fier designed in Chapter 2, and sweeping its single-ended response over a wide bandwidth. The corresponding k-factor plot shows the k value falling well below unity at lower frequencies (Figure 13.16). There would be a serious stability issue if this design were to be cascaded directly with another single-ended driver stage, and some substantial lossy elements would have to be incorporated into the matching and bias circuitry to reduce the out-of-band gain and increase the k-factor. This would degrade the in-band gain performance, which would in turn necessitate the use of a higher power driver. The alternative strategy would be to create a balanced module using two such amplifier stages. This, of course, is not a direct comparison with the single stage in that the balanced module would have 3 dB more power, but Figure 13.16 shows that the k-factor of such a configuration, using the coupler response depicted in Figure 13.13, displays impressive stability over a very wide bandwidth. This is a dramatic demonstration of the stability benefits of the balanced configuration. Figure 13.17 illustrates how the improved stability of a balanced amplifier arises. A plot of the input match when both active coupler ports are termi-nated by a short circuit shows a much more benign environment for interfacing with other components. Figure 13.17 also shows that the use of narrower-band couplers such as the branchline structure can seriously degrade this particularly valuable property of the balanced configuration.

Throughout this section it has been strongly implied that balanced amplifiers have such outstanding advantages over single-ended designs that they should be considered for narrowband applications. The virtual elimination of stability and interstage matching problems, along with the manufacturing and product flexibility presented by a standard module set, would seem to outweigh the extra board space and component requirements. The frequent use of interstage drop-in isolators in narrow-band PA designs only serves to emphasize the advantages offered by a bal-anced configuration, which provides better isolation, over much broader bandwidths, than the isolator approach.

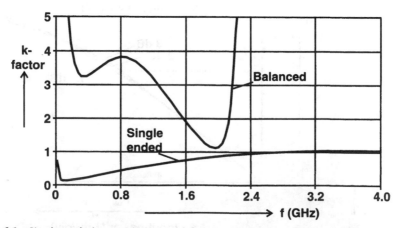

Figure 13.16 Single-ended versus balanced k-factor comparison (Class A design from Chapter 2, coupler response as in Figure 13.13).

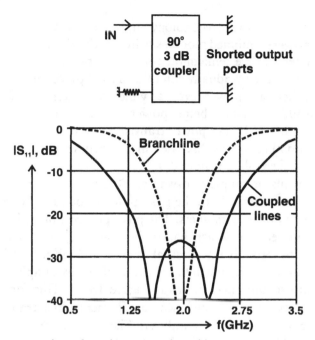

Figure 13.17 Input return loss of quadrature couplers with short-circuited output ports.

13.3 Power Combining

The role of the power combiner in RFPA design is enigmatic. Multiway power combiners at lower microwave frequencies should, in principle, be easy to build with precision, low loss and low cost. Yet the circuit designer seems all too eager to surrender this cheaper pathway to higher power levels in favor of bigger, more exotic, more expensive, higher Q transistors as soon as they become available. Obviously, if the higher power semiconductor technology simply gets better and cheaper, then there is little to be gained by not using it. The reality, however, would appear to be more convoluted. Semiconductor manufacturers seem to be prepared to go to greater lengths to produce more power in a package than amplifier manufacturers are to design low loss combiners. The resulting situation is that combiners tend to get used as a last resort, using the highest power transistors which are available to reach yet higher power levels.

There would seem to be several reasons not to follow this conventional trend. It has already been commented, in connection with the simple two-way combiner represented by a balanced configuration, that the lower periphery devices have lower Qs and can be designed to give flatter gain and phase characteristics over given bandwidth. Lower periphery devices also have significantly higher gain, due to the less demanding requirements for source or emitter inductance. Logically speaking, it should also be cheaper to purchase two devices of a given periphery, as compared with a single device with twice the periphery, due to yield considerations. Unfortunately, transistor manufacturers do not always show this in their small quantity pricing structures, presumably due to differential profit margin strategy.[2]

---

2. This, I suppose, is a euphemism for selling bigger devices below cost.

Different eras do see technology improvements which would ultimately result in multiple power-combined modules being gradually replaced by single package solutions. Higher voltage semiconductors, in particular, tend to give more power at a similar price as they gradually enter the market place. But at any given time, it would seem that there are power levels at which a greater use of power combining techniques would result in a better power amplifier than would be obtained using a smaller number of higher power transistors. Some possibilities for such large scale power combining will be discussed in due course, but initially some of the simpler building blocks will be reviewed.

Two forms of simple two-way combining have already been discussed in the previous two sections. Both the push-pull and balanced configurations can be considered to be power combiners. The push-pull combiner has the advantage of doubling the impedance which has to be matched; the balanced combiner has the advantage of broadband external match and added stability. A third type of two-way combiner, in which the two amplifiers are simply fed in phase, is known as the Wilkinson combiner, shown in Figure 13.18. This form of power combiner is essentially a pair of 2:1 impedance transformers which transform each 50Ω input up to 100Ω, where the two are paralleled. The function of the isolation resistor is to terminate any odd-mode signals; this also shows itself as a midband null in transmission between opposite combining ports. This port isolation is widely regarded as an asset in RF power combiner design, mainly through its ability to suppress odd mode instability between the combined amplifiers. In fact, the balanced quadrature hybrid will also display similar isolation between the inputs and outputs of the separate amplifiers, and over a broader bandwidth. Depending on the bandwidth required, the Wilkinson combiner can have two or more quarterwave matching sections, as shown in Figure 13.19. Although the single section would appear to be flat enough for moderate bandwidth applications, it should be noted that the rolloff from midband will accumulate with each combiner in the chain.

Such combiners have been described extensively in the literature [7] and will not be analyzed in detail here. A critical issue in power combining applications is the insertion loss of such structures, especially if realized in the cheapest and most convenient manner, using open microstrip lines. Radiation losses, and simple copper losses due to the use of thin dielectric board material (and hence thin conductor strips) at higher frequencies, can cause the losses of such structures, especially double section designs, to creep up to unacceptable levels for RFPA work. The use of a good low loss dielectric board material does not, unfortunately, guarantee low loss performance in these applications. This problem may even be a part of the trend, discussed in the introduction to this chapter, to use higher power devices in preference

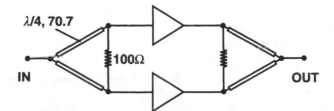

Figure 13.18   In-phase power combiner using single section Wilkinson power combiners.

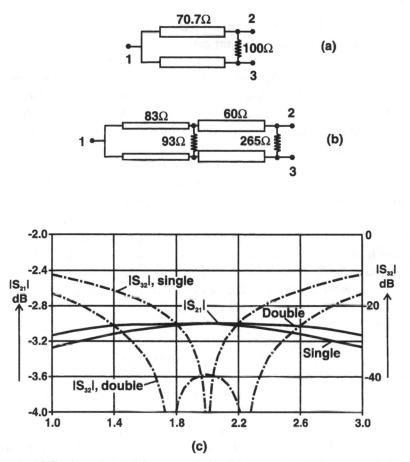

Figure 13.19   (a) Single section Wilkinson combiner; (b) two section Wilkinson combiner; and (c) simulated power transfer and isolation responses (lossless models).

to combiners. Much lower losses can be obtained using enclosed stripline, suspended stripline, or coaxial construction, but these are not so convenient to integrate with the amplifier matching circuitry.

Since the most common power combiner requirements are to combine two, or at most four, individual devices or single stage amplifiers, it would be appropriate at this point to review the relative merits of the three candidates discussed so far. Push-pull combining is not commonly used to combine the power from separate, matched amplifier modules. The advantage of common lead cancellation cannot be realized if the devices are in separate packages, and the impedance benefit is much less useful at the 50Ω level, rather than at transistor level. The final choice therefore rests between the in-phase Wilkinson and the quadrature coupler. In all important performance aspects, the quadrature coupler wins over the in-phase combiner, but the Wilkinson lends itself to easy, if lackluster, realization using simple etched microstrip lines. The availability of good, cheap, solder-in quadrature couplers may well cause a change in the traditional preference for Wilkinson type combiners in microwave PA designs.

Although there have been many other ideas for power combining techniques published over the years, most of these refer to higher, microwave frequency operation. In particular, the properties of waveguides can be used to make very effective multiway power combiners that can have almost negligible losses [8, 9], but these designs would appear to be impractical at lower frequencies due to the physical size of the waveguide. The loss performance may nevertheless become sufficiently attractive as the final power requirement reaches toward kilowatt levels. At millimeter wavelengths, quasi-optical techniques have been demonstrated, which can combine the power from hundreds of individual devices to reach power levels of over 10 watts at 30 GHz [10].

Most of the remaining reported combiner concepts are effectively derivatives of the cascaded Wilkinson type, particularly with the added generality of unequal power division [11]. One interesting derivative in this category is the so-called "traveling wave" combiner [12], which is effectively a string of unequal power dividers, each succesive splitter being set to give an equal overall power division factor. The benefit of this combiner would appear to be a more compact layout than an equivalent cascaded Wilkinson "tree," which often incurs extra losses through the need for multiple wavelengths of 50Ω lines to distribute the various inputs and outputs. Another, more straightforward, derivative of the multiple Wilkinson tree is the "radial" combiner [13, 14], which again reduces the impact of distribution line losses by adopting a more compact three-dimensional structure. The radial combiner is still quite widely used to generate multiple watts of power at mm-wave frequencies.

In pursuing the possibilities for combining a very large number of devices, it pays to consider an analogous situation in large aperture antenna design. Here, just as with microwave amplifiers, there is a clear choice between a single large aperture antenna, such as a large parabolic dish having a single feedpoint, and a phased array which may in some cases consist of hundreds of small individual elements with a correspondingly more complicated corporate feed structure. Any antenna designer will emphasize the benefits of the phased array approach. The large dish has very limited and specialized uses whereas the phased array is more flexible and requires much less structural engineering. Of course, antennas need to be steered and in this respect the analogy has an important aspect missing, although the possibilities for power management and linearization in a large array of power-combined transistors are considerable and largely unexploited. The main point is that the techniques used for feeding large antenna arrays are well established and usually differ significantly from the conventional multiple-bifurcating 3 dB power splitters which are conventionally used for high power amplifiers. Figure 13.20 shows a couple of examples of antenna-style multiway combiners. If the $\lambda/2$ sections could be realized using some form of enclosed structure, rather than lossy microstrip lines, it would seem possible to make combiners extending into tens or even hundreds of individual cells.

Although this seems an alien concept in power amplifier design, it is a highly regarded technique in the antenna world. Such a structure would occupy more boardspace than a smaller number of higher power devices, but would have much better thermal management. Gain, bandwidth, phase linearity, and soft failure would be additional benefits. As has already been commented, there would also be new possibilities for linearization and efficiency enhancement, based on some of the principles outlined in Chapter 14.

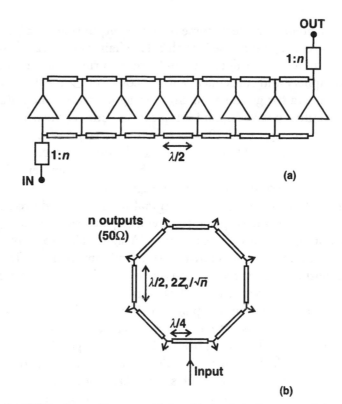

Figure 13.20   Possibilities for multiway combining: (a) simple halfwave paralleling, and (b) Bagley Polygon. (*From:* [4]. © 1992 Peter Peregrinus. Reprinted with permission.)

13.4   Multistage PA Design

Any PA stage requires a driver, and before the details of circuit design are considered, it is important to decide how much drive power, and at what linearity, any particular case requires. Clearly, in applications where nonlinear amplifiers are acceptable, and the PA stage is a saturated design, there is correspondingly no issue regarding driver linearity; the driver can be nonlinear too. In practice, however, it is quite common to revert to linear designs in order to build up PA driver chains which may extend to 40 dB or more in overall gain. This is due to the relatively low impact of driver efficiency given a PA stage with over 10 dB power gain (see Figure 3.17), plus the increased gain available from linear stages. Probably the only stage which may still benefit from higher efficiency mode operation would be the driver itself; subsequent predriver stages might as well be designed for maximum linear gain at appropriate power levels.

Driver chains become more challenging when overall linearity is important. The key issue is to control the generation of nonlinear products in the driver stages, while maintaining efficiency and power supply specifications. At first sight, this may seem a simple enough task of selecting the periphery of each driver and predriver, taking due account of the gain in front of each respective stage. Assuming the output PA stage is designed to run at a given level of backoff in order to meet a given specification, it would seem obvious that the drivers all have to run at the same level of relative backoff, but with some added margin in order to ensure that most of the nonlinear effects come from the output stage; 3 dB extra backoff is a typical rule-of-thumb often used. Such a PA chain is illustrated in Figure 13.21. The extra margin for driver linearity can be seen to run up a significant cost in terms of device periphery and consequent current consumption. All too often, this has been a traditional area for good sound engineering practice to be redlined by commercial and competitive pressures, and a typical outcome is shown in a revised block diagram shown in Figure 13.22. Here the driver stage reaches its 1 dB compression point at the same amplifier input level as the output stage.

Curiously, this is one area where the evils of commercial pressure have been known to trump sound engineering design principles. PA chains have a mysterious habit of working better than conventional wisdom would predict, even when one or more of the driver stages appear to be working at a similar relative level of gain compression, or power backoff, as the final PA stage. This appears to be a contradiction when taking a conventional power series viewpoint of amplifier performance. Envelope simulation sheds some light on the process, as illustrated qualitatively in Figure 13.23. If each stage is assumed to be a hard voltage clipper, there is no loss in overall linearity in performing the clipping function in a driver, rather than a PA stage. Provided the clipping levels of each stage are compatible with the subsequent gain, there will be no additional clipping from subsequent stages. So if gain compression characteristics were ideally sharp, it would be quite acceptable to have all the individual stages in the chain running at the same relative level of gain compression, and the final distortion characteristics would be the same as a single stage running at this

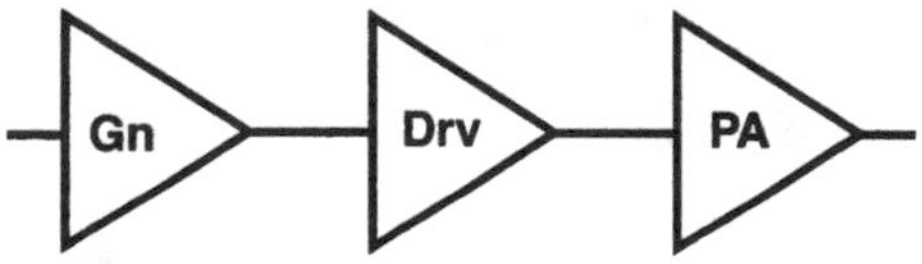

	Gn		Drv		PA
P_{OUT} (dBm)	19.1		31		40
g/c (dB)	0.1		0.1		1
gain (dB)	13		11.9		9
$P_{1 dB}$ (dBm)	23		34		40
Periphery	W/50		W/4		W

Figure 13.21   PA drive chain, "linear" driver stages.

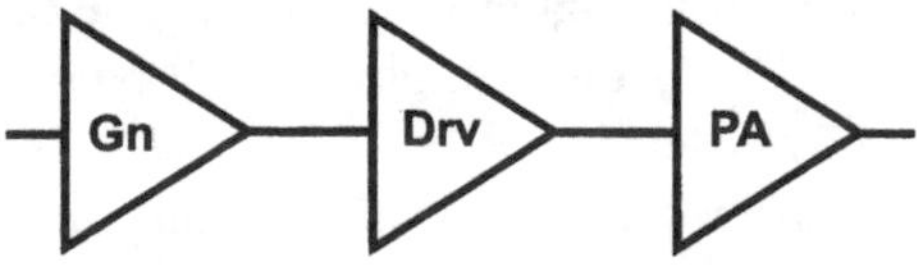

	Gn		Drv		PA
P_{OUT} (dBm)	19.1		31		40
g/c (dB)	0.1		1.0		1.0
gain (dB)	13		11.0		9
$P_{1 dB}$ (dBm)	23		31		40
Periphery	W/50		W/8		W

Figure 13.22   PA drive chain, driver stage 1 dB compressed.

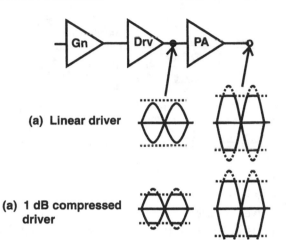

(a) Linear driver

(a) 1 dB compressed driver

Figure 13.23 (a, b) PA chain compression, assuming ideal clipping action.

level of compression. Unfortunately, this ideal picture will not be reproduced in practice because the devices will not behave as ideal clippers. But the insight is valid, and at least shows that over-zealous elimination of gain compression in driver stages may indeed lead to unnecessary increases in cost and power requirements.

Related, although not identical, effects can be observed in the AM-PM characteristics of drivers. Looking back at the measured AM-PM data in Figure 9.13, it is clear that the AM-PM phase shift can be either positive or negative, depending on the specific device characteristics, the input and output tuning, and the bias conditions. In a given situation, it is quite possible to find a fortuitous cancellation between the AM-PM characteristics of the driver and the PA stage. This cancellation only becomes possible if the driver stage itself is driven into compression, and is biased more towards cutoff. Well backed-off Class A drivers may be more linear in themselves, but do not provide the opportunity for such helpful cancellation scenarios. Such fortuitous cancellations are pleasing to observe, but have some dangers associated with their long-term reproducibility.

The above analysis has been qualitative. Commercial system simulators can now be used to explore possible cancellation effects and to obtain more quantitative and objective choices for driver chain configuration. Having obtained a basic device lineup it then remains to design the interstage matching networks. In principle, the output power matching for a driver stage can be treated in much the same manner as has been considered in some detail in previous chapters, where it was usually assumed that the target impedance would be the output 50Ω termination. In a simple single-ended design, such matching networks (see Figure 13.24) will be challenging for a number of reasons. The target matching impedance will have a reactive component, being now the input impedance of another RF transistor. This gives it a frequency-dependent characteristic, a figurative moving target. Such is not the case when matching into a 50Ω termination. There is also a serious stability issue. Even though a simple lowpass matching structure may appear to do a decent job for the required in-band response, it is likely that the out-of-band response will place the low, highly reactive PA transistor input impedance directly across the driver output at some lower frequency, an odds-on stability conflict. Typically, sta-

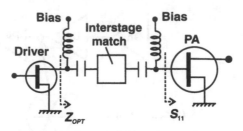

Figure 13.24   Interstage match for single-ended driver-PA configuration.

bility can only be restored by the placement of lossy elements which will increase the driver power requirements (a common technique is the insertion of resistive attenuator pads in the 1–3 dB range). There is an additional complication of bias insertion. An interstage matching network has to include both output and input bias insertion networks for the driver and PA devices, respectively, including a DC block.

The design of the interstage match between a driver and PA stage can be just as challenging as the matching of the PA output itself. Indeed, this is often a reason for poor initial results in such amplifier designs. It is not uncommon to see eleventh-hour insertion of interstage isolators, or even attenuators, to band-aid the various problems arising from the interfacing of a driver-PA stage combination. Drop-in isolators covering common communications bands have become readily available at fairly low cost, and although their use in this application may seem to some as an admission of shortcomings in the design process, the benefits they offer can sometimes outweigh such philosophical considerations.

13.5   Conclusions

The techniques discussed in this chapter allow individual transistors and amplifier modules to be combined to produce powers extending into the hundreds of watts. Environmental issues notwithstanding, it is probably a reasonable assumption that RF and microwave PA requirements will continue to demand more power at higher frequencies. Up to a point, the rapid growth of wireless communications throughout the 1990s has been sustainable with pre-existing technology. Fortunately, existing technology included some more esoteric semiconductor materials and processes such as Gallium Arsenide which came into being largely through military requirements. The emergence of high band gap semiconductors, in particular Gallium Nitride is an important new development in this field. This material has been successfully grown on both Si and SiC substrates [15] and offers substantial improvements over GaAs in terms of current density and voltage breakdown. This may well result in microwave power amplifiers breaching the kilowatt barrier, although maybe not at bargain-basement prices.

References

[1] Sevick, J., *Transmission Line Transformers*, Atlanta, GA: Noble, 1996.

[2] Riddle, A., "Ferrite and Wire Baluns with Under 1 dB Loss to 2.5 GHz," *IEEE Intl. Microw. Symposium*, MTT-S 1998, WE2E-1.

[3] Raicu, D., "Design of Planar, Single-Layer Microwave Baluns," *IEEE MTT Intl. Microw. Symposium*, MTT-S 1998, WEIF-07.

[4] Burberry, R. A., *VHF and UHF Antennas*, London, England: Peter Peregrinus, 1992.

[5] Eisele, K. M., R.S. Engelbrecht, and K. Kurokawa, "Balanced Transistor Amplifiers for Precise Wideband Microwave Applications," *IEEE Int. Solid State Circuits Conf.*, February 1965, pp. 18–19.

[6] Lange, J.,"Interdigitated Stripline Quadrature Hybrid," *IEEE Trans. Microwave Theory & Tech.*, MTT-17, No.12, 1969, pp. 1150–1151.

[7] Cohn, S. B., "A Class of Broadband Three-Port TEM-Mode Hybrids," *IEEE Trans. Microwave Theory & Tech.*, MTT-16, No. 2, pp. 110–116.

[8] Chen, M. H., "19-Way Isolated Power Divider Using TE01 Circular Waveguide Mode Transition," *Proc. IEEE Intl. Microw. Symp.*, MTT-S 1986, pp. 511–513.

[9] Tokumitsu, Y., "6 GHz 80W GaAsFET Amplifier with TM Mode Cavity Power Combiner," *IEEE Trans. Microwave Theory & Tech.*, MTT-32, No. 3, 1984, pp. 301–307.

[10] DeLisio, M. P., et al., "A Ka-Band Grid Amplifier Module with Over 10 Watts Output Power," *Int. Microwave Symp.*, Tu4A-2, Fort Worth, TX, June 2004.

[11] Choinski, T. C., "Unequal Power Division Using Several Couplers to Split and Recombine the Input," *IEEE Trans. Microwave Theory & Tech.*, MTT-32, No. 6, 1984, pp. 613–620.

[12] Bert, A., and D. Kaminsk, "The Travelling Wave Divider/Combiner," *IEEE Trans. Microwave Theory & Tech.*, MTT-28, No. 12, 1980, pp. 1468–1473.

[13] Goel, J., "K-Band GaAs FET Amplifier with 8.2W Output Power Using Radial Combiner for 16 Devices," *IEEE Trans. Microwave Theory & Tech.*, MTT-32, No. 3, 1984, pp. 317–324.

[14] Belohoubek, E., "30-Way Radial Combiner for Miniature GaAsFET Power Amplifiers," *IEEE Intl. Microw. Symp.*, MTT-S 1986, pp. 515–518.

[15] Brown, J. D., et al., "Performance of AlGaN/GaN HFETs Fabricated on 100mm Silicon Substrates for Wireless Basestation Applications," *IEEE Intl. Microw. Symp.*, WE-5C-7, Fort Worth, TX, 2004.

CHAPTER 14

Power Amplifier Linearization Techniques

Introduction

PA linearization has become a subject in its own right in recent years, and has seen a large investment of research and development funding in both the industrial and academic sectors. It has been the subject of dedicated books [1–3] and on the conference circuit has grown from one or two papers into multiple sessions, and even entire symposiums. It has also been a very active area for venture capital spending. Every few months, it seems, a new start-up company lets out a press release proclaiming their totally new, disruptive linearized PA technique which will revolutionize the industry; another heavily patented area. This chapter is presented as an update and an overview of PA linearization. More focused treatment has been restricted to certain topics in digital predistortion and feedforward, these being the techniques that are in current use and still undergoing intensive development for wireless communications applications. Other techniques such as various kinds of feedback will be described but not fully analyzed, since they have found less use in modern wireless PA applications.

Chapter 13 described techniques which improve efficiency, without attempting to quantify the impact on linearity or RF power output. Such techniques are of paramount importance in mobile systems, where battery lifetime and thermal management are critical. There are other PA applications where efficiency becomes an important, but secondary consideration, in comparison to linearity. Such applications would typically be single or multichannel basestation transmitters in ground or satellite communications systems. Commercial competition, in particular between system operators who have paid billions of dollars for the exclusive use of a small frequency segment, has created a new generation of spectral regulations which are so stringent that even the compliance of passive components cannot be assumed. Traditionally "acceptable" spectral distortion levels, often specified in the −30 to −40 dBc range, can typically be 30 or 40 dB higher than the specified requirements in modern mobile communications systems. Not only are the required levels of intermodulation distortion tens of dB lower than in single channel applications, but the envelope variations of the composite signals can have instantaneous amplitude variations occurring at frequencies in the tens of MHz range.

These linearity specifications are so challenging for an RFPA that most first and second generation systems used a channelized approach, shown in Figure 14.1. This approach has the advantage of largely eliminating the possibility of intermodulation distortion between the separate signals, and greatly simplifies the

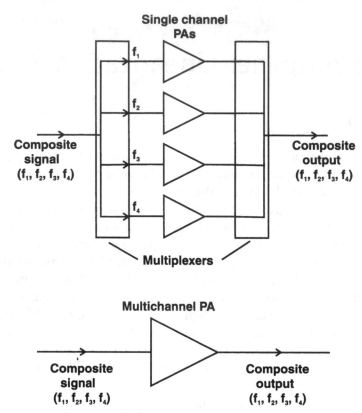

Figure 14.1 Channelized versus multichannel PA approaches.

design of the individual amplifiers. The problem with the channelized approach is the design and realization of suitable multiplexer networks. These monuments of traditional microwave engineering are usually laden with precision-machined silver-plated cavities, exotic dielectric materials, and galaxies of tuning screws which require individual adjustment. The motivation to design amplifiers that can handle the composite multichannel signal without channelization has been further compounded by the rapid growth in the numbers of remote stations and the lack of frequency agility in the passive multiplexers.

So the RFPA industry has had to tackle the challenge to design and manufacture amplifiers which have to meet the same stringent spectral distortion specifications without the facility of channelization. Such specifications can typically call for IM and ACP levels of –60 dBc. Clearly, the required linearity cannot be realistically contemplated using power backoff. If the IM3 for a two-carrier input is at –20 dBc for an amplifier running at 1 dB compression point, to get this down another 40 dB requires that the power is backed off by 20 dB; this means that a 100 watt transistor would only be good for a 1 watt output, at an efficiency measured in fractions of one percent. In the face of these numbers, linearization techniques can afford to be relatively costly in power and hardware requirements.

Rapid advances in high-speed digital signal processing (DSP) technology have brought a vast new resource, and an associated new culture, to the RFPA industry. The possibilities this offers, under the generic heading of "Digital Predistortion" (DPD, Section 14.2) are in the process of revolutionizing the traditional preserves of the analog RFPA designer. Even the more conventional, analog, and surprisingly vintage feedforward technique described in Section 14.3, is enjoying a new lease of life due to the use of digital techniques in the various control and monitoring functions.

The new developments and new requirements have also caused some fallers by the wayside. Feedback techniques, most notably those which use input and output detection in order to form a corrective drive signal to the PA, have been largely sidelined by the increasing bandwidths of modern communications signals. Even a single carrier WCDMA signal has a bandwidth in excess of 1 MHz, which poses great problems in a feedback linearization system due to the virtually irreducible delays in detecting, differencing, and amplifying a corrective signal. Techniques in this category, such as the Cartesian Loop, will thus be given a fairly brief overview in Section 14.4.

14.1   Introduction to PA Linearization

Most PA linearization techniques use the amplitude and phase of the input RF envelope as a template with which to compare the output, and so generate appropriate corrections. It is the way in which the correction is applied which forms the main distinction between the two most important linearization techniques currently in use. This is shown conceptually in Figure 14.2, where two approaches are shown; input and output correction. A linearizer which applies a corrective signal to the output of the PA will need to generate a significant amount of power to perform its function, but in so doing it will physically increase the peak power capability of the linearized PA. If the correction is applied at the input, it clearly cannot in any way increase the PA peak power. Input correction also poses an additional problem in that the composite input signal will be subjected to the nonlinearities of the PA. We will see in Section 14.2 that this can have some serious consequences, which can significantly reduce the apparent benefits of the linearization process.

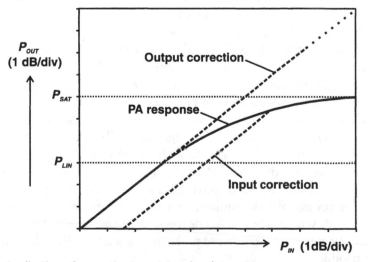

Figure 14.2   Application of corrective linearizing signals to a PA.

In order to meet IM specifications in the order of –60 dBc, much of the required action of a linearizer takes place in the lower left-hand region of Figure 14.2, where the thickness of the printed curves make the linearized and unlinearized responses appear coincident. In this extensive backed-off power range, a greatly magnified scale would show small fractions of a dB of gain compression and fractions of a degree of AM-PM distortion. *Predistortion* is the generic term given to techniques which seek to linearize a PA by making suitable modifications to the amplitude and phase of the input signal. *Feedforward*, a technique which has thus far dominated multicarrier PA linearization, applies a corrective signal at the PA output. The observation that an output corrective system such as feedforward can play an additional role in increasing the peak power capability of the target PA has long been the source of confusion, and much debate. It is actually possible to identify three sub-categories of feedforward PA systems: those which act primarily to linearize the PA at backed-off levels, and usually require the PA to be used below its peak power capability; those which act primarily in a peak power restoration mode; and those which do some of both. The second of these subcategories can be considered to be an efficiency enhancement technique in applications where the signal peak-to-average ratio is very high (see Chapter 10, Section 10.9.1).

A feedback loop can also be categorized as a form of input correction, or predistortion. A feedback system, working in a lower signal bandwidth application, seeks to generate a corrective control which is usually applied to the amplifier input signal. Given that the required predistorted signal at any specific signal input level is unique, it is a temptation for the analog RF traditionalist to speculate on why the use of digital techniques to generate the signal should be apparently immune from considerations of loop gain and stability. The key difference in the two implementations is that whereas the analog feedback loop computes the correction in real time, the digital version relies on a data store, which remembers the correction previously required for the same signal amplitude. Thus if for any reason the distortion properties of the amplifier have changed since the last calibration, the process will not work to its full capability. Memory effects in RFPAs have become a subject of great importance and close scrutiny, as further discussed in Chapter 9, and some kinds of memory effects can be incorporated in the digital correction process. This issue will be considered further in Section 14.2.

A quantitative analysis of the spectral distortion caused by the processes of gain compression and AM-PM is useful in order to obtain some quantitative feel for what a linearizer has to do. As always, a simple third degree model for the PA is helpful in this thought process. Figure 14.3 shows the levels of third order IM products in relation to the corresponding gain compression, and Figure 14.4 shows the IM levels which correspond to typical AM-PM distortion. These curves are based on a signal envelope having 3 dB peak-to-average ratio. It becomes clear that AM to PM distortion effects can contribute significantly to IM products in the precompression zone, and in order to reduce the IM levels down to –60 dBc the linearizer has to reduce gain compression down to 0.01 dB and AM-PM to well under 1°. The AM-PM curve is probably less familiar, and very significant. It shows that correction of AM-PM down to one or two degrees, over the whole dynamic signal range, will always be necessary to achieve more than about 10 dB reduction in spectral distortion products.

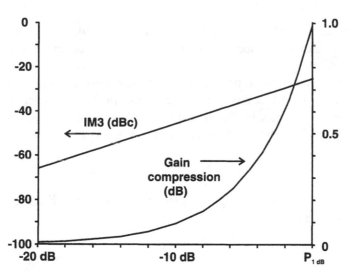

Figure 14.3   A two-carrier IM distortion and single carrier gain compression plotted as a function of drive level (third degree model).

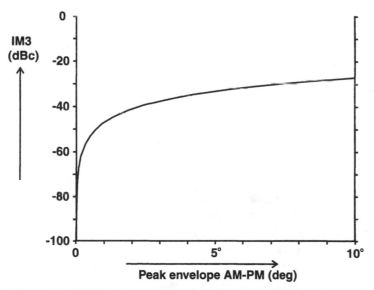

Figure 14.4   A two-carrier IM distortion as a function of peak AM-PM, based on analysis of a two-carrier signal in Chapter 9, (9.24) through (9.28).

14.2   Predistortion

14.2.1   Introduction to Predistortion Theory

Analog predistortion techniques have a long history, having been used extensively to correct the nonlinear characteristics of Traveling Wave Tubes (TWTs), a combination still in use today for high power applications in the upper GHz frequency bands [4–6]. Analog predistorters also still find use for more modest linearization requirements. For example, an extension of the useable range of an RFPA into its compression region can result in the gain of a few valuable percentage points of efficiency. Such applications include mobile, or handset, PAs, which frequently incorporate a simple analog predistortion network, often consisting of some kind of diode gain expander.

The theory behind predistortion can get quite complicated, and is not treated in detail here [1]. A few basic theoretical concepts are however worth noting. As usual, these principles can be illustrated quite well using a simple third degree model for the target PA. If the PA characteristic has the form

$$v_o = a_1 v_p + a_3 v_p^3 \tag{14.1}$$

where the PA input voltage v_p is the output of a predistorter, which in turn has a characteristic

$$v_p = b_1 v_{in} + b_3 v_{in}^3 \tag{14.2}$$

then the PA output, in terms of the predistorter input signal v_{in}, is given by

$$v_o = a_1 \left(b_1 v_{in} + b_3 v_{in}^3 \right) + a_3 \left(b_1 v_{in} + b_3 v_{in}^3 \right)^3 \tag{14.3}$$

which expands out to be

$$v_o = a_1 b_1 v_{in} + \left(a_1 b_3 + a_3 b_1 \right) v_{in}^3 + 3 a_3 b_1^2 b_2 v_{in}^5 + 3 a_3 b_3^2 b_1 v_{in}^7 + a_3 b_3^3 v_{in}^9 \tag{14.4}$$

and contains some minor surprises. Clearly, if we carefully design our predistorter so that

$$b_3 = -\frac{a_3 b_1}{a_1}$$

the original third degree distortion which was exhibited by the PA will be removed. Unfortunately, in so doing we have actually created some additional distortion terms, extending all the way up to the ninth degree. This demonstrates the first important "theorem" of predistortion:

*A predistorter/PA combination will, in general, **create** distortion products that were not present under the same signal drive conditions for the PA by itself.*

As a result, the predistorter has to generate some higher degree components in order to cancel the distortion which is caused by this process. This leads to a second theorem:

The output from a predistorter will display a spectrum of distortion products which equal, and in most cases significantly exceed, the spectral bandwidth of the uncorrected PA output under the same drive conditions.

This second theorem has some important ramifications in all predistorter design. The actual ratio of the necessary predistorter signal bandwidth to the unlinearized PA output spectrum is a much-debated number, but a factor in the 3–5 range seems to be quite typical, if the full dynamic range of the PA is to be used. This means that sampling and DSP clock rates may have to be as much as an order of magnitude faster than those used to generate the original signal. The generation of high order distortion by the predistortion process can be particularly troublesome when attempting to meet regulatory spectral distortion masks at higher frequency offsets ("Alternate Channel" power), which may have originally been met by the uncorrected PA, as indicated in Figure 14.5.

The above analysis has assumed that the linearization process is aimed entirely at the in-band, or close-to-carrier distortion. In Chapter 9 it was shown that such distortion is entirely the result of odd degree nonlinearity, so that in modeling the PA only odd degrees of nonlinearity need to be considered. The use of an odd degree nonlinearity in the predistorter, as analyzed in (14.1) through (14.4), therefore seems a logical choice. It is appropriate however to consider the effects of reintroducing the even degree nonlinearities when analyzing a predistorter-PA combination. In particular, if (14.1) is amended to include second degree distortion,

$$v_o = a_1 v_p + a_2 v_p^2 + a_3 v_p^3 \qquad (14.5)$$

and we use a different flavor of predistorter which has just a second degree nonlinearity,

$$v_p = b_1 v_{in} + b_2 v_{in}^2 \qquad (14.6)$$

the PA output becomes

$$v_o = a_1 \left(b_1 v_{in} + b_2 v_{in}^2 \right) + a_2 \left(b_1 v_{in} + b_2 v_{in}^2 \right)^2 + a_3 \left(b_1 v_{in} + b_2 v_{in}^2 \right)^3$$

which expands out to

$$v_o = a_1 b_1 v_{in} + \left(a_1 b_2 + a_2 b_1^2 \right) v_{in}^2 + \left(2a_2 b_1 b_2 + a_3 b_1 \right) v_{in}^3$$
$$+ \left(a_2 b_2^2 + 3a_3 b_1^2 b_2 \right) v_{in}^4 + 3a_3 b_2^2 b_1 v_{in}^5 + a_3 b_2^3 v_{in}^6 \qquad (14.7)$$

So if the square-law predistorter is designed such that

$$b_2 = -\frac{a_3}{2a_2}$$

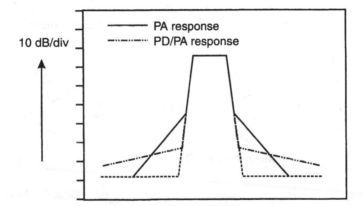

Figure 14.5 Effect of higher order IM generated by PD/PA combination.

the primary goal of canceling the harmful third degree PA nonlinearity can be achieved. The interesting aspect of this approach is that although the predistortion process again can be seen to generate several new and higher order nonlinearities, only one of them this time, a fifth degree term, is potentially harmful in its ability to generate additional close-to-carrier distortion products.

Unfortunately, the addition of second degree distortion, or "two-ness" to the input signal cannot be done within a normal narrow bandwidth of a typical communications channel. Two-ness has to be introduced either as a baseband or a second harmonic component. This basic principle has been used with some success as an analog predistorter, and will be considered again under that heading in Section 14.2.3.

14.2.2 Digital Predistortion (DPD)

It has already been noted that DPD has become the most active area for PA linearization development [7–9], and has been made possible mainly through unconnected developments in high speed digital processing (DSP) technology. Another enabling factor has been the addition of a new conceptual input socket on the RFPA assembly; a digital representation of the baseband signal which, in upconverted format, is being applied to the PA at any given time. This important addition is illustrated in Figure 14.6, and clearly represents a significant restriction in the range of potential RFPA applications. Such an input will only be available within the confines of a complete transmitter system which generates all of the signal information from scratch. Systems which use multiplexing at the RF band would not have such a baseband signal format available, and although in principle a peak detector and an ADC could be used to recover the envelope amplitude information, the phase modulation component would present greater difficulties. We will, in this section, make the same assumption that is now widely made in the linearized RFPA community, which is that such signal information is available.

Figure 14.7 shows a basic DPD PA system flow chart. If the PA behavior is quasi-static, it follows that the output amplitude has a fixed, monotonic relationship to the input signal,

$$V_o = F\{V_i(\tau)\}$$

where V_i and V_o are the magnitudes of the input and output modulated RF carrier, measured at the same instant in envelope domain time, τ. Thus a static look-up table, or LUT, can be compiled using *a priori* measurements on the PA.

Figure 14.6 Requirement for an additional input to a digitally predistorted PA.

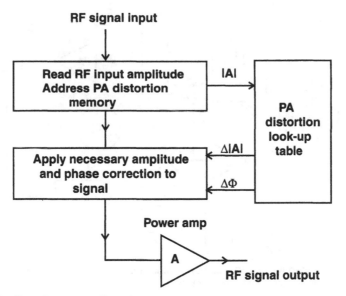

RF signal input

Read RF input amplitude
Address PA distortion
memory

|A|

PA
distortion
look-up
table

Apply necessary amplitude
and phase correction to
signal

Δ|A|

ΔΦ

Power amp

A

RF signal output

Figure 14.7 Predistortion system flow chart.

Such a system represents the first generation of DPD PA development, and has obvious limitations. It is entirely open loop, and any change in the PA characteristics will rapidly degrade the correction process. Nevertheless, implementers were not quite prepared for the lackluster performance such systems gave, even for short-term LUT calibration. Even with time and amplitude resolution beyond suspicion, 10–15 dB reduction in the close-to-carrier spectral distortion would be a representative result. The fundamental reason for these mediocre results was recognized as short-term variations in the PA characteristics, which on closer examination appeared to be present on all timescales, from nanoseconds to seconds. The term "memory effects" was coined to categorize these effects, as much as anything to pass the buck back to the PA designer, as to provide an explanation of their physical origins.

The various causes and manifestations of memory effects have been covered in Chapter 9. Subsequent generations of DPD configurations [9–11] have gradually got on top of this problem, through three main developments:

- Reduction of memory effects through more careful PA design;
- Addition of an adaption loop in the DPD system, enabling continuous monitoring of the linearization integrity, and LUT refreshing;
- Use of DPD algorithms which can correct shorter-term memory effects through the inclusion of some of the recent "history" of the signal.

The second of these developments means in effect that the RFPA system has to have its own receiver, as shown in Figure 14.8. With the ongoing developments in DSP technology, it has even become possible to compare the signals in a digitized IF format, thus avoiding some ambiguity issues when converting back to baseband. The system of Figure 14.8 now starts to look suspiciously like a feedback loop, and questions concerning its stability still arise. It seems that even using current DSP

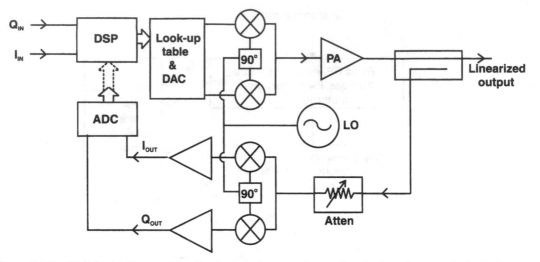

Figure 14.8 Digital adaptive predistortion system. In normal operation the loop is opened; the look-up table can be updated by occasional closing of the loop.

hardware, the LUT refreshment process is sufficiently slow that stability issues can be sidestepped.

The third development, algorithms which use previous signal values, raises a few eyebrows in the more traditional world of electrical circuits. A physical system whose output is not just a function of the input excitation at the current time, but also shows some dependency on previous input excitations, is hardly a new discovery. For example, Figure 14.9 shows a simple electrical circuit containing resistors, inductors, and capacitors. It is a matter of elementary electrical theory that the output voltage at time t cannot be expressed as a function of the input voltage excitation at time t alone. The known physical properties of capacitors and inductors lead directly to a much more complicated function, which involves both differential and integral functions of the input excitation. This in turn means that to determine the current value of the output voltage, a considerable number of previous time values of all of the voltages and currents are required. The simulator has to keep a record of these multiple value sets as it progresses in the time domain, in order to compute new updates on the output value.

It could be argued that this is a somewhat extreme and unrepresentative case of a system which displays memory effects, in comparison to the response of an RFPA. It could also be countered that the only difference is the 200 years of research into the physical behavior of inductors and capacitors that has enabled us to formulate a physically based model for electrical circuits. Physical modeling of memory effects in PAs could be straightforward enough if, as discussed in Chapter 9, they can primarily be attributed to dynamic thermal, and bias supply modulation, effects. Unfortunately there is the much less tractable intrusion of semiconductor trapping effects, which if present will make physical modeling much more of a challenge. Mainly for this latter probability in many cases, there has thus far been a strong preference to take a behavioral approach for PA memory effects, using the more general formulation of Volterra as a starting point [12].

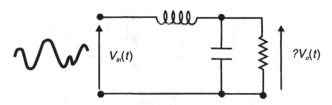

Figure 14.9 Time domain response of an LCR circuit.

Although the road to digitally predistorted Utopia has been a longer and more arduous journey than originally anticipated by many, the present status is starting to look impressive. Several stand-alone commercial DPD products are available, which come with full development support [13, 14]. Manufacturers of RF power semiconductors are starting to quote efficiency and linearity performance of their devices with predistorted test drive signals. Results using a range of RFPA technologies show that 20–30 dB of ACP reduction can be achieved, at a hardware cost in the sub-$50 range. This compares favorably in cost and performance to the only other linearization option in this class, feedforward, which will be discussed in Section 14.3.

Digital predistortion, and in particular the methodology to handle memory effects, is a heavily patented area [15–17].

14.2.3 Analog Predistortion

Despite a long history, analog predistortion techniques in RFPAs have never really reached mainstream use. Here and there, the use of a cheap and simple circuit which nudges the PA characteristic more or less in the right direction, for some part of the power range, has found its way into commercial use. But the subject has something of a lightweight status, and seems to have survived without much in the way of serious attempts to devise an *a priori* approach [1]. In the pre-DPD era, there was some more substantial effort to devise configurations which could synthesize the prescribed polynomial characteristic of a predistorter [1, 2], but such efforts would today be reasonably viewed as a very tough way of doing what can be readily achieved using DSP.

Analog PD circuits usually take the form of attenuators with an expansive insertion loss characteristic. Figure 14.10 shows a conceptual circuit which can create an expansion characteristic over a few dBs of dynamic range. At low signal levels, the diodes do not conduct, and the attenuation is determined by the series resistor. At higher drive levels, the diodes start to conduct and a lower value resistor is shunted across the first resistor, giving lower attenuation, as shown in the response curve in Figure 14.10. There have been many variations and derivatives of this basic circuit [18, 19]. Suitable placement of small value capacitors can generate an approximate phase characteristic in such circuits, as can the use of a varactor [20].

A more systematic approach is shown in Figure 14.11, in which some odd degree distortion components can be generated, and then scaled in amplitude and phase before being combined with the PA input signal. Such an arrangement has been called a "cuber," and can be arranged to work over a power range where the nonlinear element generates mainly just third degree distortion. This approach represents an important step in analog predistortion techniques, since the final PD dis-

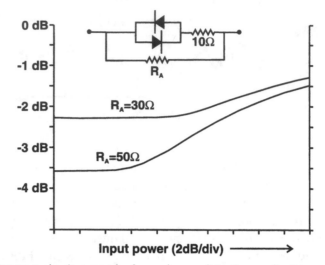

Figure 14.10   Conceptual gain expander for analog predistortion applications.

tortion is not so closely coupled to the properties of the nonlinear element. For example, a nonlinear element which has a gain compression characteristic can now be used, and one useful implementation would be to use a lower power amplifier whose characteristics are matched to those of the target PA [1]. This basic concept can be extended to synthesize a polynomial characteristic with higher degree terms. Other than the possibility of operating over much higher signal bandwidths, such polynomial synthesis has now become the exclusive preserve of DSP techniques. It is worth noting that the analog version would be unlikely to fare any better than a memoryless DPD scheme in its ability to handle PA memory effects.

A more recent development in analog predistortion, which appears to show results comparable to those obtained using memoryless DPD, is the use of second degree predistortion [21, 22]. As already noted in Section 14.2.1, second degree predistortion involves adding the PD component either at baseband, or in the form of a second harmonic. The baseband option appears to be less demanding, although the harmonic approach has been demonstrated [22]. Taking an RF carrier which has a baseband sinusoidal modulation of frequency Ω,

$$v = V \sin \Omega t \sin \omega t$$

so the second degree predistortion component will have the form

$$b_2 v^2 = b_2 \frac{V^2}{4} (1 - \cos 2\Omega t)(1 - \cos 2\omega t)$$

from which it is desirable to discard the second RF harmonic, leaving a baseband voltage which can be considered to be a variation on the DC bias applied to the signal in the amplifier, a conveniently accessible input. Figure 14.12 shows the somewhat unfamiliar-looking effect of this predistortion on a two-carrier signal. The predistorted signal now has the form

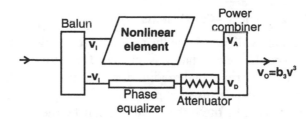

Figure 14.11 "Cuber" predistortion circuit. The linear branch is set to cancel the linear signal passing through the nonlinear element, giving a third degree output signal which can be scaled and phase-shifted as required.

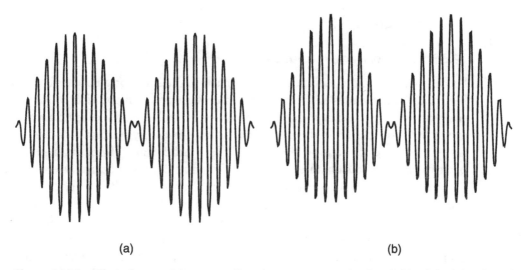

(a)   (b)

Figure 14.12 Effect of second degree predistortion on a two-carrier signal: (a) original signal, and (b) signal with baseband second degree component added.

$$v_p = vb_1 \sin \Omega t \sin \omega t + \frac{v^2}{4} b_2 \cos 2\Omega t$$

which will clearly generate a PA output component at the IM3 frequencies when subjected to a second degree PA distortion,

$$v_{pIM3} = a_2 \frac{v^3}{4} b_1 b_2 \sin 3\Omega t \cos \omega t.$$

This technique has produced some good results using an analog implementation [21, 22], which compare favorably with the best-reported results using memoryless DPD. The effectiveness will however be closely coupled to the individual PA device characteristics. The underlying theory points to a relatively unexplored area, which is the potential for linearization offered by the application of a suitably profiled bias variation to the PA stage.

14.2.4 Predistortion—Conclusions

DPD techniques will undoubtedly continue to sideline older analog methods. It should be noted, however, that most current DPD configurations require not only *a priori* knowledge of the signal modulation in digital form, but also a phase coherent sample of the RF carrier used to perform the down conversion. There are applications where neither of these requirements can be met. There are also applications such as in microwave links, where the individual signals are sufficiently widely spaced that the composite envelope variations extend into hundreds of MHz, well beyond the reach of DSP at the present time. Analog techniques, in principle, do still have the potential to perform usefully over these extended bandwidths.

14.3   Feedforward Techniques

14.3.1 Feedforward, Introduction

Feedforward is an old technique, dating back to the same original feedback patent by Black [23]. Its inventor saw it as merely a different implementation of the same basic concept as feedback except that the correction is applied to the output, rather than the input, of the amplifier. In this manner, the "time causality conflict"[1] of direct feedback is removed, and it could be fairly stated that feedforward offers the benefits of feedback without the disadvantages of instability and bandwidth limitations [24, 25].

Needless to say, this all comes at a price. In applying the feedback correction at the output of the amplifier, outside the correction loop, the correction signals need to be amplified up to the necessary higher power level. This necessitates an additional amplifier whose own distortion properties place an upper limit on the correction that can be obtained from a single feedforward loop. Great accuracy in gain and phase tracking also needs to be maintained in the different elements of the system, and the correction which can be obtained is dependent on the precision with which this tracking can be maintained over frequency, temperature, and time. Notwithstanding these difficult aspects, feedforward PA systems have been in volume production and have been key elements in the implementation of mobile communications infrastucture. Commercially available feedforward amplifier systems typically claim about 20–30 dB of ACP or IM reduction for a single feedforward loop. To the considerable chagrin of the DPD camp, a feedforward loop performs its error corrections in real time, literally at lightening speed, and as such is essentially immune to PA memory effects.

Figure 14.13 shows a basic feedforward correction loop. A delayed sample of the undistorted input signal is subtracted from a coupled sample of the PA output signal. If the amplifier has no gain or phase distortion, this subtraction will produce zero output. Any gain or phase distortion in the amplifier, in the form of compression or AM-PM effects, will result in an RF error signal at the output of the combiner. This error signal is then amplified back to the original level at which the output sample was taken, and recombined with the PA output, following a delay line in the main signal path which compensates for the delay in the error signal

---

1.   See Section 14.4.1.

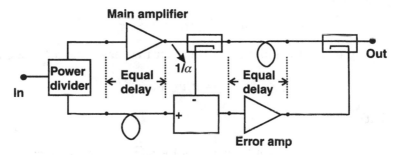

Figure 14.13   Basic feedforward error correction loop.

amplifier. The key aspects to note in the basic correction process are that both amplitude and phase errors are removed, and that the addition and subtraction process is performed on the fully formatted RF signal, rather than detected or down-converted derivatives. This means that the correction process operates over a bandwidth which is limited only by the phase and amplitude tracking capability of the various components of the system.

Although the basic operation of such a system is simple enough to understand in a qualitative manner, it pays to examine certain aspects in more quantitative detail. Particular issues which require quantitative analysis are the relative power capabilities of the main and error amplifiers and their impact on the overall efficiency, the related problem of the nonlinear contributions of the error amplifier, and the effects of imperfect gain and phase tracking. Such analysis quickly gets complicated and with the best of intentions, the wood and the trees can become somewhat indistinguishable [1–3]. A detailed analysis of all the tradeoffs is therefore not attempted here. But following many years of discussion and development efforts in this area, there appear to be a few *causes célèbres* which demand some more detailed examination:

- Does a feedforward loop correct gain compression, or just cancel IM distortion?
- Isn't power from both amplifiers "wasted" in the output coupler?
- What are the benefits (or otherwise) of putting an adaptive control on the sampled PA output in order to reduce the workload of the error amplifier?

Going boldly in the footsteps of Pandora, these topics form the subjects of the next three sub-sections.

14.3.2   Feedforward—Gain Compression

The ability of the error amplifier (EPA) in a feedforward loop to restore, and even enhance, the power output of the main PA (MPA) is not always used. Fundamentally, any FFW loop *will* seek to correct gain compression, as well as removing the other more visible spectral distortion products. The effectiveness with which it performs the task is essentially a function of the power capability of the EPA and the manner in which the loop is adjusted. In order to address this topic, it is important to recall some basic theory from Chapter 9. Taking a basic two-carrier signal,

$$v_{in} = v \cos \omega_1 t + v \cos \omega_2 t$$

and using this as the input to an amplifier that has some third degree distortion,

$$v_{out} = a_1 v_{in} + a_3 v_{in}^3$$

so that

$$v_{out} = a_1 \left(v \cos \omega_1 t + v \cos \omega_2 t \right) + a_3 \left(v \cos \omega_1 t + v \cos \omega_2 t \right)^3 \qquad (14.8)$$

from which it was determined that the two fundamental components at ω_1 and ω_2 both have output amplitudes given by

$$v_{0,\omega_1,\omega_2} = a_1 v + a_3 \frac{9}{4} v^3 \qquad (14.9)$$

and the amplitude of the upper and lower IM3 products are

$$v_{0,2\omega_1 - \omega_2, 2\omega_2 - \omega_1} = a_3 \frac{3}{4} v^3 \qquad (14.10)$$

thus the PA output can be written in the form

$$v_o = \left(a_1 v + a_3 \frac{9}{4} v^3 \right) \cos \omega_1 + \left(a_1 v + a_3 \frac{9}{4} v^3 \right) \cos \omega_2$$

$$+ a_3 \frac{3}{4} v^3 \cos(2\omega_1 - \omega_2) + a_3 \frac{3}{4} v^3 \cos(2\omega_2 - \omega_1) \qquad (14.11)$$

In Figure 14.13, the basic function of the first loop is to scale down the output by the sampling factor α, and then subtract the result from the appropriately delayed (and hence time-coherent) input signal, which results in an input to the EPA, v_e, given by

$$v_e = \left\{ \left(1 - \frac{a_1}{\alpha} \right) v + \frac{1}{\alpha} a_3 \frac{3}{4} v^3 \right\} \cos \omega_1 t + \left\{ \left(1 - \frac{a_1}{\alpha} \right) v + \frac{1}{\alpha} \frac{3}{4} a_3 v^3 \right\}$$

$$\cos \omega_2 t - \frac{1}{\alpha} \frac{9}{4} v^3 \cos(2\omega_1 - \omega_2) - \frac{1}{\alpha} \frac{9}{4} v^3 \cos(2\omega_2 - 2\omega_1) \qquad (14.12)$$

It is frequently asserted that the fundamental components are always cancelled by this subtraction process, which can be adjusted by a suitable setting of the sampling factor α (or, more conveniently, usually using an additional variable attenuator at this point in the system). Such a cancellation however will be a function of the signal amplitude v, as can be seen by examining the bracketed coefficients of the two fundamental signals in (14.12). For example, at small values of v, the α parameter would be set such that

$$\alpha = \frac{1}{a_1}$$

which is the logical and so-called "standard adjustment" for a feedforward loop, whereby the loop does not attempt any correction when the MPA is linear. With this setting, higher values of v will cause gain compression or expansion, depending on the sign of a_3. This in turn will greatly increase the input power of the fundamental components to the error amplifier, given that the third degree distortion terms are both proportional to v^3, but the fundamental components are three times higher in voltage magnitude than the IM3 components. Clearly, any particular value of v can be chosen as the condition where the α value is set to null the fundamental components in the error signal. In general, such alternative settings will enable the FFW loop to remove the IM3 distortion, but will no longer restore the fundamental components to their original expected linear values. The dynamic power transfer characteristics for two values of α are shown in Figure 14.14. By suitable resetting of the α parameter, the action of a FFW loop becomes more comparable to that of a predistorter; spectral distortion is removed, but the MPA power is not increased.

When α is set to the normal adjustment value, the FFW loop seeks to behave as a power restoration system. The EPA will essentially top-up the power shortfall in the MPA as it starts to show gain compression at higher drive levels. Although this may have some attractions in terms of fully utilizing the power capability of both amplifiers, it comes at a price. A FFW loop with this adjustment requires a higher power from the EPA to perform the prescribed task, and in most practical implementations a different α setting, as also shown in Figure 14.14, would be used. This alternative setting represents a compromise between power utilization, efficiency, and linearization limits. A full analysis of this tradeoff is a complicated problem which has to take full account of the signal environment and use a more realistic PA model [1]. But the overall conclusion, that the setting of the sampling ratio has a critical bearing on the EPA power requirement, remains intact. Figure 14.14 shows also how in this idealized analysis, the power requirement of the EPA can be significantly reduced using different α settings. For each setting, a dynamic linearized characteristic is obtained, with different EPA requirements. In practice, the EPA power levels will additionally be affected by the presence of AM-PM. A FFW loop does correct

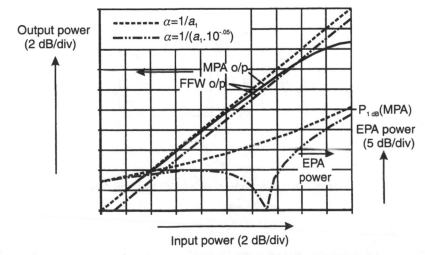

Figure 14.14 Effect of sampling ratio α on FFW power restoration and EPA power.

AM-PM distortion in the MPA, but anything more than about 10° will have a major impact on the EPA power requirement [1, 3].

It is worth noting that the IM distortion generated by the EPA is caused primarily by the residual fundamental carrier component in the expression for v_e (14.12), and will be proportional to the ninth power of the input signal magnitude. The minimization of these components is of paramount importance in the successful operation of a FFW loop, and cannot be optimized using a single fixed setting of α. This raises the question of making α adaptable to signal conditions, and this will be the subject of Section 14.3.4.

14.3.3 Feedforward—Effect of the Output Coupler

The analysis so far, and in particular the conclusion that a FFW loop can be set up such that it literally restores the power lost through gain compression of the MPA, has apparently ignored the transmission loss of the output coupler.[2] Looking again at Figure 14.13, it would appear that the addition of the corrective signal to the delayed PA output will be an imperfect and possibly wasteful process. Superficially, a high coupling factor appears desirable in order to prevent too much valuable PA output power being lost into the unused port. Conversely, if the coupling factor is made high, then the same coupling factor is presented in the EPA path and thus increases the power requirement from the EPA; this is a critical issue for the efficiency of a FFW loop. It pays, however, to take a closer look at the combining process, assuming that the coupler is a conventional quadrature component.

In the presence of signals at more than one port, the properties of couplers are not intuitively obvious, and the various power relationships have to be analyzed initially through voltage superposition. For example, Figure 14.15 shows a typical low loss, high directivity, coupler. If we assume a typical coupling value of 10 dB (β = 10), and supply a 9 watt signal into port 1, and into port 3 a phase-coherent 1 watt signal having the same frequency but in-phase quadrature to the port 1 input, it is a fundamental property of this quadrature coupler that the power output at port 4 will be 10 watts, the sum of the two input powers. With this particular power ratio,

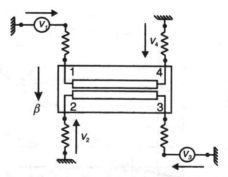

Figure 14.15 Terminated directional coupler with two sinusoidal co-phased input signals.

---

2.   The analysis required to plot the EPA power levels in Figure 14.14 did in fact assume the use of a 10 dB coupler.

which was carefully chosen, there will be no power dissipated in the terminated port 2. So the power at port 4 is

$$P_4 = P_1 + P_3, \left(\text{only for } \frac{P_1}{P_3} = 9\right)$$

and *not*, for example, the possibly more intuitive result,

$$\text{xxxxx} \quad P_3 = \left(1 - \frac{1}{10}\right)P_1 + \left(\frac{1}{10}\right)P_3 \quad \text{xxxxx}$$

More detailed analysis shows that it behaves as a perfect power combiner for signals at ports 1 and 3, if the relationship

$$\frac{P_1}{P_3} = \beta - 1$$

applies, where β is the coupling factor expressed as a power ratio. This is also assuming that the two signals are phase coherent, and have a quadrature phase relationship; this will usually be ensured by originally splitting the two signals from a single source, using another quadrature component. This relationship can be demonstrated by reconsidering the coupler in Figure 14.15, and defining input voltages v_1 and v_3 at ports 1 and 3, respectively. These voltages represent the amplitude of phase-coherent RF carriers, having the same frequency and in phase quadrature. The ports are all terminated with the appropriate characteristic impedance. The voltage at port 2 is given by

$$v_2 = \left(\frac{1}{\sqrt{\beta}}\right)v_1 - \left(1 - \frac{1}{\beta}\right)^{\frac{1}{2}} v_3 \tag{14.13}$$

and the voltage at port 3 is given by

$$v_3\left(1 - \frac{1}{\beta}\right)^{\frac{1}{2}} v_1 + \left(\frac{1}{\sqrt{\beta}}\right)v_3 \tag{14.14}$$

So setting $v_2 = 0$ in (14.13),

$$v_1 = v_3(\beta - 1)^{\frac{1}{2}}$$

hence the original result, $\frac{P_1}{P_3} = \beta - 1.$

So, for example, recalling Figure 14.14, in the case where a FFW loop is adjusted to give full MPA power restoration, it is possible to select a coupler ratio such that the ideal combining property is utilized at a selected point in the characteristic. One obvious point would be the 1 dB compression point of the main amplifier.

Suppose the main PA has a cw output power of 80 watts at the 1 dB compression point. The gain characteristic thus shows that the extrapolated linear power at this point would be approximately 100 watts (1 dB $\approx \frac{5}{4}$). So if the EPA tops up the MPA output with a 20 watt output into port 3, the full 100 watts will duly appear at port 4 as long as the coupling ratio is ($\frac{80}{20}$ +1), or about 7 dB. It is important to note that although this result assumes a lossless coupler, this loss refers entirely to the resistive and dielectric losses which result from the materials used in the coupler manufacture. For example, good quality airline couplers at this coupling level should only exhibit around 0.1 to 0.2 dB loss. The power at port 4 is fully restored even when taking full account of the $(1 - \frac{1}{\alpha})$ power transmission factor between ports 1 and 4, which would be observed if a solitary signal were applied to port 1.

The concept that the output coupler in FFW systems causes power wastage due to the port 1-4 transmission factor is a widely misquoted detraction, although the ideal power addition only applies exactly at a single power level. At backed-off power levels, which take the coupler away from its optimum power-combining point, the coupler transmission factors come back into the equation, but now at much lower power levels from the two PAs. The impact of the output coupling ratio is illustrated in Figure 14.16, which shows that a wide range of selections, all the way from 3 dB to nearly 20 dB, can make useful FFW systems for different applications. Lower coupling ratios can be used when there is an emphasis on peak power restoration, where the EPA power becomes an important contributor to the main power output. A compromise value of 10 dB is often used where maximum linearization is required, and there is no requirement to restore or supplement the MPA power.

The analysis in this section has assumed an idealized model for the MPA. In practice, the effects of any AM-PM will have a significant effect on the calculated EPA power levels, and this becomes a major issue if the AM-PM levels exceed about

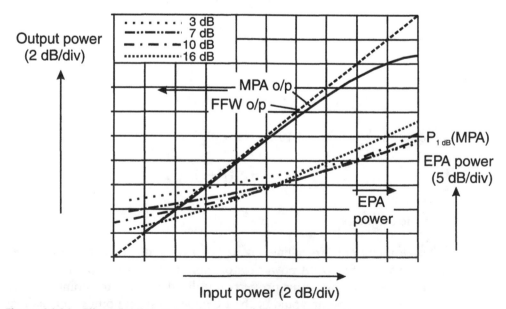

Figure 14.16   Effect of different coupling ratio choices on EPA power.

20° [1, 3]. It is also worth noting that whereas the coupling ratio can be selected to give ideal power-combining action at a prescribed PA output compression level, there is no such convenient choice available for the phase relationships of the two signals. In conditions of higher AM-PM, the corrective signal from the EPA will not track the AM-PM from the MPA in a linear fashion, and the coupler will no longer act as a perfect power combiner, at any level.

14.3.4 Feedforward—Adaptive Controls

Recalling (14.12), is is clear that the FFW will always cancel the IM distortion, regardless of the sampling coefficient (α) setting, provided that the EPA is not generating any comparable levels of IM itself, and also assuming that the loop is kept perfectly gain and phase tracked. For any particular signal environment, it is usually possible to find a setting which minimizes the error signal entering the EPA; in practice this is mainly an issue of arranging for cancellation of the amplified signal component. Provided this is done slowly enough, so that it does not interfere directly with the intended ongoing action of the FFW loop, significant improvement in linearization performance, for a given EPA power, can be achieved.

Such dynamic adaption of the FFW loop settings are frequently employed in commercial systems, and have become a good deal easier to implement using DSP techniques [26]. The debatable issue is how far such an adaption concept can be taken. Clearly, if the α parameter is varied in envelope time, to keep track of the signal envelope, such that no error signal ever gets to the error amplifier, then the FFW loop will not generate any corrective signal and no linearization action will take place. On the other hand, if the settings are varied too slowly, some potential benefit will be missed. It has been suggested that the EPA input can be minimized using a small variable attenuator at the MPA input, as indicated in Figure 14.17. The issue now is that if this adjustment is driven by a detector at the input to the EPA, what we have effectively done is to build an envelope amplitude feedback linearization loop. Such a loop is a linearization technique in its own right, and would probably be used in the first place if it were not for limitations which will be discussed in Section 14.4.

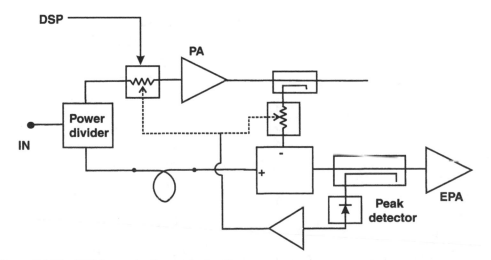

Figure 14.17 FFW loop adaptive control options.

The problem with such a feedback loop is that the time delay associated with detecting the difference signal and processing it to provide a suitable corrective drive signal to the attenuator gain control, is typically too long when measured relative to the excursions of a multicarrier RF signal envelope. This can typically result in making the wrong corrections to the signal, and creating more distortion.

Assuming that a FFW loop can have the same special benefits that are given to DPD systems, in particular being presented with an *a priori* sample of the signal in digital form, it would be possible to derive a suitable time coherent drive signal to the input attenuator, using DSP, also indicated in Figure 14.17. This, of course, is nothing more nor less than a reinvention of the digital predistorter. It is worth noting that such a hybrid combination of techniques could be beneficial. A basic memoryless DPD could present a very linear signal to the FFW loop, which would then focus on removing the memory effect distortion, possibly using a much lower power EPA than that usually required for a stand-alone FFW system.

14.3.5 Feedforward—Practical Issues, Conclusions

FFW system practicalities are mainly focused around the need to maintain accurate gain and phase tracking. These tracking errors will most likely come from gain and phase variations in the error amplifier over frequency and temperature. Unlike the main PA tracking errors, these errors occur *after* the error correction signal has been defined, and cause unretrievable cancellation errors. Some basic amplitude and phase arithmetic quickly generates some formidable tracking targets, as shown in Table 14.1.

This table shows the degree of cancellation between two nominally equal amplitude, sinusoidal signals expressed as a dB ratio between the resultant and one of the original signals. The phase errors refer to the deviation from an exact antiphase condition. It is clear that to maintain the cancellation greater than 40 dB, the output gain and phase tracking errors have to be down in the hundredths of one degree of phase and hundredths of one dB of amplitude. Maintaining even more modest tracking limits in the output cancellation loop, over frequency and temperature, remains

Table 14.1 Gain and Phase Tracking Errors in an FFW Loop

Gain Error	Phase Error	Cancellation (dB)
1 dB	0	−19
0.5 dB	0	−25
0.25 dB	0	−31
0.1 dB	0	−39
0.01 dB	0	−59
0	10°	−15
0	5°	−21
0	2.5°	−27
0	1°	−35
0	0.1°	−55

one of the most challenging aspects of feedforward amplifier design. Fortunately, in a system which is already well provided with digital and software controls, the accurate compensation of temperature effects is a fairly easy addition to the overall control infrastructure.

Frequency variations represent a tougher challenge for the designer of the RF elements of the system. Most commercial systems only operate to full specification over quite narrow RF bands, typically less than 5%. There is an additional problem that newer modulation systems generate signals with bandwidths stretching into tens of MHz, such that the gain and phase variations even within the signal bandwidth may start to degrade the performance, and will not respond to conventional tracking adaption methods. Apart from the gain and power flatness requirements, there is the issue of the tracking delay lines. It has already been noted in an earlier section that the equivalent electrical length of a multistage microwave amplifier can be several meters, but unfortunately the lumped element matching networks do not have precisely linear phase-frequency characteristics, so that a simple length of cable may not do an adequate job over the necessary bandwidth. Techniques to alleviate this problem have been reported [27]. Gain and phase tracking techniques possibly form the largest of all the patent mountains in the linearized PA area. The most popular methods are based on the use of pilot tones, as originally reported by Seidel [24, 25].

It seems that despite the difficulties and challenges, feedforward solutions have thus far dominated the MCPA mobile communications market. The linearization performance seems to have been such that users have been willing to accept low efficiencies, typically in the 10–15% range for MCPA applications. The extra cost of the various couplers, delay lines, and the EPA itself all point to a need for an alternative solution capable of higher efficiency and lower cost. Clearly, DPD in principle offers a solution which dispenses with all of these RF components, and only adds about $10 worth of silicon. It would seem that the long reign of the feedforward amplifier may be coming to an end.

14.4 Feedback Techniques

14.4.1 Introduction, Direct Feedback Techniques

The use of feedback techniques to reduce or eliminate the quirky and unpredictable behavior of electronic devices has been standard practice for just about the full duration of the electronic era. It runs into problems, however, when the time delay of the propagating signals becomes significant in comparison to the timescale on which the signals themselves are changing. The point was made eloquently by Seidel [25]:

> Feedback, in comparing input with output, glosses over the fundamental distinction that input and output are not simultaneous events and, therefore, not truly capable of direct comparison. In practice they are substantially simultaneous if device speed is far faster than the intelligence rate into the system…if we could organize a system of error control which did not force a false requirement of simultaneity, not only would the problem of fabricating zero transit time device disappear, but so would the entire problem of stability, another consequence of comparing the incomparable.

In particular, the use of direct negative feedback, such as that represented by the evergreen textbook schematic shown in Figure 14.18, has diminishing value as the signal frequency enters the GHz region. The composite gain G of an amplifier with intrinsic gain of A, having a feedback network which subtracts a voltage $(\beta.V_{out})$ from the input signal V_{in}, is given by

$$G = \frac{A}{1+\beta A} \qquad (14.15)$$

where A is usually assumed to incorporate a 180° phase shift for a single stage low frequency amplifier. At microwave frequencies, and especially in multistage cascaded chains, A will typically have a rapidly varying phase characteristic with frequency, and the phase of the feedback network division ratio β will have to be adjusted to ensure subtraction at the point of feedback application. In practice, this will usually lead to oscillation at a nearby frequency. At GHz frequencies, gain is also a more precious commodity, and it is not as permissible to let A tend to infinity with a wave of the hand. As such, direct application of classical feedback theory to microwave amplifiers is somewhat irrelevant. It is worth noting however, that the RF world is a little careless about assigning the term "feedback" to various configurations which include some amount of lossy matching to improve bandwidth at the expense of gain. Microwave circuit designers use the term "feedback" to categorize amplifier designs that use a moderately high value resistor connected from the drain to gate, or collector to base. This resistor acts mainly as a damping element, which improves stability and decreases the Q-factor for broader bandwidth applications. But these amplifiers do not usually show any significant linearity benefit from using the feedback element.

14.4.2 Indirect Feedback Techniques—Introduction

The "linearity" of an RFPA, for the present purposes, is defined to be the integrity with which it preserves the amplitude and phase variations of a modulated RF carrier. It therefore becomes clear that feedback techniques can be applied at baseband, rather than at RF carrier, frequencies. An example of such a linearization scheme is shown in Figure 14.19. A conventional RF amplifier has envelope detectors coupled to its input and output ports, and a video differential amplifier forms an error correction signal which can be fed to a gain control on the amplifier. Provided the amplifier is well below saturation, it would seem that such a loop would force the

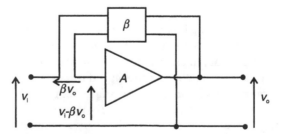

Figure 14.18 Basic direct feedback amplifier.

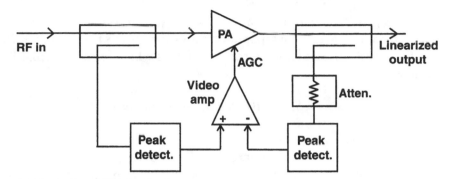

Figure 14.19 Indirect RFPA feedback system.

output envelope to replicate the input envelope, and substantial improvement in the spectral distortion could be obtained.

This simple "envelope feedback" was a mainstay of the mobile communications industry for many years. It enabled a few more valuable dBs of IM performance to be obtained from VHF and UHF solid state power amplifiers [28]. One reason why it does no more than a modest linearization job can be seen by recalling Figure 14.1. The simple envelope amplitude correction cannot increase the intrinsic power saturation of the device, so the effectiveness of the procedure will decrease markedly as the envelope swings into the compression region. At much lower power levels, the gain required from the video amplifier becomes much higher if any useful effect is to be obtained, which leads to bandwidth and stability problems. In fact, IM levels can in practice be reduced by useful amounts in the precompression range using this technique, but obviously AM-PM effects are not corrected. In fact, one of the catches in implementing this technique is to beware of creating extra AM-PM distortion through the action of the amplitude control device.

The delays in the detection and video signal processing are a fundamental limitation which becomes ever more severe as signal bandwidths increase. The problem is compounded by the fact that the detected signal envelope will have a bandwidth several times higher than the signal itself, with a rather ill-defined sprawling spectrum, and the wide video bandwidth restricts the loop delay limits for stable or effective operation. The video bandwidth problem can be somewhat alleviated by using down conversion, rather than direct detection. This also allows for the possibility of phase correction, and leads us into the most widely used technique in this class, the Cartesian Loop.

14.4.3 The Cartesian Loop

Cartesian Correction has been widely used in the design of solid state radio transmitters, and has an extensive literature [2, 29, 30]. It offers the possibility of 30 dB of IM reduction for signal modulation bandwidths that are lower than about 100 kHz. It is a transmitter, rather than a PA technique, and requires the modulating signal in its baseband form as the main input. In a modern digital system, it is most likely that the baseband signal will already be available in an I-Q format. This has some benefits in that the resulting I and Q channels can be processed in well-matched paths, each of which has a bandwidth which is comparable to the actual signal bandwidth. Figure 14.20 shows the essentials of a Cartesian loop linearization system. The separate I and Q signals are fed through differential correcting amplifiers, and into vector modulators which form the actual RF signal $S(t)$, where

$$S(t) = I(t)\cos\omega_c t + Q(t)\sin\omega_c t$$

ω_c being the RF carrier frequency.

The signal $S(t)$ is then fed into the RF power amplifier, emerging with some distortion. A small portion of this output is coupled into a downconverter which has a similar configuration to the upconverter used on the input and retrieves the now distorted I and Q signals which are then directly compared with the undistorted input baseband signals. The gain of the input differential amplifiers will force the loop into generating an output signal which closely tracks the original I and Q signals, the degree of precision depending not only on the gain and bandwidth of the video circuitry, but also on the linearity of the downconverter demodulators.

One of the advantages of the Cartesian Loop is the symmetry of gain and bandwidth in the two quadrature signal processing paths. This will reduce the phaseshifts between the AM-AM and AM-PM processes which are a prime cause of asymmetrical IM sidebands. Although the video bandwidth and stability will limit the capability to handle multicarrier signals, it should be noted that the use of a synthesized RF source makes the system frequency agile. With the widespread availability of low cost quadrature modulators and demodulators, the overall system appears to be an attractive linearized transmitter architecture. Practical systems with linearity improvements up to 45 dB, over limited bandwidths, have been reported.

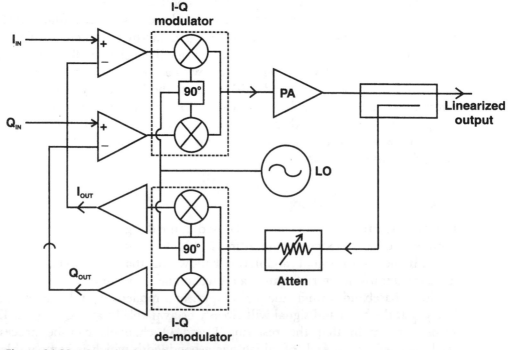

Figure 14.20   Cartesian Loop feedback system.

14.4.4 The Polar Loop

The Polar Loop [31, 32] is a complete modulator, rather than a linearizer, but it includes some linearization in the modulation process. The basic Polar Loop system is shown in Figure 14.21. The only actual input is the fully modulated signal at IF. An RFPA is used in a supply modulated ER configuration to create the AM, and a VCO source is phase-locked to the IF signal to create the PM on the RF carrier. The output from the PA is down converted, where a control loop generates an error signal which feeds the baseband drive to the ER envelope modulator. All of the comments made in this section about loop delays and video bandwidth apply to this system, along with the additional issues relating to the implementation of an ER system, which were discussed at some length in Chapter 10. This does not all add up to an enticing prospect for practical implementation. The Polar Loop has nevertheless come back into favor in the handset PA zone, especially for challenging modulation systems such as EDGE, which make excessive demands on the efficiency-linearity tradeoff in conventional Class AB approaches [33, 34]. Claims and counterclaims seem to be made concerning the feasibility of making the system work for a 300 kHz bandwidth signal in its original closed loop form.

One of the key issues with the Polar Loop is the bandwidth requirements for the amplitude and phase error amplifiers. Splitting QPSK signals into separate envelope amplitude and phase components results in a wide bandwidth for the phase channel. As with the simple amplitude envelope feedback loop considered in the last section, the bandwidth and stability limitations of the video circuitry will limit the usefulness of such systems to single carrier applications.

A problem which is quite commonly observed in feedback systems such as Cartesian or Polar Loops is that the IM specs are met comfortably for lower order products (IM3, IM5), but higher order products can be uncorrected or even enhanced, often showing gross asymmetry. This is an inevitable effect of video circuit bandwidth and phase delay limitations. Nevertheless, transmitters using Polar Loop correction have been demonstrated that have average efficiency greater than 50% and IM3 products at 50 dBc [31]. Thus, it would appear possible to operate the PA stage such that the envelope peaks are well into the compression region; this maintains

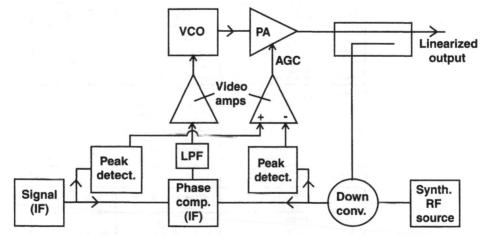

Figure 14.21   Polar Loop modulation system.

the average efficiency above 50% and the correction loop is able to reduce the IM3 levels by as much as 25 dB in comparison with the same amplifier running uncorrected.

Despite their universal impact in electronic engineering, feedback techniques are of limited use for the linearization of RFPAs. An exception is the widespread use of indirect feedback techniques, especially the Cartesian Loop, in single carrier transmitter applications which have modulation bandwidths less than 100 kHz.

14.5   Other Linearization Methods

Figure 14.22 attempts to classify linearization techniques. PA linearization can essentially be performed in three ways: modification of the input signal, modification of the output signal, or dynamic modification of the amplifier characteristics. As Figure 14.22 shows, the most heavily subscribed category appears to be the first one, given that feedback techniques are essentially closed-loop implementations of input signal predistortion. The feedforward loop is the definitive output signal modification technique, although it is somewhat debatable whether it should be classified as open or closed loop. To date, it seems that there is no mainstream technique, other than some fairly crude analog bias adaption used in low cost handset products, which modifies the amplifier gain properties on a dynamic basis. One reason for this is the two-dimensional nature of PA distortion. It is necessary to identify separate control mechanisms for the amplitude and phase corrections, and a typical RFPA design does not readily supply such orthogonal controls. This seems to be the main divide between techniques that give light correction, maybe up to 10 dB or so, and the mainstream techniques which give over 20 dB correction. The graphs in Figures 14.3 and 14.4 show how correcting even a few degrees of AM-PM becomes mandatory if more than about 10 dB reduction in IM products is required.

The other notable gap in Figure 14.22 is in the digital column for output signal correction. It is easy enough to conceive of a "Digital Feedforward" system, whereby the DSP directly generates the baseband error signal. Such a system could be given the

Modify input signal	Analog	Digital
Open Loop	Predistortion	Predistortion
Closed Loop	Envelope F/B	"Adaptive" PD
Modify output signal		
Open Loop	Feedforward	?
Closed Loop	? 2-loop Feedforward	?
Modify PA properties		
Open Loop	Bias adaption	?
Closed Loop	?	?

Figure 14.22   Summary of PA linearization techniques.

same adaptive controls as a DPD configuration, and could handle gain and phase tracking through the action of the adaptive process without the use of physical delay lines. It is possible that "smarter" use of the EPA could give some efficiency benefits in higher peak-to-average ratio applications, although there is an argument that such a system will always have lower efficiency than a full DPD approach.

It seems probable that mainstream linearization techniques will remain as FFW or DPD derivatives. For the less stringent applications, various miscellaneous techniques appear in the literature, from time to time, which claim to reduce IM distortion in an RFPA. These methods seem to catch short-lived attention, both in the literature and on the symposium circuit. Typically, these techniques are tricks which provide some degree of nonlinearity reversal over a limited power range. This limitation is what sets them aside from mainstream linearization techniques. The polynomial mathematics usually dictates that cancellation will only occur at one specific drive level, and this causes a mixed reaction among the "serious" linearized PA community. If the third order null can be set somewhere in the vicinity of the 1 dB compression point, a very useful dB or two increase in peak envelope power can be obtained from the same device, and this can translate into some serious cost savings. But if the goal is to reduce the IM or ACP by 40 dB or so, such tricks are of little real value.

One such technique is worthy of mention, which is the use of segmented, parallel-connected device elements biased to different points on the DC characteristic. This has been mainly categorized as "Derivative Superposition" [35], although this refers to specific work aimed mainly at increasing the dynamic range of small signal receiving amplifiers. Figure 14.23 indicates a possible implementation in a high power device, where a small segment can be biased into a Class C condition to give some gain expansion, which will potentially cancel the gain compression of the main segment at higher drive levels. Such techniques suffer from the problem that their success is very closely coupled to individual device sample characteristics, and as such has not found much use outside the R&D labs.

14.6 Conclusions

Linearization techniques need to be judged according to the application. Closed loop correction systems such as the Polar and Cartesian loops are worthwhile

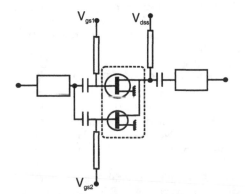

Figure 14.23 Segmented bias for linearization of a high power RF transistor.

options for single channel transmitter designers looking for optimum power and efficiency for a given device whilst maintaining spectral mask requirements. For lower signal bandwidths, these techniques can give 30 dB correction in close-to-carrier ACP levels. Multicarrier applications usually have much more stringent intermodulation specifications, and signal bandwidths that stretch into the tens of MHz range. Such requirements dictate the use of feedforward techniques, or a retreat back to multichannel system design, where closed loop techniques can be used for each individual channel. Digital predistortion techniques appear to be ready to replace feedforward systems in the mobile communications infrastructure, although DSP speed is still a limitation for signal bandwidths above about 20 MHz. Low cost, high volume PA applications such as handsets continue to make good use of simple, cheap predistortion methods. Meanwhile, the higher frequency microwave link and satellite communications system manufacturers still seek a viable linearization technique which can handle multicarrier signals spaced out over hundreds of MHz, each using high density QAM modulation formats, and with no baseband signal information available. At the present time, they are still forced into using inefficient Class A amplifiers, and must wonder whether they will yet be able to participate in the great RFPA linearization party of the last decade.

References

[1] Cripps, S. C., *Advanced Techniques in RF Power Amplifier Design*, Norwood, MA: Artech House, 2002.
[2] Kenington, P. B., *High Linearity RF Amplifier Design*, Norwood, MA: Artech House, 2000.
[3] Pothecary, N., *Feedforward Linear Power Amplifiers*, Norwood, MA: Artech House, 1999.
[4] Katz, A., et al., "Passive FET MMIC Linearizers for C, X, and Ku-Band Satellite Applications," *IEEE Intl. Microwave Symp.*, MTT-S, 1993, pp. 353–356.
[5] Kumar, M., et al., "Predistortion Linearizer Using GaAs Dual-Gate MESFET for TWTA and SSPA Used in Satellite Transponders," *IEEE Trans. Microwave Theory & Tech.*, MTT-33, No.12, 1985, pp. 1479–1499.
[6] Katz, A., and J. Matsuoka, "A Highly Efficient & Linear Integrated PA for Ka/Q Bands," *IEEE Intl. Microwave Symp.*, TH1B-2, Long Beach, CA, 2005.
[7] Stapleton, S. P., and F. C. Cotescu, "An Adaptive Predistorter for a PA Based on Adjacent Channel Emissions," *IEEE Trans. Veh. Tech.*, Vol. 41, No. 1, February 1992.
[8] Kenney, J. S., et al., "Predistortion Linearization System for High Power Amplifiers," WE5B-3, *IEEE Intl. Microwave Symp.*, Fort Worth, TX, 2004.
[9] Woo, W., and J. S. Kenney, "A Predistortion Linearization System for Low Power Amplifiers with Low Frequency Memory Effects," *IEEE Intl. Microwave Symp.*, TH2B-1, Long Beach, CA, 2005.
[10] Ding, L., et al., "A Robust Digital Baseband Pre-Distorter Constructed Using Memory Polynomials," *IEEE Trans. Comm.*, Vol. 52, No. 1, June 2004.
[11] Sparlich, R., et al., "Closed Loop Digital Predistortion With Memory Effects Using Genetic Algorithms," *IEEE Intl. Microwave Symp.*, TH2B-4, Long Beach, CA, 2005.
[12] Wood, J., and D. E. Root,(eds.), *Fundamentals of Nonlinear Behavioral Modeling for RF and Microwave Design*, Norwood, MA: Artech House, 2005.
[13] "Adaptive Predistortion Using the ISL5239," Intersil Application note AN1028, September 2002.
[14] Pauluzzi, D., and A. Wright, "Digital Adaptive Predistortion: The PMC-Sierra Approach," *IEEE Intl. Microwave Symp.*, WE5B-3, Fort Worth, TX, 2004.

[15] Braithwaite, R. N., "Digital Predistortion System and Method for Linearizing an RF Power Amplifier with Nonlinear Gain Characteristics and Memory Effects," U.S. Patent 2005190857, September 2005.

[16] Yoo-Seung, S., "Digital Predistortion and Method in Power Amplifier," U.S. Patent 2005253652, November 2005.

[17] Wright, A., "A Wideband Predistortion Linearizer for Nonlinear Amplifiers," U.S. Patent 6798843, December 2001.

[18] Yamakuchi, K., et al., "A Novel Series Diode Linearizer for Mobile Radio Power Amplifier," *Proc. 1996 IEEE Intl. Microwave Symposium*, San Fransisco, CA, pp. 831–834.

[19] Nojima, T., et al., "The Design of a Predistortion Linearization Circuit for High Level Modulation Radio Systems," *Proc. Globecom*, 1985, pp. 47.4.1–47.4.6.

[20] Gupta, N., et al., "A Varactor Diode Based Predistortion Circuit," *2004 IEEE Intl. Microwave Symp.*, WE5B-6, Fort Worth, TX.

[21] Fan, C. W., and K. M. Cheung, "Theoretical and Experimental Study of Amplifier Linearization Based on Harmonic and Baseband Signal Injection Technique," *IEEE Trans. Microwave Theory & Tech.*, Vol. 50, No. 7, 2002, pp. 1801–1806.

[22] Moazzam, M. R., and C. S. Aitchison, "A Low Third Order Intermodulation Amplifier with Harmonic Feedback Circuitry," *1996 IEEE MTT-S*, Vol. 2, 1996, pp. 827–830.

[23] Black, H. S., "Translating System," U.S. Patent 1,686,792, issued October 29, 1928, and U.S. Patent 2,102,671, issued December 1937.

[24] Seidel, H., H. R. Beurrier, and A. N. Friedman, "Error Controlled High Power Linear Amplifiers at VHF,"*Bell Syst. Tech J.*, May–June 1968.

[25] Seidel, H., "A Microwave Feedforward Experiment," *Bell Syst. Tech. J.*, Vol. 50, November 1971, pp. 2879–2916.

[26] Cavers, J. K., "Adaption Behavior of a Feedforward Amplifier Linearizer," *IEEE Trans. Veh. Tech.*, Vol. 44, No. 1, 1995, pp. 31–40.

[27] Cavers, J. K., "Wideband Linearization: Feedforward Plus DSP," *2004 IEEE Intl. Microwave Symposium*, Workshop WMC, Fort Worth, TX, 2004.

[28] Arthanayake, T., and H. B. Wood, "Linear Amplification Using Envelope Feedback," *Electron. Lett.*, Vol. 7, No. 7, 1971, pp. 145–146.

[29] Bateman, A., and D. M. Haines, "Direct Conversion Transceiver Design for Compact Low Cost Portable Mobile Radio Terminals," *39th IEEE Vehicular Techn. Conf.*, San Fransisco, CA, 1989, pp. 57–62.

[30] Johansson, M., and T. Mattso, "Linearized High Efficiency Power Amplifier for PCN," *Electron. Lett.*, Vol. 27, No. 9, 1991, pp. 762–764.

[31] Petrovic, V., and W. Gosling, "Polar Loop Transmitter," *Electron. Lett.*, Vol. 15, No. 10, 1979, pp. 286–287.

[32] Petrovic, V., and C. N. Smith,"Reduction of Intermodulation Distortion by Means of Modulation Feedback," *IEE Conf. on Radio Spectrum Conservation Techniques*, September 1983, pp. 44–49.

[33] Hietala, A., "A Quad-Band 8PSK/GMSK Polar Transceiver," *2005 Radio Freq. Integrated Circ. Symp.*, RMO1A-1, Long Beach, CA, 2005, pp. 9–13.

[34] Sowlati, T., et al., "Polar Loop Transmitter for GSM/GPRS/EDGE," *2005 Radio Freq. Integrated Circ. Symp.*, RMO1A-2, Long Beach, CA, 2005, pp. 13–16.

[35] Webster, D., J. Scott, and D. Haigh, "Control of Circuit Distortion by the Derivative Superposition Method," *IEEE Microwave & Guided Wave Lett.*, Vol. 6, No. 3, March 1996.

PA_Waves

This is a program which postulates, through user inputs, voltage and current wave-forms for RFPAs. It is NOT a circuit simulator, and should not be judged as such. PA_waves is termed a "postulator." Rather than taking a device model and a passive circuit environment, and then engaging the necessary (and gargantuan) computing resources to figure out the resulting nonlinear voltage and current waveforms, PA_waves takes a complementary approach. It allows the user to postulate current and voltage waveforms which are consistent with the device IV characteristics. It then computes the impedances required to support these current and voltage waveforms. This is a very simple piece of computation, comparatively speaking. The user needs to employ some physical reasoning in order to end up with a set of impedances (fundamental, second, third harmonic at present) which are realizable; this means eliminating negative real parts, and in the case of harmonics, ideally eliminating resistive components altogether.

PA_waves analyzes any stipulated set of RF current and voltage waveforms for:

- RF output power (expressed as a normalized dB value);
- Output efficiency;
- Harmonic impedances which are "implied" by the inputted waveforms.

The user first selects a bias point and a drive level, using the "**Ibias**" and "**Isig**" text entry boxes. These values are normalized such that pinchoff (zero bias) is **Ibias**=0, and saturation **Ibias**=1. So for a conventional Class B amp, **Ibias**=0 and **Isig**=1 is a good starting point.

The voltage can be constructed using fundamental (start off with unity for this), and second and third harmonic components (start off with these set to zero). All three voltage harmonic components can have any phase with respect to the current; a conventional Class B amp would have fundamental phase of 180° and harmonics set to zero (shorted).

A key feature of this program is the use of a realistic model for the IV "knee." The drain current has the form

$$I_d = V_g \cdot \left(1 - e^{-V_d/V_k}\right) \cdot I_{max}$$

where I_{max} is normalized to unity. The Vk ("**Vknee**") parameter is normally set to 0.1. This effectively means that the device reaches 63% of its full value when the drain voltage is 10% of the DC supply. This is about right for GaAs PHEMTs running off a 10v rail (Vk=1V).

To get started, though, just set the **Vknee** box to zero.

So with these settings, you should get a classical Class B set of waveforms. You may actually need to set **V1** to 1.001 to avoid any knee glitches. Then the efficiency should read the right value (78.5%). The power will now read close to 0 dB (maybe –0.1 dB). This is the reference power level, to which all other results will be compared using a dB value.

Now try setting **Vknee** to a realistic value of 0.1 and see the difference, both in power and efficiency. You can now see another feature; the current waveform plots show a "zero knee" trace, which shows how much the current has been modified by the knee effect. You can try reducing **V1** to pull the voltage out of the knee. But note how the power increases a bit at first and then goes back down again; efficiency degrades also.

You can try adding some third harmonic voltage to simulate Class F conditions; (**V3** values, 0.15 at 0 degrees); again, then see how much effect the knee has on power efficiency and the third harmonic impedance. The impedance plots basically show the magnitude and phase of the individual voltage and current components at each harmonic. The key thing is that they must be at least 90° apart from each other; they must never share a mutual relative quadrant, as this implies the device is absorbing power and a negative resistance would be needed in the external circuit.

The original reason for writing this program was to explore second harmonic voltage enhancement. The key thing with second harmonic is to watch the impedance. This is conveniently done by watching the polar plot of the second harmonic current and voltage vectors; they must not get within 90° of each other, since this then requires a negative output resistance. An example of a "Class J" result (see Chapter 4) is obtained with the following values (you could just load the **paw_J_ex** file)

$$\text{Vknee} = 0.077$$
$$\text{Ibias} = 0, \text{Isig} = 1$$
$$\text{Vdc} = 0.768$$
$$\text{V1} = 1, 202.5 \text{ deg; V2} = 0.27, 41 \text{ deg, V3} = 0$$
$$\text{I2} = 0, \text{I3} = 0$$

This gives 80.2% efficiency and a relative power of –2.9 dB (this is not "bad" power, it is always much less than 0 dB when knee effects are included). The key thing is, you have to gradually step the phase of the second harmonic until the impedance chart shows a right angle between voltage and current. The chart can be a little bit optimistic, and you need to watch the blue box below V2; this must be zero or negative.

One point to watch; if you let the voltage drop below zero, the program will "clamp" it at zero, but this will in turn generate additional harmonics of voltage that at the moment the program does not allow for. So *keep the voltage above zero* or you will get some invalid (and good-looking) results. *The blue* "Vmin"box *should always show a positive number*, preferably close to zero.

There is an additional feature which enables harmonic distortion of the input current waveform, such as that which may be caused by the input nonlinear capacitance, as discussed in Chapters 4 and 8. The parameters **I2** and **I3** control this feature, which adds these harmonics to the input voltage waveform, but still clips at the threshold point. They are probably best used in an empirical manner.

To run the program: click on the red "calculate" button.

When loading in Excel, a prompt will appear about running macros. Click the "enable" option.

Spectral Analysis Using Excel IQ Spreadsheets

As commented in Chapter 9, it is certainly desirable to bring some quantification to the qualitative descriptions of modulation systems given in Chapter 9 and their possible differences when sent through an RFPA. It is clear that such quantification cannot be beneficially pursued using formal symbolic analysis, and computation in some form or another is required. Bridging the gaps between a CCDF plot, a PA behavioral model, and a regulatory specification in a quantitative manner is a task which still challenges CAD manufacturers. In what follows, we do not in any way attempt to compete with the still-emerging professional CAD simulation tools in this area. But it is on occasion useful to be able to negotiate the gaps, even if it is with a piece of string rather than a six-lane highway.

The CD-ROM accompaniment to this book contains a number of spreadsheets which are prefixed by a modulation system. Two specific cases (EDGE.ACP and OFDM.ACP) are set up to demonstrate how a passable spectrum can be obtained using the Excel FFT routine, which is part of the data analysis package. These spreadsheets also include some basic PA distortion models, which enable the effects of PA distortion to be observed as ACP on the spectra. Readers can easily add their own variations by using these spreadsheets as templates.

The .ACP spreadsheet template uses a 4096 data sample, which is taken at 10 samples per symbol. This represents a very short data burst, and in order to obtain a meaningful spectrum, a suitable window function needs to be applied (hence the "I*Hanning" and "Q*Hanning" headings on columns 4 and 8 adjacent to the highlighted I and Q data in columns 3 and 5). Two points should be noted about using the Excel FFT analysis:

- It is not included in a standard Excel setup and usually needs to be loaded (follow **Tools/Add-ins** on the main menu, it is called "Fourier analysis").
- It does not update when parameters in the spreadsheet are changed. To run the FFT, for example, after you have changed the PA distortion parameters, you need to run it from the **Tools/data analysis** menu string.
- Using 4096 points, and depending on the speed of your computer, the FFT computation can take somewhere between 15 and 45 seconds.

In order to run an updated FFT, first select **Tools/data analysis/Fourier analysis**. You are then required to select the entire column of data headed "FFT input" (column 15, highlighted in yellow), which is the windowed output signal in the Excel format for a complex number I+iQ (it helps to use the shift/end/down key combination to select the whole column). You also have to tell the FFT function where to place the result data block, so select the single cell (r4c17, highlighted red) under the heading "FFT result" to denote the start of the FFT output computation. It is important to get the FFT output data in the right place, so that the charts will automatically update.

Sheet 1 represents a linear amplifier, by setting the "clip level" parameter to unity. Note that the IQ data is "raw" and not normalized to the average level, as is the case with the modulation format files used in Chapter 9. Sheet 3 shows a hard-clipped signal, the clipping level being set by the clip level parameter in cell r3c10. Sheet 2 implements a soft-clip PA function, having the form

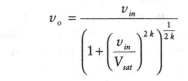

$$v_o = \frac{v_{in}}{\left(1 + \left(\frac{v_{in}}{V_{sat}}\right)^{2k}\right)^{\frac{1}{2k}}}$$

where V_{sat} (r3c10) is the saturation voltage level, and k (r4c10) is a parameter which characterizes the sharpness of the saturation, usually a value between 2 and 4 being representative. (The voltages represent magnitudes of the complex quantities represented by the IQ data; this model is quoted in 3GPP specifications and is attributed to Rapp.) Clearly, more complex (literally, sorry) functions could be implemented by intrepid readers for themselves, which would then include AM-PM effects.

The charts are to the right hand side of the sheet 1 data columns and show primarily the spectrum plots, as well as an IQ and voltage gain plot, which illustrate the PA compression characteristic. The gain plot is more useful if a smaller number of points are used and the data points are plotted as well as the lines. This then gives some indication of the statistical occurrence of the various signal levels.

Enjoy.

Bibliography

Introductory Texts on RF and Microwave Techniques

Besser, L., and R. Gilmore, *Practical RF Circuit Design for Modern Wireless Systems*, Norwood, MA: Artech House, 2003.

Lee, T., *Planar Microwave Engineering*, Cambridge, U.K.: Cambridge University Press, 2004.

Vendelin, G., A. Pavio, and U. Rohde, *Microwave Circuit Design Using Linear and Nonlinear Techniques*, New York: Wiley, 1990.

Wireless Communications

Harte, L., and S. Prokup, *Cellular and PCS/PCN Telephones and Systems*, Fuquay-Varina, NC: APDG Publishing, 1996.

Larson, L., *RF and Microwave Circuit Design for Wireless Communications*, Norwood, MA: Artech House, 1996.

Ojanpera, T., and R. Prasad, (eds.), *Wideband CDMA for Third Generation Mobile Communications*, Norwood, MA: Artech House, 1998.

Digital Modulation

Xiong, F., *Digital Modulation Techniques*, Norwood, MA: Artech House, 2000.

Nonlinear Techniques and Modeling

Maas, S., *Nonlinear Microwave Circuits*, Norwood, MA: Artech House, 1988.

Vuolevi, J., and T. Rahkonen, *Distortion in RF Power Amplifiers*, Norwood, MA: Artech House, 2003.

Wood, J., and D. E. Root, (eds.), *Fundamentals of Nonlinear Behavioral Modeling for RF and Microwave Design*, Norwood, MA: Artech House, 2005.

Power Amplifier Techniques

Cripps, S. C., *Advanced Techniques in RF Power Amplifier Design*, Norwood, MA: Artech House, 2002.

Kenington, P. B., *High Linearity RF Amplifier Design*, Norwood, MA: Artech House 2000.

Pothecary, N., *Feedforward Linear Power Amplifiers*, Norwood, MA: Artech House 1999.

Zhang, X., L. Larson, and P. M. Asbeck, *Design of Linear RF Outphasing Power Amplifiers*, Norwood, MA: Artech House, 2003.

Recommended Reading

The following books are recommended for further reading in related subject areas not fully covered in this book.

Cripps, S. C., *RF Power Amplifiers for Wireless Communications*, Norwood, MA: Artech House, 1999. Covers basic PA design at the circuit level and PA modes (referred to as *"RFPA"* throughout this book).

Harte, L., and S. Prokup, *Cellular and PCS/PCN Telephones and Systems*, Fuquay-Varina, NC: APDG Publishing, 1996. An introduction and technical overview of modern cellular communications systems.

Kenington, P. B., *High Linearity RF Amplifier Design*, Norwood, MA: Artech House, 2000. A wide-ranging treatment of PA linearization topics.

Minnis, B. J., *Designing Microwave Circuits by Exact Synthesis*, Norwood, MA: Artech House, 1996. Detailed treatment of broadband microwave circuit synthesis methods.

Ojanpera, T., and R. Prasad, (eds.), *Wideband CDMA for Third Generation Mobile Communications*, Norwood, MA: Artech House, 1998. Introduction and comprehensive treatment of 3G communications systems.

Pothecary, N, *Feedforward Linear Power Amplifiers*, Norwood, MA: Artech House, 1999. Specific coverage of feedforward techniques; emphasis on practical system issues.

Glossary

AC	Alternating current; term used generically to describe any sinusoidal signal.
ADC	Analog-to-digital converter.
AM	Amplitude modulation.
AM-AM	Term used to describe gain compression in an amplifier, whereby a given increase in input signal results in a different change in the output level.
AM-PM	Amplitude modulation to phase modulation; a distortion process in a power amplifier, whereby increasing signal amplitude causes additional output phase shift.
ACP	Alternate channel power; term used more generally to describe the spectral distortion of spread spectrum signals, which appear as continuous bands.
AGC	Automatic gain control.
BER	Bit error rate; a measure used to quantify the transmission integrity of a digital communications system.
BJT	Bipolar junction transistor.
CAD	Computer-aided design.
CDMA	Code division multiple access; a digital communications system which "chips" multiple signals in a pseudo-random fashion, making them appear like random noise unless a specific coding is applied.
CFR	Crest factor reduction. The process by which the PAR of a signal is reduced by digital means.
DAC	Digital-to-analog converter.
DQPSK	Differential Quadrature Phase Shift Keyed; a variation of QPSK.
DPA	Doherty Power Amplifier.
DPD	Digital predistortion.
DSP	Digital signal processing.
ECM	Electronic countermeasures; a generic term for military broadband microwave applications.
EDGE	Enhanced Data rates for GSM Evolution; an evolution of the GSM system, providing data rates up to 384 kbit/sec.
EER	Envelope Elimination and Restoration.
EPA	Error power amplifier; used in feedforward loop.

EPR   Ratio, usually expressed in decibels, of main PA power to EPA power; higher values imply higher system efficiency.

EVM   Error Vector Magnitude; a more comprehensive measure of amplifier distortion which incorporates both gain compression and AM-PM.

FET   Field effect transistor.

GaAs   Gallium Arsenide; one member of a group of useful semiconductor materials that are compounds between group 3 and group 5 elements in the periodic table. These materials show higher mobility and higher saturation velocity than Silicon and are therefore used in higher frequency applications.

GaN   Gallium Nitride; a newly emerging semiconductor material for high power, high frequency applications.

GSM   Global System for Mobile communication. Most extensively used worldwide digital cellular network operating in the 900 MHz and 1900 MHz bands.

HBT   Heterojunction bipolar transistor; most common variation of bipolar device used for higher frequency applications.

HEMT   High Electron Mobility Transistor; an FET device in which a "sheet" of high mobility material is created by the interaction between two epitaxial layers.

IF   Intermediate frequency.

IM, IMD   Intermodulation distortion.

LINC   Linear amplification using nonlinear components.

LDMOS   Laterally Diffused Metal Oxide Semiconductor; derivative of silicon RF MOS technology for higher frequency applications.

LO   Local oscillator.

LUT   Look-Up Table.

MMIC   Microwave Monolithic Integrated Circuit; a term now used, by convention, to describe monolithic integrated circuits which operate above 3 GHz.

MIC   Microwave Integrated Circuit; a term used in the pre-MMIC/RFIC era to describe a miniaturized chip-and-wire microwave hybrid circuit technology.

MCPA   Multicarrier Power Amplifier.

MOS   Metal Oxide Semiconductor; usually silicon-based semiconductor technology.

NADC   North Americal Digital Cellular; original digital cellular system used in the United States.

OFDM   Orthogonal frequency division multiplexing, used especially in the 802.11 and 802.16 standards.

PAE   Power Added Efficiency; efficiency definition for an amplifier that accounts for the RF input drive.

PAR   Peak-to-average power of a signal.

PEP   Peak Envelope Power.

PUF   Power Utilization Factor. A term introduced in this book to define the "efficacy" of a PA design. Defined as the ratio of power delivered in a given situation to the power delivered by the same device with the same supply voltage in Class A mode.

PHEMT   Pseudomorphic High Electron Mobility Transistor; a higher frequency variant on the HEMT.

QPSK   Quadrature Phase Shift Keyed; a generic term for a variety of modulation systems which carry information only in the phase, not the amplitude, of the RF carrier.

RFIC   Radio Frequency Integrated Circuit; a monolithically integrated device operating in the "RF" range, typically up to 3GHz.

RFPA   Radio Frequency Power Amplifier.

RRC   Raised Root Cosine; a filter of the Nyquist type, which allows a QPSK signal to be bandlimited without losing any modulation information.

SiC   Silicon Carbide; a semiconductor material which offers high voltage and high velocity saturation but low mobility. Potentially useful as a high power RF device technology in the low GHz frequency range.

SCSS   Short Circuited Shunt Stub; a distributed matching element used at microwave frequencies.

SPICE   A general purpose time domain nonlinear simulator, available in many implementations including shareware.

SSB   Single Sideband; a variation on AM developed in the 1950s, in which only one modulation sideband of an AM signal is transmitted.

3GPP   Third Generation Partnership Project.

WCDMA Wideband CDMA; a 3G system (e.g., UMTS, "CDMA2000") based on CDMA.

WiFi   Wireless Fidelity; a technical nickname for high speed data links used for the service available from WLAN networks, usually using the 802.11 standards. Typically short range, now widely available in airports, shopping malls, and bookshops.

WiMax   Worldwide Interoperability for Microwave Access; derivative of WiFi using the 802.16 standards which potentially will offer WiFi service over much greater range, giving comparable coverage to (but competitive with) the mobile telephone networks.

About the Author

Steve C. Cripps obtained his Ph.D. degree from Cambridge University in 1974. From 1974 to 1980, he worked for Plessey Research on GaAs-FET device and microwave hybrid circuit development. Dr. Cripps joined the solid state division of Watkins-Johnson in Palo Alto, California, in 1981 and since that time held various engineering and management positions at WJ, Loral, and Celeritek. His technical activities during that period focused mainly on broadband solid state power amplifier design for ECM applications. He has published several papers on microwave power amplifier design, including a design methodology that has been widely adopted in the industry.

Since 1990, Dr. Cripps has been an independent consultant, and his technical activities have shifted from military to commercial applications, which include MMIC power amplifier products for wireless communications. In 1996 he returned to England, where his focus shifted to high power linearized PAs for cellular, WLAN, and satcom applications. He teaches courses on PA design, linearization, and RF design, both in Europe and in the United States.

Index

Recent Titles in the Artech House Microwave Library

Microwave Circuit Modeling Using Electromagnetic Field Simulation, Daniel G. Swanson, Jr. and Wolfgang J. R. Hoefer

Microwave Component Mechanics, Harri Eskelinen and Pekka Eskelinen

Microwave Differential Circuit Design Using Mixed-Mode S-Parameters, William R. Eisenstadt, Robert Stengel, and Bruce M. Thompson

Microwave Engineers' Handbook, Two Volumes, Theodore Saad, editor

Microwave Filters, Impedance-Matching Networks, and Coupling Structures, George L. Matthaei, Leo Young, and E.M.T. Jones

Microwave Materials and Fabrication Techniques, Second Edition, Thomas S. Laverghetta

Microwave Mixers, Second Edition, Stephen A. Maas

Microwave Radio Transmission Design Guide, Trevor Manning

Microwaves and Wireless Simplified, Third Edition, Thomas S. Laverghetta

Modern Microwave Circuits, Noyan Kinayman and M. I. Aksun

Modern Microwave Measurements and Techniques, Second Edition, Thomas S. Laverghetta

Neural Networks for RF and Microwave Design, Q. J. Zhang and K. C. Gupta

Noise in Linear and Nonlinear Circuits, Stephen A. Maas

Nonlinear Microwave and RF Circuits, Second Edition, Stephen A. Maas

QMATCH: Lumped-Element Impedance Matching, Software and User's Guide, Pieter L. D. Abrie

Practical Analog and Digital Filter Design, Les Thede

Practical Microstrip Design and Applications, Günter Kompa

Practical RF Circuit Design for Modern Wireless Systems, Volume I: Passive Circuits and Systems, Les Besser and Rowan Gilmore

Practical RF Circuit Design for Modern Wireless Systems, Volume II: Active Circuits and Systems, Rowan Gilmore and Les Besser

Production Testing of RF and System-on-a-Chip Devices for Wireless Communications, Keith B. Schaub and Joe Kelly

Radio Frequency Integrated Circuit Design, John Rogers and Calvin Plett

RF Design Guide: Systems, Circuits, and Equations, Peter Vizmuller

RF Measurements of Die and Packages, Scott A. Wartenberg

The RF and Microwave Circuit Design Handbook, Stephen A. Maas

RF and Microwave Coupled-Line Circuits, Rajesh Mongia, Inder Bahl, and
  Prakash Bhartia

RF and Microwave Oscillator Design, Michal Odyniec, editor

RF Power Amplifiers for Wireless Communications, Second Edition, Steve C. Cripps

RF Systems, Components, and Circuits Handbook, Ferril A. Losee

Stability Analysis of Nonlinear Microwave Circuits, Almudena Suárez and
  Raymond Quéré

*TRAVIS 2.0: Transmission Line Visualization Software and User's Guide, Version
  2.0,* Robert G. Kaires and Barton T. Hickman

Understanding Microwave Heating Cavities, Tse V. Chow Ting Chan
  and Howard C. Reader

For further information on these and other Artech House titles, including
previously considered out-of-print books now available through our In-Print-
Forever® (IPF®) program, contact:

Artech House Publishers	Artech House Books
685 Canton Street	46 Gillingham Street
Norwood, MA 02062	London SW1V 1AH UK
Phone: 781-769-9750	Phone: +44 (0)20 7596 8750
Fax: 781-769-6334	Fax: +44 (0)20 7630 0166
e-mail: artech@artechhouse.com	e-mail: artech-uk@artechhouse.com

Find us on the World Wide Web at: www.artechhouse.com